LARGE-SCALE ECOLOGY
AND CONSERVATION
BIOLOGY

LARGE-SCALE ECOLOGY AND CONSERVATION BIOLOGY

THE 35TH SYMPOSIUM
OF THE BRITISH ECOLOGICAL SOCIETY
WITH THE SOCIETY
FOR CONSERVATION BIOLOGY
UNIVERSITY OF SOUTHAMPTON
1993

EDITED BY

P. J. EDWARDS
Geobotanisches Institut, ETH, Zurich

R. M. MAY
Department of Zoology, University of Oxford,
and Imperial College, London

N. R. WEBB
NERC Institute of Terrestrial Ecology,
Furzebrook Research Station, Wareham

OXFORD
BLACKWELL SCIENTIFIC PUBLICATIONS
LONDON EDINBURGH BOSTON
MELBOURNE PARIS BERLIN VIENNA

© 1994 the British Ecological Society
and published for them by
Blackwell Scientific Publications
Editorial Offices:
Osney Mead, Oxford OX2 OEL
25 John Street, London WC1N 2BL
23 Ainslie Place, Edinburgh EH3 6AJ
238 Main Street, Cambridge
    Massachusetts 02142, USA
54 University Street, Carlton
    Victoria 3053, Australia

Other Editorial Offices:
Librairie Arnette SA
1, rue de Lille
75007 Paris
France

Blackwell Wissenschafts-Verlag GmbH
Düsseldorfer Str. 38
D-10707 Berlin
Germany

Blackwell MZV
Feldgasse 13
A-1238 Wien
Austria

First published 1994

Set by Excel Typesetters Company, Hong Kong
Printed and bound in Great Britain
at the University Press, Cambridge

DISTRIBUTORS

Marston Book Services Ltd
PO Box 87
Oxford OX2 ODT
(*Orders*: Tel: 0865 791155
        Fax: 0865 791927
        Telex: 837515)

USA
Blackwell Scientific Publications, Inc.
238 Main Street
Cambridge, MA 02142
(*Orders*: Tel: 800 759-6102
            617 876-7000)

Canada
Oxford University Press
70 Wynford Drive
Don Mills
Ontario M3C 1J9
(*Orders*: Tel: 416 441-2941)

Australia
Blackwell Scientific Publications Pty Ltd
54 University Street
Carlton, Victoria 3053
(*Orders*: Tel: 03 347-5552)

A catalogue record for this title
is available from the British Library

ISBN 0-632-03832-2 (hbk)
ISBN 0-86542-801-8 (pbk)

CONTENTS

PREFACE

What will be the ecological consequences of rapid climatic change? What rate of exploitation of deep sea fish stocks is sustainable? Can we predict the dynamics of epidemic diseases such as AIDS? Where should we direct our conservation efforts, given the gathering rate of loss of biodiversity?

Never before has ecology been of such crucial importance in tackling the major social and economic problems that the world faces. However, the growing recognition that there is an essential ecological component to many of these problems, itself represents a crisis for ecological science. One of the fundamental issues is whether we can make ecological predictions at a scale appropriate to the problems that need to be addressed. Most ecological theory has developed from studies conducted at a small scale within patches of a single habitat. Indeed as we examine in progressively more detail the minutiae of ecological processes, there is a strong tendency to ignore possible complicating factors operating over a larger scale. Thus, ecologists commonly exclude from their deliberations (sometimes, but not always, intentionally) processes such as immigration and emigration, the climatic influence of neighbouring patches, or long distance transport of nutrients or genetic material, because they add an additional layer of complexity to the systems they study.

A related issue concerns our ability to acquire adequate data to tackle ecological questions over a large scale. The kind of detailed information (for example, about population structure or community composition) which provides the essential parameters for conventional ecological models is usually not available in large-scale studies. Very often we must rely upon surrogate data obtained by techniques such as remote sensing. Given the urgent need to answer questions about such things as the probability of extinction or the optimal design of refugia, it is important to know just how much (or rather, how little) information is essential; for the purist it may go against the grain, but a 'rough and ready' answer is often better than no answer at all.

The 35th annual symposium of the British Ecological Society took the theme: 'Large-scale ecology and conservation biology'. The rationale was a recognition that a science which neglects the reality and distinctive character of large-scale ecological processes provides a poor basis for tackling many of the more urgent problems in natural resources management. The organizers therefore saw a close connection between the aims of conservation biology and the need for a large-scale approach to ecology.

The symposium had three main objectives. The first was to examine the nature of large-scale ecological processes, and the adequacy of ecological concepts and models to describe and understand these processes. Secondly, the symposium was concerned with the practical problems of working at a large scale, and with the special tools (for example, remote sensing and geographic information systems) that may be needed. Finally, the social, economic and political issues associated with the application of ecological ideas in decision making and policy were also considered.

The opening papers in this volume deal with aspects of scale in ecology. In Chapter 1, May argues that an understanding of ecological processes and the ability to make predictions depend crucially upon studying the system at an appropriate scale. For example, the diversity of coral reef fishes is largely unpredictable on isolated reefs, and predictability increases with spatial scale. In contrast, the dynamics of measle epidemics in England and Wales only become apparent when the aggregated data are decomposed into data for individual cities. Angel (Chapter 4) develops this theme, demonstrating that to understand ecological processes in the ocean we must consider a very wide range of spatial and temporal scales. Davis and her colleagues (Chapter 2) remind us of the importance of considering long time scales in terrestrial communities. They describe large-scale mosaics in forest vegetation in Michigan USA, quite unlike those described in the literature, which owe their existence to events 3200 years ago when hemlock extended its range into the region.

How adequate are ecological models to make predictions of value in conservation biology? Gates *et al.* (Chapter 7) show that purely statistical models can be used to predict successfully the changing abundance of farmland birds in Britain from environmental variables such as climate and land use, though they are unreliable in predicting long-term changes. Kirkwood *et al.* (Chapter 9) examine models for estimating the sustainable yield (in particular of fish) from life history characteristics. In essence, the faster the growth of a species and the higher the level of natural mortality, the higher is the yield; thus long-lived species such as whales have a lower potential yield than sardines. However, particularly short-lived species may be exceptions to this general rule. They can provide higher yields than those predicted by these models, but are also more vulnerable to extinction from a combination of high exploitation and catastrophic events.

Webb and Thomas (Chapter 6) discuss a problem inherent in metapopulation models – that of identifying habitat patches in a manner relevant to a particular species. They take the example of heathland in southern England, an apparently simple and uniform type of habitat, and show that what seems to be a stable archipelago of heathland fragments may be anything but stable when seen through the eyes of three insect species. Harrison (Chapter 5) argues that, in fact, few well-studied species fit the 'classic' metapopulation model, and that in applying metapopulation ideas in conservation there is little choice but to consider each case individually. As the example of the northern spotted owl reveals, success may only be possible for species whose ecology is very well known. In contrast, Lawton *et al.* (Chapter 3) suggest that a knowledge of only a few relevant variables allows us to understand variation in geographical range size. Furthermore, consideration of metapopulation equilibrium dynamics should enable us to make reasonable estimates of extinction thresholds simply by measuring patch occupancy; conservation managers do not need to study every species as if it were unique.

The practical problems of studying large-scale ecological patterns are perhaps at their most acute in tropical rainforest. Campbell (Chapter 8) argues that much is still to be learnt from traditional forest inventories, even though there is no such thing as a representative small stand of species-rich tropical forest. For example, indices of similarity (Sorensen's index) between adjacent plots in lowland terra firme forest in the Amazon ranged from 0.1 to 0.21. In contrast, Newbould (Chapter 12) describes how in species-poor communities, simple measures of the performance of a dominant species, in this case *Phragmites communis*, can be valuable in developing management plans for an extensive area.

One of the important tools available to the large-scale ecologist, especially if not concerned with precise information about species composition, is remote sensing. This has been as an effective means of assessing the extent and types of blanket peatland in Britain, and does permit considerable resolution of the main types of plant community (Reid *et al.* Chapter 10). The great potential of remote sensing techniques combined with geographical information systems is vividly revealed in the paper by Rogers and Williams (Chapter 11) on the distribution of the tsetse fly in Africa.

The last few papers concern social, economic and political issues associated with the application of ecological science. One of these is where to direct our conservation efforts, given the inevitability that many species will become extinct within the next few decades. Mace (Chapter 13) describes the development of objective criteria for determining the threat of extinction. It is significant that the application of these criteria to a sample of 779 species of vertebrates suggests that 43% are threatened (defined as at least a 10% probability of extinction within 100 years), a much higher level than had previously been recognized.

Wilcove (Chapter 14) illustrates how ecological research can influence public policy, taking the bitter controversy over protecting the northern spotted owl in the Pacific Northwest of the USA as an example. More than any other natural resource controversy in recent years, this one was shaped by scientific research, though any victories were achieved through unrelenting pressures on legislators and land managers via lawsuits and media campaigns. Marsh (Chapter 15) argues that we require more adequate methods of evaluating economic activity which can attribute values to consequent ecological change. To do so involves a longer term perspective than current economics, and makes assumptions about costs and benefits to future generations which are difficult to validate: however, to ignore such issues means getting the economics wrong.

In the final chapter, Porritt considers in more general terms some of the processes by which the results of ecological research are converted into specific policy. He suggests that these processes remain opaque. Scientists concentrate on the elucidation of increasingly specialized parts of the whole, whereas administrators and politicians prefer to work in terms of simplified wholes. Scientists are understandably reluctant to infer too much from their partial evidence, and politicians are for the most part quite incapable of grappling with anything other than generalizations and half-truths. Clearly, more effective communication is not the least of the challenges for conservation biology.

The symposium was co-sponsored by the Society for Conservation Biology.

P.J. Edwards
R.M. May
N.R. Webb

1. THE EFFECTS OF SPATIAL SCALE
ON ECOLOGICAL QUESTIONS
AND ANSWERS

ROBERT M. MAY

Department of Zoology, South Parks Road, University of Oxford, Oxford OX1 3PS, UK and Imperial College, London, UK

SUMMARY

For understandable reasons, a great deal of ecological research focuses on single species or interactions between two species, usually on a time scale of a few years or less, and on spatial scales that are often smaller than the characteristic distance over which an individual member of the species moves in its lifetime. Many such studies are entirely appropriate to the questions being asked, but others derive more from the financial and time constraints of grants (often reinforced by current fashions for Popperian 'falsifiable hypotheses', which themselves owe more to philosophical musings than any real appreciation of how physical scientists actually work), than from careful assessment of the spatial scales that govern the system in question. This introductory chapter begins with some examples where the answers to ecological questions – and ultimately the under-standing of ecological systems – depend on whether or not the system is studied at an appropriate scale. It also indicates how spatial structure, on a variety of scales, can be self-organized by organisms, by simple and fully deterministic movement rules. Against this background, I offer some speculations about the increasing need for ecologists in general, and conservation biologists in particular, to deal with larger spatial scales than most of us are used to, or happy with.

INTRODUCTION

Recent advances in the study of ecological and evolutionary questions reach all the way down to the molecular biology of the genes which code for the self-assembly of living things, and all the way up to the interplay between biological and physical processes on a global scale. More than ever, ecological studies must be pursued at many different levels, and on many different spatial and temporal scales, from the way evolutionary

I

accident ('chance') combines with environmental and biomechanical constraints ('necessity') to shape the life history of individual organisms,
to the way plate tectonics and changing global climate shape entire
biogeographical realms.

For a variety of reasons, many of our universities and other institutions
do not easily accommodate work that stretches across traditional dis-
ciplinary boundaries, or that involves gathering data over a long time or a
large area. For one thing, although departmental and other organizational
boundaries are themselves usually the result of past evolutionary ac-
cidents, they are too often seen as absolute and inevitable; this can hinder
new initiatives. For another thing, the time constraints of PhD theses or
the funding cycles of research grants understandably militate against long-
term studies. Thus Weatherhead (1986) reviewed some 308 studies dealing
with ecological questions, and found the mean duration of such studies
was 2.5 years. More recently Tilman (1989) analysed some 749 empirical
studies reported in *Ecology* over the previous decade: only 13 of these
(1.7%) were field studies that lasted 5 years or more; of the 180 papers
describing manipulative experiments in field or laboratory, again only 13
(7%) lasted 5 years or more whereas 72 (40%) lasted less than 1 year. For
further discussion, see Elliott (1994).

By the same token, the spatial scale of most empirical investigations
will tend to be constrained partly as a corollary of time constraints, and
partly because 'hypothesis-falsifying' manipulative experiments are more
easily performed on relatively small spatial scales. So it is probably not
surprising that when Kareiva and Anderson (1989) analysed some 97
manipulative field experiments reported in *Ecology* between January 1980
and August 1986, they found 43 (44%) of the studies to have a characteristic physical dimension of less than 1 m, and 73 (75%) to be below
10 m. While there are many interesting and practically important questions
that can sensibly be pursued on these small scales, many others cannot.
I fear that, in recent years, too many ecologists have yielded to the
temptation of finding a problem that can be studied on a conveniently
small spatial and temporal scale, rather than striving first to identify the
important problems, and then to ask what is the appropriate spatial scale
on which to study them (and how to do this if the scale is large).

Be all this as it may, the present book deals with ecological questions
that inherently involve large spatial scales. Many of these questions are
of fundamental interest, but many also have practical ramifications. In
particular, given current concerns about likely changes in global climate
and about the accelerating effects of habitat destruction upon the global
diversity of species, much of the book focuses on the messages for

conservation biology that emerge from a better understandin
scale ecological processes.

The remainder of this introductory chapter is organized :
First, I outline some explicit ecological examples where an und
of what is going on is crucially dependent on studying the system on an
appropriate spatial scale. Second, I sketch some theoretical issues relating
to spatial scales and the persistence of populations. The chapter ends with
a brief survey of ways in which these empirical facts and theoretical
questions bear upon practical issues of conservation of species.

HOW SPATIAL SCALE CAN MATTER: SOME FACTUAL EXAMPLES

(handwritten annotation: measured by no. species, etc. and complexity simultaneous processes, at what spatial scale?)

Coral reefs

(handwritten annotation: Is ...)

Coral reefs in the sea, and tropical rainforests on land, are commonly
thought of as the most complex and species-rich communities on Earth.
Why this should be so remains a central question in ecology, with no
agreed answer.

Jackson has recently published a review of ideas about coral reef
communities. His conclusion serves as a text for my chapter: 'Failure to
consider the effects of [spatial] scale has been a major source of confusion
in theories to explain coral reef diversity' (Jackson 1991, p. 475).

Jackson shows how earlier generations of reef ecologists sought to
understand large-scale patterns in species diversity, such as why some
regions are more species-rich than others (fewer than 50 species of coral
in the tropical eastern Pacific and eastern Atlantic, around 100 in the
tropical western Atlantic, and roughly 600 in the tropical Indo-western
Pacific). This work emphasized physical differences among regions (differences in the distribution and extent of habitat suited to reef development; prevailing climate; etc.), and the intensity of various biological
interactions (competition, predation, disease).

However, these ideas of niche diversification as a factor structuring
coral reef communities and promoting diversity have been widely discredited over the past 20–30 years. Partly this results from Connell's
(1978; see also Huston 1985) studies of $1 m^2$ quadrats at Heron Island in
Australia. These studies – and other subsequent quadrat studies at other
reef sites – have demonstrated great fluctuations from year to year and
from site to site, both in numbers of coral species and in their abundance.
Numbers of coral reef fishes on small, isolated patch reefs or in other
small-scale study sites are also unpredictable, owing to large fluctuations

in larval recruitment (Doherty & Williams 1988; Sale 1988). Such instability and unpredictability on quadrat scales led Connell (1978) to propound his influential 'intermediate disturbance hypothesis': other things being equal, species diversity should be highest at some intermediate level of environmental disturbance; at low levels of disturbance, the better competitors get the chance to assert themselves, excluding other species; and at high levels of disturbance, many species will simply be absent most of the time.*

In reviewing this history of ideas, Jackson emphasizes how notions of community structure – or lack thereof – depend upon the scale at which the question is considered. Jackson contrasts the more recent 'quadrat view of diversity' with the 'landscape view of diversity', and he also reviews newer work on landscape scales. His conclusion (Jackson 1991, pp. 480–481) is that:

> predictability of reef coral distribution, abundance, and diversity increases with spatial scale. If we look only at individual quadrats, we see ecological anarchy. If we examine many quadrats within the same habitat, we observe a series of successional states that result from interspecific differences in coral life histories and vulnerability to disturbance. If we look across habitats, we see gradients in predictable associations commonly termed zonation. These patterns appear to result from interspecific differences in resource use as well as from effects of life histories and disturbance . . . species differ greatly in the ways they exploit resources and dominate habitats, and these patterns have persisted for at least half a million years.

Measles: endemic oscillations

Before vaccination campaigns were begun in the early 1960s, measles was endemic in developed countries, where in most cases it was – and is – a 'notifiable disease'. There are thus good records, dating back to the turn of the century, of the number of cases of measles reported each week for

---

* To my knowledge, this idea was first clearly set out by Horn at the symposium in memory of Robert MacArthur in 1973. The idea is expounded in the text and in an illustrative figure (based on a model of forest succession): 'intermediate disturbances produce higher diversity than either very high or very low levels' (Horn 1975, p. 209). Connell's (1975) paper from the same symposium is devoted to 'Some mechanisms producing structure in natural communities', and it makes no mention of degree of disturbance as a factor. I think it unfortunate that Connell's elegant and influential paper in *Science* in 1978 makes no reference to Horn.

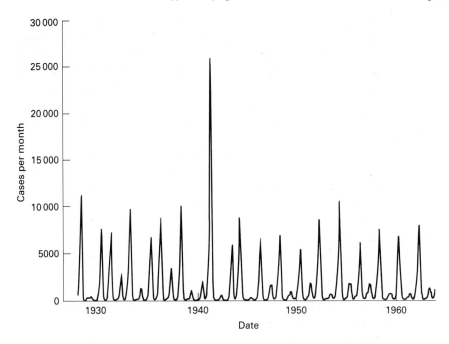

FIG. 1.1. Cases of measles reported in New York City, on a monthly basis, from the late 1920s until the advent of vaccination in the early 1960s. After Sugihara *et al.* (1991).

England plus Wales, and in the USA reported monthly for New York City and for Baltimore. These time series show yearly oscillations (associated mainly with schools opening and closing). As shown in Fig. 1.1, they also show a discernible but irregular 2-year 'interepidemic cycle', with the total number of cases tending to alternate high and low in successive years.

As discussed much more fully elsewhere (Anderson & May 1991, Chapter 6), these time series probably provide the best example of a Lotka–Volterra prey–predator cycle to be found in the ecological literature (broadly defined). The 'prey' is the population of susceptible hosts, continually being augmented by newly born susceptibles, and being depleted as individuals acquire measles infection (recovering into a lifelong immune state). The 'predators' are those infected-and-infectious hosts, who are serving as vehicles for measles reproduction and transmission. The corresponding characteristic lifetime of 'prey' is the average age at which hosts are infected (around 5 years in the UK and in the USA, before the advent of vaccination), and of 'predators' is the average duration of measles infectiousness (around 1 week or 1/50 years); hence the characteristic Lotka–Volterra period is around 2π (5 times $1/50)^{1/2}$, or roughly 2 years.

So far, so good. Unfortunately, the simplest models for this process predict very slowly damped oscillations, with the oscillatory period estimated above, and a damping time of a century or so. So what kept the oscillations persisting, until vaccination altogether changed this dynamical system? One idea (Bartlett 1957) is that fluctuations in birth rates are enough to keep the 'prey–predator' cycle going. On this view, the observed 'interepidemic cycles' are regular periodicity plus noise. An alternative idea is that the annual periodicity (associated with schools opening and closing) interacts with the Lotka–Volterra cycle to give a 'forced pendulum', which vibrates chaotically (Yorke & London 1973; Olsen & Schaffer 1990; Sugihara & May 1990; Bolker & Grenfell 1993). On this view, both the regularities and the irregularities in the observed 'interepidemic cycle' are deterministic chaos (and hence the irregularities are predictable in the short term, but – owing to sensitivity to initial conditions – not in the long term). These measles data are therefore of great interest, because they serve as a test-bed for new ideas about how to distinguish an apparently random time series that is generated by deterministic chaos from a 'really random' time series (as generated, for instance, by a roulette wheel).

I make no attempt here to develop these ideas about new methods of 'non-linear forecasting'. Suffice to say, there are several related approaches (for a brief review, see Sugihara & May 1990). They all tend to suggest the measles data from New York City and Baltimore are 'deterministic chaos', that is, a deterministic prey–predator cycle forced by the annual school cycle; the irregularities are as much a part of the deterministic, but chaotic, dynamics as are the annual and 2-year prey–predator cycles. But the measles data for England plus Wales show up as 'cycle plus noise'; the irregularities appear to be just noise superimposed on the regular cycles. This is most unsatisfactory. Measles should be one dynamical system or the other, independent of geography.

This problem may recently have been resolved (Sugihara et al. 1991), by taking the aggregated data for England plus Wales and decomposing it into data for individual large cities: Birmingham, Liverpool, Sheffield, etc. When this is done, the conclusion is, as for New York City and Baltimore, that the dynamics are deterministic chaos. London shows up as intermediate between the smaller cities and the aggregated England plus Wales: not quite cycles plus noise, but not low-dimensional chaos either. This has been a quick and impressionistic account of a complex and still-developing subject (for a more full discussion and illustrations, see Sugihara *et al.* 1991).

However, the essential point is clear. If the data are analy appropriate scale (here, epidemiologically homogeneous citie one answer (deterministic chaos resulting from interplay between annual school cycles and intrinsic prey–predator cycles). But if we aggregate too coarsely (England plus Wales), the fine structure of the data from individual cities is smudged out, and we get a different answer (rough interepidemic cycles, plus external noise). In this explicit example the answer we get depends on the spatial scale on which data are gathered or aggregated.

Huffaker's oranges and other metapopulations

As discussed more fully elsewhere in this book (Harrison; Lawton *et al.*; Wilcove), many species are thought to persist as 'metapopulations'. Such metapopulations are made up of many subpopulations, distributed among a mosaic of patches: at any one time, some patches are empty (but liable to colonization), while others are occupied (but liable to local extinction). In these circumstances, the lights of individual patches wink on and off unpredictably, but the overall average level of illumination – the overall density of the metapopulation – may remain relatively steady. As Harrison observes, these ideas are essentially a population-level version of the theory of island biogeography developed for communities by MacArthur and Wilson (1967).

In particular, I think that relatively specialized prey–predator associations (broadly defined to embrace plant–herbivore, host–parasite, and host–parasitoid along with more conventional prey–predator interactions) typically require spatial patchiness for their persistence. In homogeneous settings, such associations are liable to extinction via ever-diverging cycles of boom and bust of prey populations, interacting with similar but lagged cycles for the predators. But in a heterogeneous, patchy world, the associations can persist as a shifting mosaic of empty patches, prey only patches where prey populations are doing well, and patches with prey and predator where predators are currently flourishing. The dynamics of such systems obviously depend on the rates of prey and predator movement among patches and on the demographic details of the prey–predator interactions within any one patch, but they can also depend in a complicated and non-linear way on the number of patches and on the overall prey density. Halving the average prey density, or halving the number of patches (or, equivalently, the geographical range of the prey), can produce qualitative changes (May & Beddington 1981). The classic laboratory illustration of these ideas is Huffaker's (1958) experiment involving predatory mites (*Typhlodromus occidentalis*) seeking prey (the mite *Eotetranychus sesmaculatus*) in a universe of island oranges separated by seas of grease. If the patches/oranges were too few in number, or had too little separation, the system rapidly oscillated to extinction. More oranges, more widely separated, enabled the system to last longer before eventually crashing. But Huffaker never did have an orange-universe big enough to enable long-term persistence. Nor are the available data about the demography and movement parameters of prey and predator sufficient to permit an estimate of how big the system would have needed to be, to endure indefinitely.

When we turn to field studies, we have only anecdotes. Silver-backed jackal populations of the Serengeti appear to have exhibited an increase in average density, accompanied by a contraction in average territory size, in the late 1970s and early 1980s, followed by a crash that may be associated with an infectious disease (probably canine distemper); see Moehlman (1983). Wild dogs in the Serengeti are also known to have suffered from periodic outbreaks of disease during the past 30 years (again, probably distemper); see Sinclair and Norton-Griffiths (1979). These and others (e.g. the black-footed ferrets of Meeteese; May 1986) could be examples of canid host/viral parasite associations, where long-term persistence depends on a complex interplay among host densities, contact rates and transmission of infection, and other factors.

The basic feature common to these and many other possible instances of 'prey–predator' metapopulations is that the spatial scale required for the overall system to persist indefinitely may be very large indeed. This spatial scale – related to the effective number of patches needed for persistence – depends in a complicated and ill-understood way on the patterns of movement of prey and predator individuals among patches, and on the dynamics of prey–predator interactions in any one patch. In general, however, the spatial scale required for metapopulation persistence is very much larger than the spatial scale characteristic of the lifetime movement of individual prey or predators.

SPATIAL SCALE AND THEORETICAL MODELS

Given that there are virtually no field or laboratory studies where we have enough understanding and/or data to infer the spatial scale needed for a metapopulation (whose constituent subunits are inherently unstable and impermanent) to persist, it is useful to ask what can be learned from theoretical models.

Beginning about 20–30 years ago, there is today a burgeoning literature dealing with metapopulation models. Most of this assumes a world consisting of N distinct patches, and then asks questions either broadly about overall occupancy and persistence in terms of patch colonization and extinction rates, or more specifically about the relations between within-patch dynamics and the overall metapopulation dynamics. Much of this work deals with single species, but extensions to competing species or prey–predatory associations are increasing. Earlier studies mainly used 'mean field' approximations to generate deterministic equations for the proportion of patches occupied. More recent extensions deal with the complexities introduced by demographic and environmental stochasticity. Some of this literature is reviewed in the chapters by Harrison, Wilcove, and Lawton *et al.*, and I shall not dwell on it.

Common to all this earlier work is the assumption – usually implicit – that the distinct patches, among which the subunits of the metapopulation are distributed, are themselves simply determined by external environmental factors. They are, as it were, a 'given' in the problem. This will, of course, often – perhaps almost always – be the case. However, it has recently been recognized that simple and fully deterministic rules about the movement of individuals can, under some circumstances, lead to self-organized spatial heterogeneity even in a world which is inherently homogeneous and where there is no overt tendency for individuals to form clumps. Such self-organized spatial structure can take the form of spiral waves, or spatial chaos, or even apparently static lattices of 'good patches' and 'bad patches', even though the underlying substrate is purely homogeneous. I focus on this recent work, partly because it provides some new perspectives on the dynamics of metapopulations, and partly because it affords some disconcerting insights into the spatial scales required for metapopulation persistence in relation to the scales characteristic of individuals.

Self-organized spatial structures and metapopulation persistence

Spatial structure and patchiness has long been recognized as probably the most important factor underlying the long-term persistence of host–parasitoid associations in nature (Hassell 1978). In a homogeneous setting, parasitoids that search independently and randomly lead to diverging oscillations in host and parasitoid populations, and thence to extinction of both (Nicholson & Bailey 1935). Although various refinements to the interactions between hosts and parasitoids are in principle capable of stabilizing the association in spatially homogeneous situations, such refinements are not generally thought to be sufficiently influential to explain the persistence of host–parasitoid associations in the natural world (May & Hassell 1988; Hawkins & Sheehan 1994). Rather, it is thought that spatial heterogeneity, coupled with sufficient variability in the parasitoid attack rates experienced by hosts in different places, is the main mechanism.

The underlying spatial heterogeneity may be inherent, given by environmental factors, or it may arise because hosts and/or parasitoids have some prescribed tendency to distribute themselves non-randomly among patches, giving rise to an 'overdispersed' or clumped distribution. We can explore such models, but because the assumptions about the mechanisms generating the required heterogeneity are arbitrary, our conclusions about the spatial scales required for metapopulation persistence are correspondingly arbitrary, and thus not very illuminating.

Suppose, however, that we start with a square array of $n \times n$ patches – leaves, bushes, etc. – which are all identical. Hosts and parasitoids are initially distributed randomly among these patches. The parasitoids search independently and randomly. At the start of each new adult generation, some fixed and density-independent fraction of hosts and of parasitoids (μ_H and μ_P, respectively) are divided evenly among the eight immediately neighbouring patches; the remaining fraction of hosts and parasitoids stays home. We thus have a system of purely deterministic rules of local movement, with no in-built heterogeneity or clumping mechanisms. As discussed in detail elsewhere (Hassell *et al.* 1991; Comins *et al.* 1992; Solé & Valls 1992), this system can exhibit self-organized spatial heterogeneity, which may be orderly – spiral waves, or apparently static 'crystal lattices' – or apparently random 'spatial chaos'. It can also extinguish itself via the familiar Nicholson–Bailey diverging oscillations. The outcome depends only on the movement parameters μ_H and μ_P, on the overall spatial scale of the system in relation to its constituent elements (leaves, bushes, etc.) as set by n^2, and on the basic reproductive rate of the hosts, λ. Conversely, if the dispersing fractions of hosts and parasitoids move globally rather than locally (to any other patch rather than the immediate neighbours), then the entire system rapidly oscillates to extinction.

The spatially self-organizing features of this model are further underlined by extending the analysis to models in which there are no 'cells' or patches, but simply a continuous and homogeneous spatial arena, with deterministic local movement (described by diffusion equations). The essential conclusions remain intact (Godfray, personal communication).

Regardless of whether the homogeneous substrate is discrete patches or a continuum, numerical simulations suggest that – for a wide range of values of the parameters μ_H, μ_P and λ – the system will fluctuate to extinction unless its overall area is around 1000 to 10000 times the characteristic range within which individual hosts and parasitoids move during their lifetime (i.e. in the spatially discrete case, unless n exceeds 30–100). This is a fairly robust conclusion for this class of models. But, if it is more generally true, it is an unhappy conclusion: it suggests that, for populations which owe their persistence to this kind of self-organized spatial structure, the spatial scale of conservation areas must be very large in relation to the lifetime ranges of individuals (larger by factors of at least 10^{3-4}, or more).

Metapopulations and habitat destruction

If a species is persisting by virtue of local extinction and recolonization among a mosaic of patches, what fraction of these patches can be destroyed without extinguishing the species? In Chapter 3, Lawton *et al.* present a new, and empirically oriented, approach to this question for single species. An analysis of the sometimes surprising effects of habitat destruction upon the balance among competing species has recently been given by Nee and May (1992), and will not be recapitulated here. Instead, I sketch some new work on the effects of patch removal upon metapopulations of prey and predators.

Suppose that, in the pristine state, we have N_0 patches which support metapopulations of a prey and a predator species. Working with a 'mean field' approximation, we assume that, at time t, the number of empty, prey only, and prey plus predator patches are $X(t)$, $Y(t)$, and $Z(t)$, respectively:

$$X(t) + Y(t) + Z(t) = N_0. \tag{1}$$

We further assume that empty patches are colonized, converting them to prey only patches, at a rate αXY (the probability that an empty patch, of which there are X, is colonized is proportional to the total number of prey only patches, Y, dispersing colonizers). Similarly, prey only patches move to prey plus predator patches at a rate βYZ. Prey only patches have an intrinsic death rate dY (which will often be low, but finite), while predators overexploit prey to extinguish Z-type patches at a rate δZ. Assuming a fixed number of patches, N, we have the two-dimensional dynamical system

$$dY/dt = \alpha XY - \beta YZ - dY, \qquad (2)$$

$$dZ/dt = \beta YZ - \delta Z. \qquad (3)$$

This system has an equilibrium point at

$$Y^* = \delta/\beta, \qquad (4)$$

$$Z^* = [N - N_c][\alpha/(\alpha + \beta)], \qquad (5)$$

and X^* is given by the generalized version of equation (1), $X + Y + Z = N$. Here N_c is defined as

$$N_c = \frac{\delta}{\beta} + \frac{d}{\alpha}. \qquad (6)$$

This solution exists, and is globally stable, if and only if $N > N_c$. This critical patch number, N_c, is seen from equation (6) to depend on the 'death/birth' ratio for prey plus predator patches, δ/β, augmented by the corresponding 'death/birth' ratio for prey only patches, d/α. If $N < N_c$, then the predator population cannot maintain itself ($Z^* \to 0$), and we revert to the familiar single-species situation.

Re-expressing this in terms of the effects of patch destruction or removal, we see that as N decreases from its pristine value (N_0), the number of patches occupied by predators declines, as does the number empty. The number of patches containing only prey remains constant in this simplest model. Once the remaining number of patches falls below the critical number, N_c, the predators are extinguished. Thereafter, the number of prey occupied patches declines, until the prey are also extinguished (for $N < d/\alpha$). Figure 1.2 illustrates this, showing X^*, Y^*, Z^* as functions of $h = N/N_0$; h is the fraction of the original patches that remain.

Alternatively, Fig. 1.3a shows the proportions of the *remaining* patches that are empty, prey only, and prey plus predator ($x = X/hN_0$, $y = Y/hN_0$, $z = Z/hN_0$), again as functions of h, the number of remaining patches expressed as a fraction of the pristine numbers ($h = N/N_0$). From this perspective, the proportion of prey only patches seems to increase as patches are removed; of course, the proportion of remaining patches occupied by predators decreases, while the proportion empty may increase or decrease (depending on whether d is larger or smaller than δ). In the extreme case when prey only patches have no intrinsic death rates ($d \to 0$), we have Fig. 1.3b. Here the proportion of remaining patches occupied by prey increases as patches are destroyed, until we attain the critical destruction fraction ($h_c = N_c N_0$) beyond which predators are extinguished, and all remaining patches are occupied by prey; none is

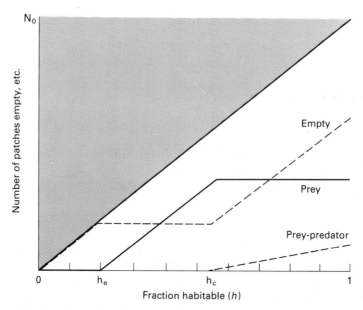

FIG. 1.2. Showing the number of patches empty, occupied by prey only, and occupied by prey and predators, as a function of the fraction of the patches remaining undestroyed, h; $h = N/N_0$, where N is the number of patches remaining and N_0 is the pristine number. Specifically, this illustrative figure is based on equations (2) and (3), with the relevant combinations of the parameters defined in the text having the values: $\delta/(\beta N_0) = 0.35$, $d/(\alpha N_0) = 0.20$, and $\alpha/\beta = 1/2$ (whence $h_c = 0.55$).

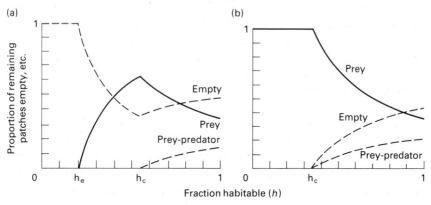

FIG. 1.3. (a) As for Fig. 1.2, except now the proportion of the remaining (undestroyed) patches that are empty, prey only, and prey plus predator, is shown, again as a function of the fraction of patches that remain habitable, h. The parameters have the same values as in Fig. 1.2, and the general features are as discussed in the text. (b) As discussed in the text, this figure illustrates the extreme case of the more general (a) which arises in the limit when the intrinsic death rate of prey only patches tends to zero, $d = 0$. As in Fig. 1.2 and (a), the remaining parameter combinations have the values $\delta/(\beta N_0) = 0.35$ and $\alpha/\beta = 1/2$ (whence $h_c = 0.35$). In this extreme case, all patches are occupied by prey, once the fraction remaining habitable falls below the critical value, h_c, required for predators to persist.

e are familiar with the phenomenon of prey species increasing
ce, deer on the east coast of the USA and elsewhere) as
re removed, but I believe there is a less general awareness of
...unding possibilities that can be superimposed by the changing
dynamics of metapopulations as a result of habitat destruction. In particular, the proportion of remaining habitats occupied by prey species will
usually increase as habitat is destroyed (and predator populations are
consequently reduced), even if no other changes occur to affect the vital
rates of prey and predator species.

In short, simple models for the effects of habitat destruction and patch
removal suggest consequences for the dynamics of metapopulations – be
they single species, competitors, or prey–predator – which are not always
intuitively obvious. These are, in essence, problems in scale. They remind
us again that the spatial scales needed for structured populations to
maintain themselves are not easily deduced from the life histories of
individual animals or plants.

SPATIAL SCALE AND CONSERVATION BIOLOGY

I conclude with a list of questions relevant to conservation biology, which
arise from the preceding discussion.

1   *How big should a reserve be?* Earlier application of the theory of
island biogeography to the design of reserves and conservation areas
aimed at producing broad guidelines, such as one big reserve is more
effective than several smaller ones adding to the same total area. Later
work continues to tease out the particular circumstances that all too often
invalidate such generalities. As we move from broad questions about
species totals, to more focused questions about the area needed to pre-
serve a particular species of plant or animal, things do not get easier. As
we have seen, both in empirical examples and in theoretical models,
many species persist by virtue of a shifting balance within a mosaic of
 patches: at one extreme, we have local extinction and recolonization; but
in other settings (e.g. host–parasitoid systems) we do not need this
extreme, requiring only a sufficient degree of heterogeneity in density
from patch to patch. Under these circumstances, the dimension of the
area needed for a population to maintain itself is not easily inferred from
the life history or movement scale of individuals, nor is this area related
in any simple way to considerations of demographic or environmental
stochasticity. In the particular case of host–parasitoid models where
persistence derives from patterns of self-organized heterogeneity, a 'conservation area' characteristically needs to be at least a thousand times

larger than the area typically covered by individuals in their lifetime. I
believe we need more analyses of this kind, relating the landscape scale
for survival of the metapopulation to the more intuitively appreciated
scale of individual movements.

2 *Migration and spatial scale.* Obviously a wider range of questions
arises for the conservation of species that exhibit systematic patterns
of large-scale migration, as do many birds or Serengeti mammals. For
temperate-zone birds that overwinter in the tropics, we have all the
problems of spatial scale and effects of fragmentation just mentioned, but
now at both ends of their range. Small wonder that there is still debate as
to the basic cause of the decline of populations of neotropical migrant
birds (Terborgh 1989).

3 *Restricted versus widespread species.* For conservation applications,
the kinds of questions I have discussed seem more directly relevant to
species whose natural ranges are restricted. Such species, of course,
include many on the 'Red Lists' of conservation biologists. Bibby *et al.*
(1992), for example, have analysed the geographical ranges of birds,
focusing on those endemic species with breeding ranges smaller than
$50\,000\,km^2$ (Terborgh & Winter 1983). These account for 27% of all bird
species, but 77% of all threatened birds (according to the IUCN classification of 'vulnerable' or 'endangered'). Using this analysis, Bibby *et al.*
(1992) propose a list of 221 'Endemic Bird Areas', three quarters of
which are in the tropics, adding up to some 6.5 million km^2 (or about 5%
of the Earth's land surface). These would serve to protect 70% of all
endemic bird species.

    This is trail-blazing work, relating directly to large-scale ecology and
conservation biology. Even so, awkward questions remain. As emphasized by Leck (1993), many endemics are common and able to handle
human disruption within their restricted ranges: the white-eye family
(Zosteropidae) is an example, with 78% of its species classed as endemics,
but relatively few endangered. Conversely, few species in the falcon
family (Falconidae) have restricted ranges – only 6% by Bibby *et al.*'s
criterion – yet a high proportion are endangered.

    In summary, questions of the scale of a species' range are only one
part of any assessment of conservation status. As Rabinowitz *et al.* (1986)
emphasize, and as Lawton *et al.* remind us in this book, range scale is one
dimension in multifactorial questions about what constitutes commonness
and rarity. Unfortunately, so much of our attention is, and has been,
focused on the ecology and behaviour of individuals or small groups,
typically on time scales of 3 years or less and spatial scales of 10 m or less.
The pressing concerns of conservation biology are on longer time scales,

and vastly greater spatial scales. Data gathered, and theories developed, to deal with individual interactions within and between species do not always provide the kind of information or intuition we need if we are to optimize our efforts to save what we can from the flames.

ACKNOWLEDGEMENTS

I am grateful to Bryan Grenfell, John Lawton, Paul Harvey, Sean Nee and others for helpful discussions. This work was supported in part by The Royal Society, and by the NERC Centre for Population Biology at Imperial College, Silwood Park.

REFERENCES

Anderson, R.M. & May, R.M. (1991). *Infectious Diseases of Humans: Dynamics and Control*. Oxford University Press, Oxford.

Bartlett, M.S. (1957). Measles periodicity and community size. *Journal of the Royal Statistical Society Series A*, **120**, 48–70.

Bibby, C.J. et al. (1992). *Putting Biodiversity on the Map: Priority Areas for Global Conservation*. International Council for Bird Preservation, Tring.

Bolker, B.M. & Grenfell, B.T. (1993). Chaos and biological complexity in measles dynamics. *Proceedings of the Royal Society of London, Series B*, **251**, 75–81.

Comins, H.N., Hassell, M.P. & May, R.M. (1992). The spatial dynamics of host–parasitoid systems. *Journal of Animal Ecology*, **61**, 735–748.

Connell, J.H. (1975). Some mechanisms producing structure in natural communities. *Ecology and Evolution of Communities* (Ed. by M.L. Cody & J.M. Diamond), pp. 460–490. Harvard University Press, Cambridge, MA.

Connell, J.H. (1978). Diversity in tropical rain forests and coral reefs. *Science*, **199**, 1302–1310.

Doherty, P.J. & Williams, D.W. (1988). Are populations of coral reef fishes equilibrial assemblages? *Proceedings of the VI International Coral Reef Symposium* (Ed. by D.J. Barnes), pp. 131–139. Symposium Committee, Townsville, Australia.

Elliott, J.M. (1994). *Qualitative Ecology and the Brown Trout*. Oxford University Press, Oxford.

Hassell, M.P. (1978). *The Dynamics of Arthropod Predator–Prey Associations*. Princeton University Press, Princeton, NJ.

Hassell, M.P., Comins, H.N. & May, R.M. (1991). Spatial structure and chaos in insect population dynamics. *Nature*, **353**, 255–258.

Hawkins, B.A. & Sheehan, W. (1994). *Parasitoid Community Ecology*. Oxford University Press, Oxford.

Horn, H.S. (1975). Markovian properties of forest succession. *Ecology and Evolution of Communities* (Ed. by M.L. Cody & J.M. Diamond), pp. 196–211. Harvard University Press, Cambridge, MA.

Huffaker, C.B. (1958). Experimental studies on predation: dispersion factors and predator–prey oscillations. *Hilgardia*, **27**, 343–383.

Huston, M.A. (1985). Patterns of species diversity on coral reefs. *Annual Review of Ecology & Systematics*, **16**, 149–177.

Jackson, J.B.C. (1991). Adaption and diversity of reef corals. *Bioscience*, **41**, 475–482.

Kareiva, P. & Anderson, M. (1989). Spatial aspects of species interactions: the wedding of models and experiments. *Community Ecology* (Ed. by A. Hastings), pp. 35–50. Springer-Verlag, New York.

Leck, C. (1993). Bird conservation. *TREE*, **8**, 37.

MacArthur, R.H. & Wilson, E.O. (1967). *The Theory of Island Biogeography*. Princeton University Press, Princeton.

May, R.M. (1986). The cautionary tale of the black-footed ferret. *Nature*, **320**, 13–14.

May, R.M. & Beddington, J.R. (1981). Notes on the management of locally abundant populations of mammals. *Problems in Management of Locally Abundant Wild Mammals*. (Ed. by P.A. Jewel, S. Holt & D. Hart), pp. 205–216. Academic Press, New York.

May, R.M. & Hassell, M.P. (1988). Population dynamics and biological control. *Philosophical Transactions of the Royal Society of London, Series B*, **318**, 129–169.

Moehlman, P.D. (1983). Socioecology of silver-backed and golden jackals. *Recent Advances in the Study of Mammalian Behaviour* (Ed. by J.F. Eisenberg & D.G. Kleiman), pp. 423–438. American Society of Mammals (Special Publication No. 7), New York.

Nee, S. & May, R.M. (1992). Patch removal favour inferior competitors. *Journal of Animal Ecology*, 61, 37–40.

Nicholson, A.J. & Bailey, V.A. (1935). The balance of animal populations, Part I. *Proceedings of the Zoological Society of London*, **1**, 551–598.

Olsen, L.F. & Schaffer, W.M. (1990). Chaos versus noisy periodicity: alternative hypothesis for childhood epidemics. *Science*, **249**, 499–504.

Rabinowitz, D., Cairns, S. & Dillon, T. (1986). Seven forms of rarity. *Conservation Biology* (Ed. by M.E. Soulé), pp. 182–204. Sinauer, Sunderland, MA.

Sale, P.F. (1988). What coral reefs can teach us about ecology. *Proceedings of the VI International Coral Reef Symposium* (Ed. by D.J. Barnes), pp. 19–27. Symposium Committee, Townsville, Australia.

Sinclair, A.R.E. & Norton-Griffiths, M. (1979). *Serengeti: Dynamics of an Ecosystem*. University of Chicago Press, Chicago.

Solé, R.V. & Valls, J. (1992). Spiral waves, chaos and multiple attractors in lattice models of interacting populations. *Physical Letters, Series A*, **166**, 123–128.

Sugihara, G. & May, R.M. (1990). Nonlinear forecasting as a way of distinguishing chaos from measurement error in time series. *Nature*, **344**, 734–741.

Sugihara, G., Grenfell, B.T. & May, R.M. (1991). Distinguishing error from chaos in ecological time series. *Philosophical Transactions of the Royal Society of London, Series B*, **330**, 235–251.

Terborgh, J. (1989). *Where Have All The Birds Gone?* Princeton University Press, Princeton.

Terborgh, J. & Winter, B. (1983). A method for siting parks and reserves with special reference to Colombia and Ecuador. *Biology Conservation*, **27**, 45–58.

Tilman, D. (1989). Ecological experimentation: strengths and conceptional problems. *Long-term Studies in Ecology* (Ed. by G.E. Likens), pp. 136–157. Springer, New York.

Weatherhead, P.J. (1986). How unusual are unusual events? *American Naturalist*, **128**, 150–154.

Yorke, J.A. & London, W.P. (1973). Recurrent outbreaks of measles, chickenpox and mumps, II. Systematic differences in contact rates and stochastic effects. *American Journal of Epidemiology*, **98**, 469–482.

2. HISTORICAL DEVELOPMENT OF ALTERNATE COMMUNITIES IN A HEMLOCK–HARDWOOD FOREST IN NORTHERN MICHIGAN, USA

MARGARET B. DAVIS, SHINYA SUGITA,
RANDY R. CALCOTE, JAMES B. FERRARI
AND LEE E. FRELICH*

*Department of Ecology, Evolution and Behavior,
University of Minnesota, St Paul, MN 55108, USA*

SUMMARY

Hemlock–hardwood forests of Sylvania, Michigan, USA, are composed of 1–20 ha patches dominated either by hemlock (*Tsuga canadensis*) or by sugar maple and basswood (*Acer saccharum* and *Tilia americana*). The mosaic pattern began about 3200 years ago when hemlock extended its geographical range into the region. Local stand histories reconstructed from fossil pollen preserved in small hollows 5–15 m in diameter show that hemlock invasion of the preceding white pine–red maple–oak (*Pinus strobus–Acer rubrum–Quercus rubra*) forest was patchy. Where hemlock invaded, hemlock-dominated stands were established which have persisted ever since. In the intervening patches sugar maple and basswood became dominant. The two stand types have both differentiated from a very different kind of forest – a forest type that no longer exists at Sylvania.

Hemlock patches and hardwood patches each have distinctive soil humus, nutrient availability, microclimate and ground flora. Local dominance by either hemlock or sugar maple creates a local environment in which recruitment by the competing species is reduced, a positive feedback that encourages the persistence of patches. We do not observe invasion of the alternate stand type today, although hemlock invasion of the pine–oak–red maple forest apparently occurred rapidly 3200 years ago as climate and fire regime changed to favour the invading species and to increase the competitive advantage of resident sugar maple, basswood and yellow birch relative to pine, oak and red maple. Thus the mosaic is

---

* Present address: Department of Forest Resources, University of Minnesota, St Paul, MN 55108, USA.

quite different from forest mosaics described in the literature that depend upon patchy disturbances or patchy distributions of physical features of the environment.

The mosaic is seen only in remnants of the original old-growth forest, because logging has converted both stand types to hardwoods. Restoration of the mosaic pattern could prove difficult, because the climatic and biotic environments that permitted hemlock stands to become established are no longer present.

INTRODUCTION

Large-scale vegetation patterns on the landscape often result from long-term, large-scale processes. These may change so slowly that understanding requires the long time series provided by palaeoecological records. Many large-scale vegetation patterns result from differences in the physical environment, for example substrate texture or chemistry, or climatic gradients related to topography (Curtis 1959; Whittaker 1960; Bormann *et al.* 1970). Only a few of the many examples in the literature have been followed through time, but in every case where palaeoecological information at the appropriate spatial scale is available, it shows that patterns change with time: community composition changes and boundaries appear or disappear (Brubaker 1975; Graumlich & Davis 1993; Spear *et al.* 1994). As climate and vegetation change, minor differences in substrate can become critical, creating a boundary, or they may no longer limit species and the boundary can disappear. In the case of landscape patterns resulting from natural disturbance (Heinselman 1973; Bormann & Likens 1979; West *et al.* 1981; Sprugel 1984; Pickett & White 1985), palaeoecological records show that both rate and extent of disturbance can change with time (Clark 1990; Swetnam 1993). A change in disturbance regime can alter the distribution of vegetation at the landscape scale (Grimm 1984).

A landscape vegetation pattern that depends on species interactions

---

Fig. 2.1. (*Opposite.*) (a) Infra-red leaf-off aerial photograph, showing locations of sites mentioned in text. Dark areas are coniferous trees, light areas are hardwoods. Plots A and C are outlined in white. Arrow points to a hollow in the Clark–Crooked Lake hardwoods. Outline map indicates location of Sylvania Wilderness in USA. (b) Infra-red aerial photograph of the northwest corner of the Sylvania reserve, showing the locations of Snapjack Lake hardwoods and Long Lake hemlocks. The boundary of the reserve shows clearly because the extensive hemlock stands exist only in the old-growth forests within the boundary; second growth forests outside the boundary are dominated by stands of hardwood species.

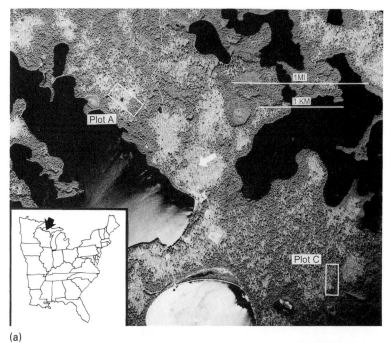

(a)

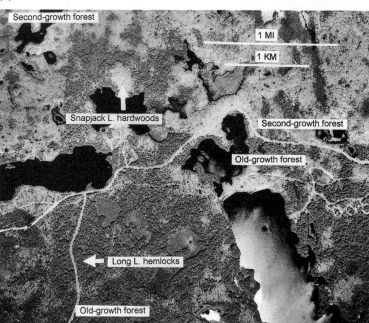

(b)

and feedbacks that influence recruitment is exemplified by the mosaic of
hemlock (*Tsuga canadensis*) and hardwood communities in forests of
northern Michigan, USA, where patches 1–20 ha in size form a striking
pattern on the landscape (Fig. 2.1). Frelich *et al.* (1993) term the pattern
a 'competition' mosaic, because the two dominant species, hemlock
and sugar maple (*Acer saccharum*), each alter their local environment,
creating conditions in the immediate vicinity of trees that favour self-
recruitment and discourage recruitment of seedlings of the other dominant
species. Positive feedbacks of this kind are termed 'two-sided switches' by
Wilson and Agnew (1992). Sugar maple seedlings are disadvantaged by
the dense shade and low nutrient conditions under hemlocks (Rogers
1978; Ferrari 1993). The small seeds of hemlock produce tiny first-year
seedlings that cannot penetrate the thick leaf mat under maples to reach a

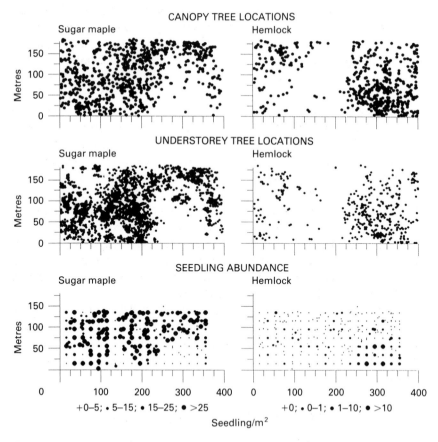

FIG. 2.2. Map of plot A, showing locations of canopy sugar maple and hemlock trees and
subcanopy saplings, and seedling densities.

reliable source of moisture (Davis & Hart 1961); hemlocks survive best on hemlock or pine rotted nurse logs or on bare mineral soil (Goder 1955; Rogers 1978).

The strong spatial association of subcanopy trees, saplings and seedlings with canopy trees of the same species (Fig. 2.2) means that replacement probabilities within each patch are highest for the dominant species (Frelich *et al.* 1993). A spatial Markov model (MOSAIC) simulates tree-by-tree replacement by deriving replacement probabilities for canopy trees from the abundances of species within 10 m of the dead tree (Frelich *et al.* 1993). Simulations show that the 'neighbourhood' effect on replacement probabilities is sufficient to create patches within a forest of randomly distributed hemlock and sugar maple trees. In simulations, patches formed within a few hundred years, growing larger and more concentrated with time.

These results explain why the mosaic pattern persists. However, the origin of the mosaic remains problematic. Regional palaeoecological records indicate that hemlocks invaded Sylvania only a few thousand years ago (Fig. 2.3) (Davis *et al.* 1986; Davis 1987). As hemlock recruitment does not now occur in stands dominated by sugar maple, it is difficult to explain how existing forests could have been invaded, and thus how the mosaic originated. Pastor and Broschart (1990) noted that large hemlock stands are often adjacent to bogs. Frelich *et al.* (1993) suggested that hemlock might have invaded wet sites first, establishing hemlock stands, with subsequent invasion of drier sites. Stearns (1990) has invoked disturbance to promote hemlock seedling establishment, and Spies and Barnes (1985a) mention fire as a factor in hemlock stand establishment. Stearns postulates that many large hemlock stands were established during the Little Ice Age (AD 1450–1850) after catastrophic windstorms followed by fire exposed mineral soil. Analysis of stand shapes supports an alternative disturbance scenario, that hardwood stands are superimposed on a matrix of hemlock following disturbance, and that hemlock are currently invading hardwood stands (Pastor & Broschart 1990).

In this chapter we will show how a spatially precise palaeoecological record can be used to test hypotheses regarding the origin of the mosaic of hemlock and hardwood forest communities, providing new information about factors affecting stand invasion and persistence of the forest mosaic through time.

STUDY SITE AND METHODS

The study site is the Sylvania Wilderness, a 8500-ha US Forest Service reserve in northern Michigan (46°5'N, 89°15'W). Sylvania includes a

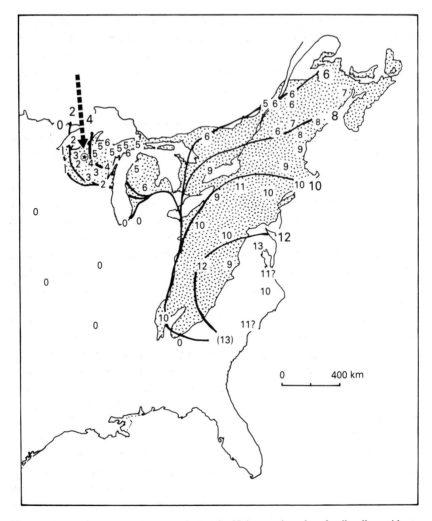

FIG. 2.3. Migration map for hemlock during the Holocene, based on fossil pollen evidence. Isochrones represent the approximate position of the frontier of the continuous species population at 2000-year intervals from 12 000 years ago (12) to the present (0). Small numbers represent the radiocarbon age (in thousands of years) at specific sites of the earliest sharply rising quantities of hemlock pollen. Arrow indicates location of Sylvania. Redrawn from Davis (1990).

number of large lakes and about 6000 ha of old-growth upland forest (Frelich & Lorimer 1991). These forest stands have never been clearcut, although large white pines (*Pinus strobus*) were selectively logged around the turn of the century. Aerial photographs (Fig. 2.1) display the patchy nature of the forests, which consist of a mosaic of stands dominated by hemlock, and stands dominated by hardwoods – primarily sugar maple and basswood (*Tilia americana*). Yellow birch (*Betula alleghaniensis*) is a constituent of both communities, but is more common in hemlock stands. Boundaries can be sharp or diffuse, and there are many mixed stands (Spies & Barnes 1985a,b; Pastor & Broschart 1990).

Sugar maple and hemlock, the two dominants, are shade-tolerant species with advance regeneration in the understorey, while basswood and yellow birch are recruited (from sprouts or seeds, respectively) following disturbances that create large gaps. All four species can persist for several hundred years in the canopy (Frelich & Lorimer 1991). White pine occurs as scattered, infrequent trees and oak (*Quercus rubra*) is very rare. These species require disturbance by wind or fire and relatively high light conditions for establishment (Curtis 1959). Windstorms that remove about 10% of the canopy (usually the larger and older trees) are the most common source of disturbance; catastrophic storms have a return time greater than 1000 years, and fire greater than 2000 years (Frelich & Lorimer 1991; Frelich 1992).

The forest mosaic pattern has been studied on the landscape scale by Pastor and Broschart (1990), who used aerial photographs and a geographical information system to map the stands and to compare them with various physical features of the landscape. Stem maps of permanent plots totalling 27 ha document the distributions of trees and understorey in more detail (Leebens-Mack 1989; Ferrari 1993; Frelich *et al.* 1993) (Fig. 2.2).

The landscape is glacial moraine, deposited about 12 000 years ago (Attig *et al.* 1985). The soils are sandy loams of variable texture with a cover of windblown sand and silt. The soils are fragiorthods (Frelich *et al.* 1993). Within plot A (Fig. 2.2) where soils have been studied in most detail, substrate texture shows no significant difference between the hemlock stand and the hardwood stand. Mor humus occurs where hemlock is abundant, and mull humus under hardwoods (Ferrari 1993), but the profiles throughout the plot are podsolic with a fragipan.

The terrain is hummocky, with numerous kettleholes. Shallow forest 'hollows' or swales, with organic sediment, are very numerous, often occurring at densities of one per hectare. Sediment from the hollows ranges in thickness from 0.3 to 1.5 m, encompassing up to 12 000 years' deposition. Although sediment accumulation was slow, beginning later in some hollows than others, sedimentation appears to have been continuous, and pollen preservation is good. Sediment has been analysed at intervals of 0.2–5.0 cm representing time intervals from 20 to 500 years.

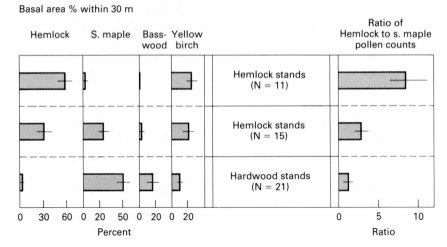

Fig. 2.4. Average tree abundances in forest samples within 30 m of small hollows, and pollen ratios in surface samples of sediment. Samples are grouped by stand type to show the differences in ratios among hemlock stands, mixed stands and sugar maple stands.

The age scale is based on Atomic Mass Spectrometer dates derived from macrofossils of terrestrial plants.

The forest canopy is continuous over many of the smaller hollows, and comparisons of pollen in surface sediments with nearby forest show a strong correlation between pollen abundances and tree abundances within a 30–50 m radius (Calcote 1992). Theory also predicts that the 50% source area should be about 50 m for pollen from the species that are dominant in Sylvania (Sugita 1993, and unpublished).

Pollen assemblages in small hollows can be used to identify the type of forest growing within the nearest hectare. The ratio of hemlock pollen to sugar maple pollen is particularly diagnostic for the stand types now present at Sylvania, with characteristic values (95% confidence intervals) of 6–11 for hemlock stands, and 1–2 for hardwoods. Intermediate values are found in mixed stands (*n* = 15) (Fig. 2.4) (Calcote 1992). The ratio

---

Fig. 2.5. (*Opposite.*) Aerial photograph of a large hardwood patch, and location of a transect of samples of surface humus collected at 50-m intervals across the patch. Pollen percentages of sugar maple and hemlock pollen are plotted against the position of the samples on the transect, and the ratio of hemlock to sugar maple pollen is projected against shading that indicates the expected ratio for hemlock stands (dark grey) and hardwoods (light grey).

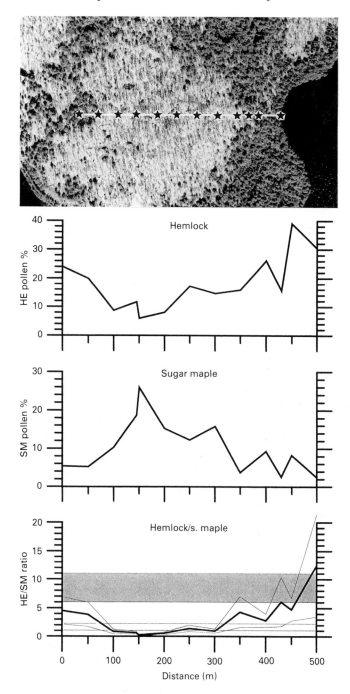

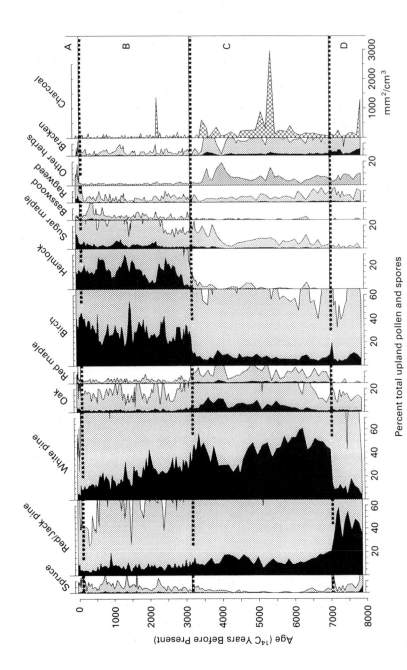

Fig. 2.6. Pollen percentages plotted against sediment age for the past 8000 years in a hollow in plot C, located on a border between a hemlock stand and a hardwood stand. The site receives pollen from both communities. Pollen zones are designated by the letters A–D.

works better than multivariate techniques because it screens out long distance pollen such as pine and oak, as well as local pollen from yellow birch, which grows in all three stand types but at locally varying densities. For example, the ratio of hemlock:sugar maple pollen in humus samples collected along a transect across a large hardwood patch correctly identified mixed, hardwood and hemlock forest (Fig. 2.5). The ratio is diagnostic despite hemlock pollen dispersal into the patch from hemlock forest beyond its borders.

Charcoal fragments have been counted using an IMAGE-1 image-analysis system. Even microscopic charcoal in small hollows can be used to indicate local fire: charcoal in surface humus sampled at various distances from a point source within the forest (a lightning-charred hemlock snag) showed a sharp decline in abundance of charcoal of both microscopic (<250 μm) and macroscopic (>450 μm) charcoal beyond 10 m from the source. From these results we deduce that peak densities of charcoal in the sediment represent local fires that burned within a few tens of metres of the hollow, while background levels reflect the general frequency of fires within the nearest kilometre.

RESULTS

Origin of hemlock–hardwood forest

Small forest hollows at Sylvania all record a transition from mid-Holocene pollen assemblages dominated by pollen of white pine, oak and red maple (*Acer rubrum*) to the hemlock–hardwood pollen assemblages of the late Holocene. The transition occurred around 3000 years BP when pollen from first birch, then hemlock and subsequently sugar maple and basswood increased in abundance. A pollen diagram (Fig. 2.6) from a hollow located between a hemlock and a hardwood stand in plot C shows the transition clearly, marked on the diagram as the boundary between zones C and B. Zone A is the cultural horizon, deposited during the last 100 years.

Forests prior to hemlock invasion

Ten small hollow pollen assemblages that date from 3500 years BP, before hemlock invaded Sylvania, have been compared with a suite of 62 surface samples from hollows in Michigan and Wisconsin, using detrended correspondence analysis (Calcote 1992). The fossil assemblages cluster near surface pollen assemblages from forest hollows in western Wisconsin,

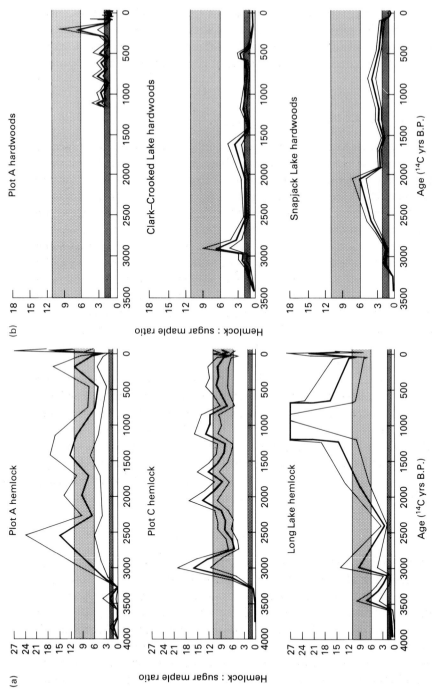

Fig. 2.7. Ratios of hemlock pollen to sugar maple pollen plotted against sediment age in cores from three hollows within hemlock stands (a) and from three hollows within hardwood stands (b). The ranges that are characteristic of hemlock stands (6–11) and of hardwood stands (1–2) are indicated by shading. Radiocarbon age of the fossil material is indicated on the abscissa.

beyond the range limit of hemlock, and from hollows on droughty soils within the geographical range of hemlock. Forests surrounding the analogous surface samples are dominated by white pine, oak, and red maple, implying that these tree species dominated the forests at Sylvania 3500 years ago.

Pollen records from hemlock stands and from hardwood stands

Figure 2.7 shows ratios of hemlock pollen to sugar maple pollen plotted against sediment age in cores from small hollows near the centres of three hemlock stands and from near the centres of three hardwood stands. The locations of these stands are indicated in Fig. 2.1a and b. Within each of the three hemlock stands, the ratio of hemlock to sugar maple pollen rose within a few hundred years to values indicating hemlock dominance. Invasion occurred 3100 years ago at all three stands, although hemlock was apparently present several hundred years earlier in the Long Lake stand, which is about 5 km NW of plots A and C. Plots A and C are 3 km apart. Continuing high values of the hemlock:sugar maple pollen ratio at all three sites indicate that hemlock remained dominant in these forest stands to the present time. A single sample at the Long Lake site shows a lower ratio indicating a mixed stand at about 2400 years BP, but this is the only exception.

At the two hardwood sites with long records, hemlock pollen appeared as background pollen at the same date, 3100 years BP, but the hemlock: sugar maple pollen ratio remained low, indicating that these stands were not invaded by hemlock (Fig. 2.7). During the course of the last 3000 years, pollen from white pine, red maple and oak declined, indicating the decreasing abundances of these species, and their replacement by sugar maple, basswood and yellow birch. Sugar maple has become especially abundant within the last few centuries (Fig. 2.6).

Charcoal concentrations

Charcoal counts from hollows in stands in plots A and C show that the overall abundance of charcoal was much higher in older sediment, deposited before hemlock invasion, than in sediments deposited after hemlock invasion. Charcoal peaks, indicating local fires, suggest a return interval of several hundred years, similar to presettlement forests dominated by pine. Overall abundance of charcoal declined after hemlock invaded 3100 years ago, and there is only one peak, in plot C, between 3000 and 2000 years BP (Fig. 2.6), suggesting a fire return interval similar to modern hardwood–hemlock forest (Frelich 1992). The charcoal records from the Long Lake hemlock stand and the Snapjack Lake hardwoods, 5–6 km away, suggest that in NW Sylvania fires were infrequent throughout the past 6000 years.

DISCUSSION: PALAEORECORD OF THE HEMLOCK–HARDWOOD MOSAIC

Hemlock invasion and mosaic formation

The hemlock stands within the mosaic cannot be older than 3200 years, because the pollen record shows that hemlock first appeared within Sylvania forests 3200 years ago. Birch (presumably yellow birch) also became more abundant at Sylvania at about the same time, and the frequency of fires decreased. Sugar maple and basswood, which were present in low abundance, expanded soon after hemlock invaded the forest (Fig. 2.6).

The forests that were invaded by hemlock were dominated by white pine, red maple and red oak, not by the hardwoods that characterize the modern mosaic. Pollen assemblages at the time of hemlock invasion were very different from surface samples at Sylvania. An example of old-growth white-pine dominated forest that may be similar to the forests hemlock invaded grows on sandy substrate about 20 km east of Sylvania. Basal area in a 0.8-ha plot was 55% white pine, 27% red maple and 2% hemlock. On sites similar to this, hemlock and sugar maple are often found in the understorey, succeeding white pine eventually if there are long intervals between fires. However, white pine and oak forests are fire-prone, with an average fire return time of 150–340 years (Frelich 1992).

Hemlock invasion 3200 years ago was patchy. The history of three large hemlock stands shows that these locations were sites of rapid increases of hemlock abundance starting 3200–3100 years ago; the stands have been dominated by hemlock ever since. Pollen records from three hardwood stands show that these sites were never invaded by hemlock. At locations that were not invaded by hemlock, sugar maple and basswood gradually became more abundant starting about 3200 years BP, displacing white pine, red maple and oak.

The palaeoecological record resolves the dilemma posed by observations in the extant forest, which suggest that hemlock would have difficulty invading hardwood forest, while sugar maple would have difficulty invading hemlock forest. The mosaic came into existence starting about 3200 years BP. With reduced fire frequency and changes of climate that apparently favoured hemlocks and hardwoods, the resident white pine, red maple and oak were outcompeted by hemlock and yellow birch at some locations, and by sugar maple, basswood and yellow birch at others. Neither hemlock nor hardwoods can be regarded as a background with patches superimposed upon it (Pastor & Broschart 1990). Instead the mosaic came into existence as hemlock invaded certain sites, creating hemlock stands, while the intervening sites differentiated into hardwood forests. This does not explain the mixed hemlock–hardwood stands, which make up one fourth of the area (Pastor & Broschart 1990); the history of these stands has been more dynamic and will constitute the subject of a subsequent paper.

Regional changes of climate

Palaeoecological records from outside Sylvania suggest that climatic conditions were gradually changing throughout the period when the mosaic was differentiating. A reduction in seasonality due to Milankovitch forcing (COHMAP 1988) might have reduced summer temperatures and evapotranspiration, effectively increasing soil moisture in late summer, a change that would have favoured hemlock and sugar maple. The expansion of population sizes and geographical ranges of trees that grow in mesic habitats (Webb *et al.* 1983), a rise in lake levels (Winkler *et al.* 1986) and expansion of bogs (Futyma & Miller 1986; Miller & Futyma 1987), all suggest more mesic conditions in northern Michigan and Wisconsin during the late Holocene (Bartlein *et al.* 1984).

It seems likely that the climatic changes were continuous, changing to become yet more favourable for hemlock and sugar maple throughout the last 4000 years. Sugar maple appears to have increased especially during the last several hundred years, perhaps reflecting particularly favourable climate during this period. Hemlock invaded Sylvania forests about 3200–3100 years BP in the course of extending its range limit westward and southwestward from a limit 50 km to the east where hemlock had became established 5000 years BP (Fig. 2.3) (Davis *et al.* 1986; Davis 1987). The westward extension of hemlock range through Michigan and northern Wisconsin was apparently in equilibrium with climate, as it occurred unusually slowly, at an average rate of about 5 km/century. The present western range limit is about 120 km west and southwest of Sylvania.

Disturbance regime

Charcoal concentrations in the sediment suggest that in the vicinity of plots A and C, fires were more frequent during the millennia prior to hemlock invasion than afterward (Fig. 2.6). The decrease in the frequency of fires fits with the climatic trend described above. However, it also seems plausible that the change in litter type and the more humid micro-climate under the densely shaded hemlock stands acted to reduce the probability of fire (Davis *et al.* 1991). As both hemlock and sugar maple are fire-sensitive species, especially as seedlings and young saplings, the reduction in fire frequency would have created a positive feedback, increasing their competitive advantage relative to white pine and oak. Reduction in disturbance frequency would also have favoured shade-tolerant species over pine, oak and red maple.

Additional hypotheses on mosaic formation and maintenance

The fossil record provided by small hollows allows us to test additional hypotheses. Work is now in progress to inspect large numbers of pollen assemblages from the pre-hemlock period, to test the idea that a pre-existing vegetational mosaic, perhaps resulting from patchy distributions of fires (Frissel 1973; Heinselman 1973), governed the locations of stands that were invaded by hemlock and eventually became hemlock-dominated stands. Ecological theory suggests that some communities are more invasible than others (Elton 1958; Orians 1986). Today hemlock seedlings are much more common in stands dominated by red maple and white pine than in stands dominated by sugar maple. Sugar maple might have become abundant in infrequently burned forests, while white pine grew at sites that had been burned in recent centuries. Biological differences on this scale could have governed the locations of sites initially invaded by hemlock.

Physical differences in the environment may also have been important. We have not been able to demonstrate physical differences in slope, aspect, or soil texture that correlate with the presence of hemlocks or hardwoods, although at the landscape scale hardwood stands are often on hilltops and hemlock stands are often adjacent to bogs or lake shores (Pastor & Broschart 1990; Frelich *et al.* 1993). However, there are patches of silty soil in Sylvania, and in these locations hardwoods occur on the fine-grained soils while hemlocks grow on adjacent coarse substrates. It seems likely that under the limiting climatic conditions that existed 3000 years ago, even subtle differences in substrate texture or depth to water table could have been important, governing the locations that eventually became hemlock stands. Certainly the palaeoecological studies cited in the introduction suggest that changing climate can make substrate dif-ferences that are unimportant under one climatic regime critical under another. Topography can also be an important influence on fire frequency (Grimm 1984).

Some hemlock and hardwood stands may have had very different origins. Our sample size is small (six out of many hundreds of patches that could be studied). Our results do not rule out the possibility that some hemlock or hardwood stands originated following disturbances, or differentiated from mixed stands. Preliminary results from sites 2.5 km to the south suggest this may be the case. Histories could have been different on different substrates, or they may have been different adjacent to the fire-prone outwash plain along the southern boundary of the Sylvania forest reserve. Many more forest hollows could be investigated to test these possibilities.

THEORETICAL IMPLICATIONS, AND IMPORTANCE FOR CONSERVATION BIOLOGY

Positive feedbacks involving the two dominant species, hemlock and sugar maple, produce a forest mosaic of two alternative communities. In contrast to a disturbance mosaic, both communities are dominated by shade-tolerant, late-successional species. Wilson and Agnew (1992) review many examples of positive feedback 'switches', suggesting that positive feedbacks are more common than generally recognized. They point out that unlike successional changes that facilitate replacement of residents, positive feedbacks tend to perpetuate the resident community.

In most cases of alternate communities with positive feedbacks, one can only speculate about the forces that initiated the pattern in the first place (examples in Wilson & Agnew, 1992). But at Sylvania, where palaeo-records exist at a fine spatial scale, information can be obtained from many sites within the mosaic, providing an unusual opportunity to observe the formation of the mosaic before the positive feedbacks, which depend on dominance by sugar maple and hemlock, were fully operative. For example, the fossil record suggests that the extreme dominance of sugar maple in some hardwood stands is a feature of the past several hundred years, when some factor, possibly climate, changed to give sugar maple a competitive advantage relative to other hardwood species, and apparently relative to hemlock growing in mixed stands. Hemlock recruitment might have had a different pattern prior to that time. However, as far as we can tell, the overall pattern of hemlock and hardwood patches has been similar to today throughout the past 3000 years, at least in the areas where our studies have been conducted. At the time of hemlock invasion 3100 years BP, the forest community may have been more uniform, or at least was patchy on a larger spatial scale, and it was apparently readily invasible by hemlock.

From the standpoint of conservation biology, understanding this system is important because human disturbance destroys the mosaic, changing size and species composition of patches, and lowering their fractal dimension (Mladenoff *et al.* 1993). Clearcutting 50 years ago, which extended right up to the boundary of the reserve, converted both hemlock-dominated stands and hardwood-dominated stands to hardwoods (Hix & Barnes 1984). The disappearance of the hemlock stands and the consequent loss of forest diversity show clearly in Fig. 2.1b, an aerial photograph of the NW corner of the Sylvania tract, showing the forest change at the boundary. The failure of hemlock to regenerate following logging has been attributed to browsing by deer, which have increased in density in recent decades (Anderson & Loucks 1979; Frelich & Lorimer 1985). However, several authors have observed that reproduction is strongly dependent on years of higher than average rainfall; periods of dry climate during the present century may have disadvantaged hemlock (Graham 1941; Hough & Forbes 1943; Mladenoff & Stearns 1993). We suggest that a third factor could be important: the rapid regrowth of sugar maple, and its dominance in the canopy of second growth stands, has produced a forest with a positive feedback that favours continued maple dominance and discourages recruitment of hemlock. It is an interesting question whether in the past the forest system responded to catastrophic disturbance in the same manner as it has responded in the present century to logging. The pollen record from a hollow on the border between the hemlock and hardwood stands in plot C contains evidence for past disturbance by fire and wind; the pollen record shows that the hemlock border retreated several tens of metres three times during the past 3000 years, as hemlocks were replaced by hardwoods (Davis *et al.* 1991), but the record shows that the hemlock stand boundary advanced again gradually to regain the lost territory over the next 500–800 years.

Manipulations might possibly restore hemlock stands on a shorter time scale – decades or even a century. Manipulations would have to be quite drastic to simulate the conditions recorded in the fossil record; they would involve removal or reduction of sugar maple and alteration of seedbed conditions to make them favourable for hemlock. The palaeorecord from Sylvania shows that old-growth forests represent ecosystems that have undergone a long history. Harvesting timber from these forests may mean destroying a system that biological succession cannot reproduce in several centuries, under the climatic regime of the present.

ACKNOWLEDGEMENTS

This work was supported by National Science Foundation grants BSR 8615196, BSR 8916503 and DEB 9221375, and by the Mellon Foundation. We gratefully acknowledge the co-operation of the United States Forest Service. Christine Douglas, Julie Blackett and Anthony Macdonald counted pollen and charcoal from some of the hollows, and Thomas Daugherty counted the charcoal in surface humus samples. Seedling counts from some of the sampling points shown in Fig. 2.2 are by James Leebens-Mack (1989).

REFERENCES

Anderson, R.C. & Loucks, O.L. (1979). White-tail deer (*Odocoileus virginianus*) influence on structure and composition of *Tsuga canadensis* forests. *Journal of Applied Ecology*, **16**, 855–861.

Attig, J.W., Clayton, L. & Mickelson, D.M. (1985). Correlation of late Wisconsin glacial phases in the western Great Lakes area. *Geological Society of America Bulletin*, **96**, 1585–1593.

Bartlein, P.J., Webb III, T. & Fleri, E. (1984). Holocene climatic change in the northern midwest: Pollen-derived estimates. *Quaternary Research*, **22**, 361–374.

Bormann, F.H. & Likens, G.E. (1979). *Pattern and Process in a Forested Ecosystem*. Springer, New York.

Bormann, F.H., Siccama, T.G., Likens, G.E. & Whittaker, R.H. (1970). The Hubbard Brook ecosystem study: Composition and dynamics of the tree stratum. *Ecological Monographs*, **40(4)**, 373–388.

Brubaker, L.B. (1975). Postglacial forest patterns associated with till and outwash in northcentral Upper Michigan. *Quaternary Research*, **5**, 499–527.

Calcote, R. (1992). Differentiating forest communities using pollen from forest hollows. *Bulletin of the Ecological Society of America*, **73(2)**, 132.

Clark, J.S. (1990). Fire and climate change during the last 750 yr in northwestern Minnesota. *Ecological Monographs*, **60(2)**, 135–159.

COHMAP Members (1988). Climatic changes of the last 18000 years: Observations and model simulations. *Science*, **241**, 1043–1052.

Curtis, J.T. (1959). *The Vegetation of Wisconsin*. University of Wisconsin Press, Madison, WI.

Davis, G. & Hart, A. (1961). *Effects of Seedbed Preparation on Natural Reproduction of Spruce and Hemlock under Dense Shade*. U.S. Forest Service. N.E. For. Exp. Station Pap. 160.

Davis, M.B. (1987). Invasions of forest communities during the Holocene: Beech and hemlock in the Great Lakes region. *Colonization, Succession and Stability* (Ed. by A.J. Gray, M.J. Crawley & P.J. Edwards), pp. 373–393. Blackwell Scientific Publications, Oxford.

Davis, M.B. (1990). Climatic change and the survival of forest species. *The Earth in Transition: Patterns and Processes of Biotic Impoverishment* (Ed. by G.M. Woodwell), pp. 99–110. Cambridge University Press, Cambridge.

Davis, M.B., Woods, K.D., Webb, S.L. & Futyma, R.P. (1986). Dispersal versus climate: Expansion of *Fagus* and *Tsuga* into the Upper Great Lakes region. *Vegetatio*, **67**, 93–103.

Davis, M.B., Sugita, S., Calcote, R.R. & Frelich, L. (1991). Effects of invasion by *Tsuga canadensis* on a North American forest ecosystem. *Response of Forest Ecosystems to Environmental Changes* (Ed. by A. Teller, P. Mathy & J.N.R. Jeffers), pp. 34–44. Elsevier Applied Science, London.

Elton, C.S. (1958). *The Ecology of Invasions by Animals and Plants.* Chapman & Hall, New York.

Ferrari, J.B. (1993). *Spatial Patterns of Litterfall, Nitrogen Cycling and Understory Vegetation in a Hemlock–Hardwood Forest.* PhD Thesis, University of Minnesota, USA.

Frelich, L.E. (1992). The relationship of natural disturbances to white pine stand development. *White Pine Symposium: History, Ecology, Policy and Management, Duluth, MN, Proceedings*, NR-BU-6044. (Ed. by R.A. Stine & M.J. Baughman), pp. 27–37. Minnesota Extension Service, St Paul, Minnesota, USA.

Frelich, L.E. & Lorimer, C.G. (1985). Current and predicted long-term effects of deer browsing in hemlock forests in Michigan, USA. *Biological Conservation*, **34**, 99–120.

Frelich, L.E. & Lorimer, C.G. (1991). Natural disturbance regimes in hemlock–hardwood forests of the Upper Great Lakes region. *Ecological Monographs*, **61(2)**, 145–164.

Frelich, L.E., Calcote, R.R., Davis, M.B. & Pastor, J. (1993). Patch formation and maintenance in an old-growth hemlock–hardwood forest. *Ecology*, **74(2)**, 513–527.

Frissel, S.S., Jr. (1973). The importance of fire as a natural ecological factor in Itasca State Park, Minnesota. *Quaternary Research*, **3**, 397–407.

Futyma, R.G. & Miller, N.G. (1986). Stratigraphy and genesis of the Lake Sixteen peatland, northern Michigan. *Canadian Journal of Botany*, **64(12)**, 3008–3019.

Goder, H.A. (1955). *A phytosociological study of* Tsuga canadensis *at the termination of its range in Wisconsin.* PhD Thesis, University of Wisconsin.

Graham, S.A. (1941). The question of hemlock establishment. *Journal of Forestry*, **39**, 567–569.

Graumlich, L.J. & Davis, M.B. (1993). Holocene variation in spatial scales of vegetation pattern in the Upper Great Lakes. *Ecology*, **74(3)**, 826–839.

Grimm, E.C. (1984). Fire and other factors controlling the Big Woods vegetation of Minnesota in the mid-nineteenth century. *Ecological Monographs*, **54**, 291–311.

Heinselman, M.L. (1973). Fire in the virgin forests of the Boundary Waters Canoe Area, Minnesota. *Quaternary Research*, **3(3)**, 329–382.

Hix, D.M. & Barnes, B.V. (1984). Effects of clear-cutting on the vegetation and soil of an eastern hemlock-dominated ecosystem, western Upper Michigan. *Canadian Journal of Forestry Research*, **14**, 914–923.

Hough, A.F. & Forbes, R.D. (1943). The ecology and silvics of forests in the high plateaux of Pennsylvania. *Ecological Monographs*, **13**, 298–320.

Leebens-Mack, J. (1989). *Spatial patterns in seed rain density and seedling recruitment of northern mixed hardwood species in Sylvania Wilderness Area, Michigan.* MS Thesis, University of Minnesota.

Miller, N.G. & Futyma, R.P. (1987). Palaeohydrological implications of Holocene peatland development in northern Michigan. *Quaternary Research*, **27**, 297–311.

Mladenoff, D.J. & Stearns, F. (1993). Eastern hemlock regeneration and deer browsing in the northern Great Lakes region: A re-examination and model simulation. *Conservation Biology* (in press).

Mladenoff, D.J., White, M.A., Pastor, J. & Crow, T.R. (1993). Comparing spatial pattern in unaltered old-growth and disturbed forest landscapes. *Ecological Applications*, **3(2)**, 294–306.

Orians, G.H. (1986). Site characteristics favoring invasions. *Ecology of Biological Invasions in North America and Hawaii* (Ed. by H.A. Mooney & J.A. Drake), pp. 133–148. Springer, New York.

Pastor, J. & Broschart, M. (1990). The spatial pattern of a northern conifer–hardwood landscape. *Landscape Ecology*, **4(1)**, 55–68.

Pickett, S.T.A. & White, P.S. (1985). *The Ecology of Natural Disturbance and Patch Dynamics.* Academic Press, Orlando, Florida.

Rogers, R.S. (1978). Forests dominated by hemlock (*Tsuga canadensis*): Distribution as related to site and postsettlement history. *Canadian Journal of Botany*, **56(1)**, 843–854.

Spear, R.W., Davis, M.B. & Shane, L.C.K. (1994). Late Quaternary history of low- and mid-elevation vegetation in the White Mountains of New Hampshire. *Ecological Monographs* (in press).

Spies, T.A. & Barnes, B. (1985a). A multifactor ecological classification of the northern hardwood and conifer ecosystems of Sylvania recreation area, upper peninsula, Michigan. *Canadian Journal of Forestry Research*, **15**, 949–960.

Spies, T.A. & Barnes, B. (1985b). Ecological species groups of upland northern hardwood–hemlock forest ecosystems of the Sylvania recreation area, upper peninsula, Michigan. *Canadian Journal of Forestry Research*, **15**, 961–972.

Sprugel, D.G. (1984). Density, biomass, productivity, and nutrient cycling changes during stand development in wave-generated balsam fir forests. *Ecological Monographs*, **54**, 165–186.

Stearns, F. (1990). Forest history and management in the northern midwest. *Management of Dynamic Ecosystems* (Ed. by J.M. Sweeney), pp. 107–122. North Cent. Sect., The Wildl. Soc., West Lafayette, IN.

Sugita, S. (1993). A model of pollen source area for an entire lake surface. *Quaternary Research*, **39**, 239–244.

Swetnam, T.W. (1993). Five history and climate change in giant sequoia graves. *Science*, **262(5135)**, 885–889.

Webb III, T., Cushing, E.J. & Wright, H.E. (1983). Holocene changes in the vegetation of the midwest. *Late-Quaternary Environments of the United States. Vol. 2, The Holocene* (Ed. by H.E. Wright, Jr), pp. 142–165. University of Minnesota Press.

West, D.C., Shugart, H.H. & Botkin, D.B. (1981). *Forest Succession.* Springer, New York.

Whittaker, R.H. (1960). Vegetation of the Siskiyon Mountains, Oregon and California. *Ecological Monographs*, **30**, 279–338.

Wilson, J.B. & Agnew, A.D.Q. (1992). Positive feedback switches in plant communities. *Advances in Ecological Research*, **23**, 263–336.

Winkler, M.G., Swain, A.M. & Kutzbach, J.E. (1986). Middle Holocene dry period in the northern midwestern United States: Lake levels and pollen stratigraphy. *Quaternary Research*, **25**, 235–250.

3. ANIMAL DISTRIBUTIONS: PATTERNS AND PROCESSES

JOHN H. LAWTON*, SEAN NEE†,
ANDREW J. LETCHER† AND PAUL H. HARVEY*†

*NERC Centre for Population Biology, Imperial College,
Silwood Park, Ascot, Berks SL5 7PY, UK and
†AFRC Unit of Ecology and Behaviour, Department of Zoology,
South Parks Road, Oxford OX1 3PS, UK

SUMMARY

Data of improved quality are revealing the range distributions of all the species belonging to particular taxonomic groups. Those range distributions show patterns which require interpretation in terms of both ecological and evolutionary processes. Using distributional data on birds and mammals from various parts of the world, we use a combination of new and standard techniques to unravel contemporary and historical causes for range limitation. We also address, at both theoretical and empirical levels, the relationships between local abundance and regional distribution, and explore the implications of these relationships for conservation.

INTRODUCTION

Recent advances in a number of areas of ecological investigation have helped further our understanding of why many organisms, particularly animals, have the geographical ranges that we currently observe. That understanding promises to help inform decisions about species conservation and management. It is not our intention in this chapter to review everything that is known about the determinants of range size. Indeed, such a task would be beyond the scope of any single volume. Rather, we tackle three general problems with the intention of providing some up-to-date information on current research programmes which, to us, seem to hold promise for future progress. First, we review recent literature which illustrates and helps explain a recurrent relationship between geographic range size and local population densities. Second, we report new results on possible reasons for the apparently widespread correlation between range size and latitude, often known as Rapoport's rule, and on the purported relationship between range size and body size. Third, we cast an eye to the future with a discussion of how models developed in the context of epidemiology can be imported almost wholesale into ecology as a basis for the study of metapopulation dynamics. Examples already demonstrate that metapopulation analysis provides a useful framework for informing conservation biologists of the consequences of particular land management plans. We argue that a reasonably informed analysis may not always require the particular details of population demography and individual dispersal patterns that seem intuitively to be necessary for such an exercise.

RANGE AND ABUNDANCE

The general problem

The abundance of species and the size of the area over which they are recorded (i.e. the size of their geographic range) are not independent (Brown 1984). Within particular taxa, a number of studies show that species occurring over large geographic areas tend to have greater local abundances at sites where they occur than do geographically more restricted species. A summary is provided by Gaston and Lawton (1990). Examples include studies on plants, birds, mammals, fish and a variety of invertebrates from molluscs and mites to zooplankton and insects. Knowledge of the phenomenon has a long pedigree; in 1922 the botanist J.C. Willis noted that localized species were often rare within their limited geographic ranges (Ricklefs 1989).

There is usually a considerable amount of unexplained variation in plots showing the relationship between range size and local abundance, so that an individual species can be widespread but rare everywhere, or locally common despite having only a small total geographic range. Nevertheless, the average trends are important for conservation because they warn that generally, and other things being equal, geographically restricted taxa tend also not to have large local populations, making them doubly vulnerable.

Despite its theoretical and practical interest, the positive correlation between size of range and local abundance is not well understood. There are some inverse correlations (to which we return below), and a minority of studies report no correlation (Arita *et al.* 1990; Gaston & Lawton 1990), although given the nature of much of the data, the occasional failure to find a correlation comes as no great surprise and may not be very informative. There are also technical (but non-trivial) problems in defining exactly what is meant by size of geographic range, so that different usages by different authors may both obscure real patterns and generate spurious ones. (For a thoughtful discussion on the problems of defining geographic ranges see Gaston 1991. We will use some of the ideas discussed by Gaston later.)

Particular concern has been expressed that positive correlations between range and local abundance are statistical artefacts, because species with low average population densities are less likely to be encountered in broad geographic surveys (McArdle 1990 is a good discussion). This could generate a spurious positive correlation between local abundance and the number of sites where the species has been recorded (Brown 1984; Gaston & Lawton 1990 and references therein); this problem appears to have been discovered independently and recently by Wright (1991). Sampling problems undoubtedly contribute to the observed correlation, but for many well-studied taxa such as birds, where both ranges and abundances are very accurately recorded, it seems extremely unlikely that the relationship is entirely spurious.

Explanations for the pattern

Brown's hypothesis

Theoretically, there are a number of ways in which a positive correlation between size of geographic range and local abundance might be generated. The simplest is that species able to exploit a wide range of resources (species with 'broad niches') become both widespread and locally abundant (Brown 1984). This hypothesis awaits critical evaluation, but such data as there are sometimes, but not always, support Brown's arguments.

Birds breeding on several types of Finnish mires (habitat generalists) have larger geographic ranges than habitat specialists although, contrary to Brown's hypothesis, they are not more abundant locally (Kuoki & Hyrinen 1991). Species of North American mammals with large ranges also occur in more habitats than species with small ranges (Pagel *et al.* 1991; unfortunately this study does not report local abundances). However, there is more to having a broad ecological niche than being a habitat generalist, so neither of these studies can be regarded as definitive.

More worrying for Brown's hypothesis are Pianka's (1986) very detailed studies of niche breadths (based on habitat and food types) and local abundances of desert lizards on three continents. Pianka finds that the most abundant lizards have moderate niche breadths; rare species include the entire spectrum from extreme specialists to broad generalists. These data are hard to reconcile with Brown's hypothesis, but lack information on geographic ranges.

Data of a very different kind on the fate of species' introductions do appear to be consistent with Brown's hypothesis, although they were not designed specifically to test it. Herbivorous insects released as biological control agents against alien weeds in a new country are more likely to establish the more widespread and abundant they are in their native environment (Crawley 1987). Similarly, introduced birds on the Hawaiian Islands (these islands have received more vertebrate introductions than anywhere else on Earth!) are also more likely to establish successfully if they have large geographic ranges in their native environment (Moulton & Pimm 1986; Pimm 1991). In other words, large ranges, high abundances and 'invasion ability' are linked characteristics of species, consistent with Brown's hypothesis.

An alternative test is provided by Gaston and Lawton (1990), who point out that if local abundances are measured in unusual habitats (i.e. the 'reference habitat' differs markedly from the spectrum of habitats in the geographic region of interest) then Brown's hypothesis predicts a negative relationship between size of range, and local abundance in the unusual reference habitat. For two bird assemblages this is what we see. There are negative relationships between local abundances on Handa island (an oceanic island on the extreme northwestern fringe of Europe) and the extent of species' geographic ranges, both in Britain and in Europe as a whole (where the majority of habitats differ markedly from those on Handa). Similar results were obtained by Ford (1990) who found a weak negative relationship between abundances of birds measured in woodland study plots on the eastern edge of Australia and continent-wide geographic distributions of species. The dominant habitats in Australia are deserts, not woodland. Further discussion and elaboration of these ideas are provided by Novotny (1991).

In summary, a variety of independent tests are broadly consistent with Brown's hypothesis. What is lacking is supporting evidence based on direct measurements of niche breadth that take into account not only habitat use but environmental tolerance, food types, feeding sites and so on. But even armed with such data, niche breadths are notoriously difficult to measure and to interpret in an objective way (Colwell & Futuyma 1971; Krebs 1989); critical, direct tests of Brown's hypothesis are therefore likely to remain elusive for some time.

Metapopulation models

Several metapopulation dynamic models (see p. 52 for a definition) generate positive correlations between geographic range measured as the number (or proportion) of patches occupied (area of occupancy in the sense of Gaston 1991) and population density within patches (Hanski 1991; Nee *et al.* 1991; Gyllenberg & Hanski 1992). A rather different, statistical model based on the consequences of patch selection by individual organisms is provided by Maurer (1990) but yields similar results. All these models seem most appropriate as descriptions of population dynamics and distributions on local and intermediate scales (Hanski 1991 provides further discussions); they may be less satisfactory as explanations for processes operating on continental scales.

It would be valuable to have independent field tests of this family of models, although the work involved will be considerable. Their implications for conservation are profound, and depressing. Very generally, if local abundance and proportion of sites occupied are positively correlated for metapopulation dynamic reasons, then reducing the number of sites at which a species is found (for example by habitat destruction) may reduce population densities at remaining sites; and reducing population densities within sites (by hunting or habitat degradation) may lead inevitably to a reduction in the number of sites occupied, even if those sites are protected. Other models and empirical data yield similar messages (e.g. Wilcove *et al.* 1986). We return to these models and problems in the final section of this chapter.

Textures of abundances within geographic ranges

One obvious reason why the correlation between size of geographic range and local abundance must inevitably be weak is that species are not evenly distributed throughout their range (Brown 1984; Carter 1988; Wiens 1989). All bird watchers know to their cost that maps in field-guides merely define each species' extent of occurrence (*sensu* Gaston 1991), and that species differ markedly in abundance within, and may be entirely absent from, large areas delimited by field-guide range maps, either because suitable habitats are lacking, or because not all suitable habitats are currently occupied. For instance, water voles *Arvicola terrestris*, small aquatic mammals that form colonies close to water, were absent from 45% of sites surveyed on river banks in the North York Moors National Park in England because the habitat was unsuitable; and of the sites that were suitable, about 30% nevertheless lacked voles because they were too isolated, and/or because of predation by mink *Mustela vison* (Lawton & Woodroffe 1991).

Very crudely, densities tend to be greatest near the centre of the range, and decline towards the boundaries (Hengeveld & Haeck 1981,

1989; for some exceptions among North American birds, see Brown (1984) argues that two phenomena are involved. ~cies tend to inhabit a progressively smaller proportion of local ~atches towards the edge of their range (Gaston's area of occupancy declines); and, as predicted from metapopulation models (see previous section), average population densities within occupied patches also decline. It would be valuable to have more examples of these patterns from a variety of taxa, along the lines pioneered by Hengeveld and Haeck (1981, 1982), Rapoport (1982) and Carter (1988). In practice, textures of distribution and abundance are often more complex than a gradual decline from the centre to the edge of the range, with multimodal patterns of abundance being common and perhaps even the norm (Taylor & Taylor 1979; Brown 1984; Root 1988; Wiens 1989, provide examples).

Because average abundances vary across species' ranges, it follows that one or more of the key demographic rates (birth, death, immigration and emigration) also change across the range; that is, a species' population dynamics must be very different near the centre compared with the edge of its range (Richards 1961; Huffaker & Messanger 1964). Thorough examples for two species of insects are provided by Randall (1982) and by Rogers and Randolph (1986) and for two species of kangaroos by Caughley *et al.* (1988). At some point close to the range boundary, intrin-sic rates of population increase, r, must on average be zero (Andrewartha & Birch 1954). Beyond the point where r = zero, 'sink populations' (Pulliam 1988) with negative average r may be sustained by immigration from 'source populations' deeper within the geographic range, where overall population performance is better.

These insights from local demography carry several important messages for conservation. First, attempts to restore populations by reintroductions into historical, but currently unoccupied parts of species ranges are more likely to succeed into the core of the historic range (where r is positive) than on the periphery or beyond it (where r is negative). This is exactly what has been observed in translocations (reintroductions) of birds and mammals (76% of 133 translocations into the core succeeded, compared with failures in 48% of 54 translocations to the periphery or beyond; Griffith *et al.* 1989). Second, wholesale persecution or habitat destruction may leave isolated populations of high conservation importance in marginal habitats (the 'where we find them now is not where they want to be' phenomenon). Relict populations of takahe *Porphyrio mantelli* in the Murcheson Mountains of New Zealand and of red kites *Milvus milvus* in south-central Wales are good examples. By definition, such populations have very low rates of increase (well below that which the species can achieve in better habitat), making their conservation even more difficult. *In extremis*, some populations may persist only because of immigration from the core; destruction of the source would make the long-term conservation of such sink populations virtually impossible (Harrison *et al.* 1988). Third, widespread changes in the environment (e.g. pollution, hunting, or for migrants, problems on the wintering grounds) that lead to a general decline in population abundance via an increasing death rate or falling birth rate, should result in overall range contraction even in the complete absence of habitat destruction. If the original range had a single, well-defined centre, we expect range contraction towards that core; if there were originally multiple modes, we expect range contraction and fragmentation into the former hot spots.

Interestingly, and for reasons that are not entirely clear to us, there have been rather few attempts to document and link patterns of population decline with changes in species distributions. Hengeveld (1989) suggests that declines in European populations of *Abies* spp. were accompanied by range fragmentation. But as Wilcove and Terborgh (1984) show for population declines of some North American birds, this is by no means always the case. The highly endangered Kirtland's warbler *Dendroica kirtlandii*, for example, withdrew to the historical centre of its range, leaving peripheral areas virtually empty, as populations collapsed by 60% between 1961 and 1971.

As we have already pointed out, fragmentation or contraction to a single core area presumably depends at least in part upon textures of population abundance within the original range (uni- versus multi-modal). Patterns of decline must also be influenced by the pace and extent of habitat destruction in a species' former range. It is difficult to know which type of range decline poses the greater threat for conservation efforts. Fragmentation and isolation will exacerbate population declines in remaining areas; contraction towards a single core area may avoid these problems but puts all the conservation eggs into one geographical basket.

Ranges and abundances in evolutionary time

Population density and size of geographic range are usually thought of as attributes determined by processes operating in ecological time; that is, we expect them to be dynamic and variable over time periods of, say, 10–100 years. But there are also poorly understood, intriguing hints of effects operating in evolutionary time. They suggest that both range and abundance are evolved, persistent, species' characteristics. The idea is implicit in Brown's hypothesis (see above), which links local abundance and size of range to a complex, elusive, albeit obviously evolved characteristic of species, namely niche breadth.

One group of studies centres on the 'taxon cycle' for birds on West Indian islands (Ricklefs & Cox 1978; Ricklefs 1989). Among passerines, putatively older taxa occur on fewer islands, have more restricted habitat distributions and tend to have reduced population densities. (It would be intriguing to revisit these analyses using independently derived, molecular criteria for species ages.)

A second group of studies also involves birds, and finds quite unexpected phylogenetic correlates of patterns in species' abundances which, if real, may imply constraints on the size of bird populations that extend back, literally, over millions of years. Briefly, summarizing a complex literature, we expect a rough, inverse correlation between body size and population density in animals (mice are commoner than elephants). But if we examine plots relating body size to abundance for species within individual tribes of birds, there are significantly more positive relationships than expected (large-bodied species are commoner than small-bodied species) in taxonomically more ancient tribes (measured in various ways, but based on molecular phylogenies) (Cotgreave & Harvey 1991; Nee 1991; T.M. Blackburn *et al.*, unpublished). Why current abundances should be correlated with phylogeny in this way is a mystery, although possible explanations have been suggested in the cited papers.

Finally, ranges also show associations with phylogeny. The Gulf and Atlantic coastal plain of North America contains one of the most diverse and best preserved molluscan faunas of the late Cretaceous. Jablonski (1987) demonstrates that individual species of bivalves and gastropods from these fossil assemblages achieved characteristic range sizes relatively early in their history; once evolved, species' range sizes changed relatively little. Moreover, pairs of closely related species have statistically similar range sizes, which in Jablonski's own words are 'in effect heritable at the species level'. In a similarly pioneering analysis, Ricklefs and Latham (1992) show that disjunct taxa within extant genera of herbaceous perennial plants relict in temperate eastern Asia and eastern North America have significantly correlated range sizes. These results imply stasis in genus-level attributes determining distributions that have been stable for at least 10 my. Woody taxa, in contrast, do not show the pattern. Possible reasons for these differences are discussed by Ricklefs and Latham.

Taken together, these remarkable studies imply that evolutionary history leaves a signal on distribution and abundance detectable through the noise of contemporary ecological events. Quite what this might mean for conservation is unclear, but we may draw two tentative conclusions, prefaced by the remark that humankind is currently changing the distributions and abundances of many (possibly most) organisms on Earth at a pace that has no antecedents in evolutionary time.

Our first conclusion is that although general effects of phylogeny on size of range and abundance are by no means proven, if they turn out to be at least reasonably common (and there is not much time to find out!), then it implies that some species – those phylogenetically predisposed to rarity – will be more than usually difficult to conserve. Second, current assaults on the distributions and abundances of organisms mean that forces acting in ecological time come to dominate large-scale ecology. Stripped of technical jargon this means that the Earth increasingly resembles a large botanical garden-come-zoological park.

RAPOPORT'S RULE

The rule and evidence for it

The latitudinal gradient in species richness, first noted by Wallace (1878), has received a great deal of attention but remains poorly understood (Simpson 1964; Schall & Pianka 1978; Brown & Maurer 1989; Stevens 1989); at least 14 different hypotheses have been proposed (Stevens 1989; Pagel *et al.* 1991). Less well known, but receiving increasingly more attention, is the latitudinal gradient in geographic range size where species at the polar end of a continent have larger ranges than those at the equatorial end. This has become known as Rapoport's rule (Stevens 1989), and has now been demonstrated for a wide variety of different taxa including mammals from the northern hemisphere, birds, fishes, reptiles, amphibians, crustaceans, molluscs, and trees (Rapoport 1982; Stevens 1989; Pagel *et al.* 1991; France 1992; Letcher & Harvey, in press), although Australian mammals and marine teleosts (Rohde *et al.* 1993) seem to provide exceptions (Smith *et al.*, in press). We report below tests of various explanations for the existence of Rapoport's rule in mammals from the northern hemisphere.

The climatic variability hypothesis

One possible explanation for Rapoport's rule is the climatic variability hypothesis (Allee *et al.* 1949; Dobzhansky 1950; Stevens 1989). The argument runs that an individual animal at the polar end of a continent experiences a much wider range of climatic conditions (both seasonally and diurnally) than one at the equatorial end. Accordingly, unlike tropical species, the more temperate species cannot specialize on a narrow set of climatic conditions. (This argument is clearly very similar to Brown's hypothesis, that species with broad niches have large ranges.) This idea has been tested for data on Palaearctic mammals (Letcher & Harvey, in press). In a stepwise multiple regression of climatic and physiographic variables on range size, annual temperature range was the best predictor of range size and when daily temperature range was also included in the regression, latitude explained no additional variation. although latitude was a better predictor of range size than either annual or daily temperature range alone, the two climatic variables together rendered latitude redundant; species with larger ranges occupied areas with greater annual but lower daily temperature ranges (annual temperature range is the much larger contributor, accounting for 28% of the variation in range size, whereas daily temperature range accounts for 6%; combined they account for 31%, which is the same as the correlation with latitude). This is a satisfying result because meridian lines do not cause ecological patterns, so some other factor, for which latitude is a surrogate measure, must be determining range size.

However, habitats may also increase in size with increasing latitude. Could it be that species are habitat limited and that Rapoport's rule follows from the fact that habitats tend to get larger as we move away from the equator? In an elegant statistical analysis, Pagel *et al.* (1991) demonstrated that this is not a sufficient explanation: more northerly species actually occupied more habitats than expected, that is, they were more generalist. Nevertheless, the climatic analyses reported in the previous paragraph might suggest that more northerly species would also occupy a greater proportion of the habitats available to them (as well as occupying more habitats), since they would be less restricted by annual climatic variation within those habitats. We tested this idea using the data on Palaearctic mammals (Letcher & Harvey, in press). Vegetation maps, obtained from Corbet (1966), divide the continent into 13 distinct habitat zones. It is possible that two species using the same habitat type in a similar manner may have very different range sizes. For instance, one species may occur in the taiga coniferous forest which stretches over a large area of the continent; that species consequently has a large range. Another species may also occur in coniferous forest, but in a smaller patch of forest separated from the taiga by unsuitable habitat. The second species therefore has a much smaller range than the first, possibly because it has never had the opportunity to invade taiga coniferous forest. For this reason each continuous area of habitat was treated as an independent habitat 'patch', thereby dividing the Palaearctic into 74 patches of habitat.

The total area of those patches that were at least partially included in a species' range was summed together, and correlated with range size. More northerly species occupied a greater proportion of the area of those habitat patches in which they were found.

In summary, the data from North American and Palaearctic mammals combined suggest that more northerly species have larger ranges, possibly because they are able to withstand greater ranges of annual temperature. Those more northerly species also occupy a greater number of habitats, as well as a higher proportion of the available area within the habitat patches in which they are found. Habitat destruction increases as one approaches the equator and, in North America and the Palaearctic, the range size of species decreases. This conjunction makes a bleak prospect.

RANGE SIZE AND BODY SIZE

Given the relationship between range size and latitude and between range size and local abundance, what might we expect the relationship between range size and body size to be? Given the general positive relationship between range size and local abundance (see above), we might expect smaller bodied species to have larger ranges because smaller bodied species are often more locally abundant (Damuth 1987). In contrast, as Boyce (1979) has emphasized, larger bodied species are able to survive under more variable and extreme climatic conditions, so (given the findings above) we might expect larger bodied species to have larger geographic ranges. In fact the limited empirical data currently analysed suggest a weak positive relationship between range size and body size, in apparent accord with the second of the above two predictions: larger bodied species have, on average, larger ranges in North American birds and in North American mammals (Brown & Maurer 1987, 1989; Pagel *et al.* 1991). However, although the correlation is positive, it tends to be caused by a triangular relationship in which large-bodied species all have large ranges, while small-bodied species may have either large or small ranges. One suggested explanation for this pattern is simply that large-bodied species with small ranges are highly likely to go extinct as they also have low population densities (Brown & Maurer 1987). However, Palaearctic mammals show a slightly different pattern: no species have small ranges. This difference arises in part because the Palaearctic data set contains no mammals with ranges extending below 20°N latitude whereas the North American data set includes many tropical mammals in Central America. When these species whose ranges extend below 20° are excluded from the American data set, most but not all of the small-bodied species with small ranges are excluded, in particular many bats.

The most plausible explanation that we can offer for the difference between Palaearctic and North American mammals results from the different topographies of the two areas. The main mountain ranges in America run north to south, whereas in the Palaearctic they run east to west. For a given latitude America will therefore have more, smaller habitats than the Palaearctic. Thus some small mammals will be constrained to having small ranges as their available habitat patches are small. Large species with small habitats and small ranges will go extinct for the reason outlined above. This explanation accords with the additional observation that, on average, species' ranges occupy a larger percentage of the Palaearctic surface area (8.4%) than that of North America (7.8%), even though the Palaearctic is 2.5 times larger in land surface area than is North America (Letcher & Harvey, in press). Even if the topographic explanation is true, we do not know if the topography has exerted its effect by evolutionary or ecological processes. An answer to this question is required before we can draw implications for conservation.

METAPOPULATION THEORY AND EPIDEMIOLOGY

Two popular theoretical frameworks for thinking about the distribution of animals across patches in a region, especially from a conservation point of view, are metapopulation theory (for individual populations) and island biogeography (for communities). In fact, these are two ends of a continuum of models; here we focus on the metapopulation end for reasons that will become obvious as we progress. (We have already touched on some applications of metapopulation dynamic theory above.) In its simplest form, a metapopulation exists as local populations of a species that occupy patches of suitable habitat which are interspersed among other patches of suitable habitat that are not occupied by the species. Individuals from occupied patches can colonize empty patches and the local population in an occupied patch can go extinct. There is an equilibrium frequency of patch occupancy determined by the balance between colonization and extinction rates. Lande (1988) demonstrated the potential power of metapopulation theory for exploring the consequences of changes in the amount of suitable habitat on population persistence. As we will discuss further below, Lande used the framework to calculate the threshold fraction of habitable patches that must be left in a region if the species is to persist in that region.

Metapopulation theory has been highly developed in population-level studies of infectious diseases, under the name of epidemiology. Here, the patch is a host organism, the species of interest is a parasite (virus, bacterium, protozoan or helminth), colonization is infection and local extinction is the death or recovery of the host. We will now argue that it is worthwhile for conservation biologists to adopt the theoretical framework that epidemiologists have developed for thinking about eradication thresholds and optimal eradication programmes (Anderson & May 1991). Essentially, the approach is, first, to derive a quantitative estimate of the eradication threshold *without* constructing an explicit mathematical model and then, second, to use explicit mathematical models to develop a qualitative understanding of how biological features, which may be considered important in any particular case, render this an over- or underestimate.

We will make the assumption of 'weak homogeneous mixing', as it is known in epidemiology (Anderson & May 1991, p. 17). Under this assumption, all patches have intrinsically similar properties with respect to colonization and extinction rates. This allows us to suppose that the rate at which a full patch colonizes empty patches is an increasing function of the fraction of such patches that are susceptible to colonization. In epidemiology this rate is typically assumed to be a linear function (in the simplest models), but, in general, it may be non-linear (see below).

It is useful to introduce some symbols. Let x be the fraction of patches in the metapopulation that is susceptible to colonization (i.e. currently empty). A local population in an inhabited patch may be assumed to colonize $f(x)$ other patches before extinction. h is the fraction of patches that is actually habitable, although a habitable patch may or may not be occupied at any particular time. In the pristine world, all patches are available and $h = 1$. As patches are destroyed, h falls below 1. In epidemiology, such intervention consists of vaccination. In metapopulations, intervention may be 'logging', 'mining', etc. We will use the word 'tarmacked' to correspond to 'immunized', since we find evocative the image of a parking lot replacing a patch of habitat. The eradication threshold is the critical value of h, h_c, such that if less than a fraction h_c is left habitable, then the metapopulation will go extinct. We wish to estimate h_c.

This is easy to do if the metapopulation is at dynamical equilibrium between colonization and extinction and if the above assumptions hold. At equilibrium, $f(x^*) = 1$, so $x^* = f^{-1}(1)$. As in epidemiology (Anderson & May 1991, p. 155), under the assumption of weak homogeneous mixing, the equilibrium susceptible fraction does not depend on what fraction of the patches has been tarmacked. This result – at first sight surprising – arises because equilibrium is achieved when the fraction of patches that are susceptible is just large enough that a patch will colonize one other before local extinction occurs. If habitable patches are of two types, colonized or susceptible to colonization, the fraction of patches that is colonized is, then, $h - x^*$. This immediately gives us x^* as our estimate of h_c, as in epidemiology.

In short, the critical fraction h_c can be estimated (under the approximation of weak homogeneous mixing) from knowledge of the fraction of patches originally occupied, in the pristine equilibrium state (or at any later equilibrium). Nothing more is needed. As is demonstrably the case in epidemiological contexts (for review of data, see Anderson & May 1991), seemingly important details of movement rates and demographic parameters cancel out, not entering into the final relation between h_c and x^*.

As an example of the use of such an estimate, consider the northern spotted owl *Strix occidentalis caurina*, which has been the subject of intense conservation interest. Following Lande (1988), the patches defining the metapopulation structure are taken to be breeding territories. The fraction of susceptible patches, that is, unoccupied breeding territories, is observed to be 0.21 ± 0.02 (in Lande's symbols, this number is calculated from $h(1 - p)$ on p. 605). Taking this as our estimate of h_c we have a result which agrees exactly with that of Lande (namely $h_c \approx 0.21$), the latter being based on a more explicit, and much more complex, exercise in mathematical modelling and parameter estimation.

There are a number of insights that are available on the basis of the present approach. First, within the constraints of the assumption of weak homogenous mixing, we can see that it makes no difference to the estimate of h_c whether colonizing propagules actively search for susceptible patches, as Lande supposed, or passively rain down and land on a susceptible patch by chance. Active search, as modelled by Lande, leads to non-linear patch colonization functions. Passive colonization generates linear functions. In either case, the full panoply of eradication threshold theory in epidemiology, based on the fundamental concept of R_0, can be employed. In addition to the nature of dispersal, the details of the owls' demography, included in Lande's model, are also clearly irrelevant to the evaluation of h_c. It must *not* be concluded that the modelling exercise was pointless as it was also used to establish that the metapopulation was at equilibrium. Lande's more biologically rich model also provides a crucial test of the simplifying assumptions that we have made here.

More importantly, perhaps, we can now use this first-step (weak homogenous mixing) analysis as a point of departure for assessing how violations of our assumptions qualitatively affect the estimate of the eradication threshold. To do this, we again borrow from established epidemiological work. Suppose that there are ecological successional processes, analogous to immunity following infection, that render habitable patches 'immune' to colonization for a period of time. The effect of this is to make x an underestimate of h_c. This is simply because, at equilibrium, the fraction occupied is not $h - x^*$, it is $h - x^* - y^*$, where y^* is the equilibrium fraction of transiently uninhabitable patches. Or consider the effects of spatial heterogeneity in colonization rates: patches may occur in clusters with varying colonization rates within the clusters, for example, x will now be an overestimate of h_c. The intuitive reason why this is found in epidemiological models, which is also valid here, is explained in Anderson and May (1991, p. 307): if all but a fraction h_c of patches are randomly chosen for tarmacking, then, from the point of view of eradication, too few patches are being tarmacked in regions of high colonization rates, and too many in regions of low colonization rates. If patches are not randomly tarmacked, but, in fact, patches in regions with high colonization rates are more likely to be tarmacked than those in regions with low colonization rates, then x may be an underestimate of h_c. This last assertion is simply the metapopulation version of a surprising result from the analysis of optimal eradication vaccination programmes (Anderson & May 1991, chapter 12).

Given the current rapid rate of habitat destruction, and the growing numbers of threatened species, it is unlikely that it will be possible to construct management models as detailed as those for spotted owls for all, or even 1%, of endangered species. The best that conservation biologists will often be able to do is provide quantitative estimates for the eradication threshold, as given by the assumption of weak homogeneous mixing, and then indicate whether the directions in which reality departs from the assumptions make this an over- or underestimate. Ecologists in general seem unaware that many of the difficult problems of eradication threshold theory and practice have already received much attention, in that epidemiology provides a powerful conceptual framework which can be imported, *in toto*, into conservation biology. However, a major conceptual problem for conservation biology, which must be solved before we can apply these methods, is the determination of what constitutes a patch: to identify breeding territories as patches in the case of the northern spotted owl was an important insight.

CONCLUDING REMARKS

Our aim in this contribution has been to focus on general principles. Our review of the repeated relationship between local abundance and geographical range shows how the same interpretative framework is likely to be relevant to the large majority of species studied. Similarly Rapoport's rule, occurring independently in so many different taxa, begs a single and probably quite straightforward explanation; a successful test for one taxonomic group (be it an Order of mammals or the whole Class) is likely to point the way to explanations for the rule occurring in other taxa. It is both fortunate and exciting that the repeated patterns we have discussed imply that a knowledge of variation in only a few relevant variables (be they measures of physiology, climate or habitat) will be necessary for us to understand reasons for variation in geographic range size. Similarly, considerations of metapopulation equilibrium dynamics indicate that we shall be able to make reasonable estimates of extinction thresholds simply by measuring patch occupancy. Conservation managers do not have to study each and every species as though it were unique. There are some general rules and guidelines that constrain what can and will happen to the distribution and abundance of threatened organisms in the messy, real world.

REFERENCES

Allee, W.C., Emerson, A.E., Park, O., Park, T. & Schmidt, K.P. (1949). *Principles of Animal Ecology*. Saunders, Philadelphia.

Anderson, R.M. & May, R.M. (1991). *Infectious Diseases of Humans*. Oxford University Press, Oxford.

Andrewartha, H.G. & Birch, L.C. (1954). *The Distribution and Abundance of Animals*. Chicago University Press, Chicago.

Arita, H.J., Robinson, J.G. & Redford, K.H. (1990). Rarity in neotropical forest mammals and its ecological correlates. *Conservation Biology*, 4, 181–192.

Boyce, M.S. (1979). Seasonality and patterns of natural selection for life histories. *American Naturalist*, 114, 569–583.

Brown, J.H. (1984). On the relationship between abundance and distribution of species. *American Naturalist*, 124, 255–279.

Brown, J.H. & Maurer, B.A. (1987). Evolution of species assemblages: effects of energetic constraints and species dynamics on the diversification of the North American avifauna. *American Naturalist*, 130, 1–17.

Brown, J.H. & Maurer, B.A. (1989). Macroecology: the division of food and space among species on continents. *Science*, 243, 1145–1150.

Carter, R.N. (1988). Distribution limits from a demographic viewpoint. *Plant Population Ecology* (Ed. by A.J. Davy, M.J. Hutchings & A.R. Watkinson), pp. 165–184. Blackwell Scientific Publications, Oxford.

Caughley, G., Grice, D., Barker, R. & Brown, B. (1988). The edge of range. *Journal of Animal Ecology*, 57, 771–785.

Colwell, R.K. & Futuyma, D. (1971). On the measurement of niche breadth and overlap. *Ecology*, 52, 567–576.

Corbet, G.B. (1966). *The Terrestrial Mammals of Western Europe*. G.T. Foulis, London.

Cotgreave, P. & Harvey, P.H. (1991). Bird community structure. *Nature*, 353, 123.

Crawley, M.J. (1987). What makes a community invasible? *Colonization, Succession and Stability. 26th Symposium of the British Ecological Society* (Ed. by A.J. Gray, M.J. Crawley & P.J. Edwards), pp. 429–453. Blackwell Scientific Publications, Oxford.

Damuth, J. (1987). Interspecific allometry of population density in mammals and other animals. *Biological Journal of the Linnean Society*, **31**, 193–246.

Dobzhansky, T. (1950). Evolution in the tropics. *Scientific American*, **39**, 209–221.

Ford, H.A. (1990). Relationships between distribution, abundance and foraging niche breadth in Australian land birds. *Ornis Scandinavica*. **21**, 133–138.

France, R. (1992). The North American latitudinal gradient in species richness and geographical range of freshwater crayfish and amphipods. *American Naturalist*, **139**, 342–354.

Gaston, K.J. (1991). How large is species' geographic range? *Oikos*, **61**, 434–438.

Gaston, K.J. & Lawton, J.H. (1990). Effects of scale and habitat on the relationship between regional distribution and local abundance. *Oikos*, **58**, 329–335.

Griffith, B., Scott, J.M., Carpenter, J.W. & Reed, C. (1989). Translocation as a species conservation tool: status and strategy. *Science*, **245**, 477–480.

Gyllenberg, M. & Hanski, I. (1992). Single-species metapopulation dynamics: a structured model. *Theoretical Population Biology*, **42**, 35–66.

Hanski, I. (1991). Single-species metapopulation dynamics: concepts, models and observations. *Biological Journal of the Linnean Society*, **42**, 17–38.

Harrison, S., Murphy, D.D. & Ehrlich, P.R. (1988). Distribution of the bay checkerspot butterfly (*Euphydryas editha bayensis*): evidence for a metapopulation model. *American Naturalist*, **132**, 360–382.

Hengeveld, R. (1989). *Dynamics of Biological Invasions*. Chapman & Hall, London.

Hengeveld, R. & Haeck, J. (1981). The distribution of abundance II. Models and implications. *Proceedings of the Koninklijke Nederlandse Akademie van Wetenschappen Series C*, **84**, 257–284.

Hengeveld, R. & Haeck, J. (1982). The distribution of abundance I. Measurements. *Journal of Biogeography*, **9**, 303–306.

Huffaker, C.B. & Messanger, P.S. (1964). The concept and significance of natural control. *Biological Control of Insect Pests and Weeds* (Ed. by P. De Bach), pp. 74–117. Chapman & Hall, London.

Jablonski, D. (1987). Heritability at the species level: analysis of geographic ranges of Cretaceous molluscs. *Science*, **238**, 360–363.

Krebs, C.J. (1989). *Ecological Methodology*. Harper & Row, New York.

Kuoki, J. & Hyrinen, U. (1991). On the relationship between distribution and abundance in birds breeding in Finnish mires: the effect of habitat specialisation. *Ornis Fennica*, **69**, 170–177.

Lande, R. (1988). Genetics and demography in biological conservation. *Science*, **241**, 1455–1459.

Lawton, J.H. & Woodroffe, G.L. (1991). Habitat and distribution of water voles: why are there gaps in a species' range? *Journal of Animal Ecology*, **60**, 79–91.

Letcher, A.J. & Harvey, P.H. (in press). Variation in geographical range size among mammals of the Palearctic. *American Naturalist*.

Maurer, B.A. (1990). The relationship between distribution and abundance in a patchy environment. *Oikos*, **58**, 181–189.

McArdle, B.H. (1990). When are rare species not there? *Oikos*, **57**, 276–277.

Moulton, M.P. & Pimm, S.L. (1986). Species introductions to Hawaii. *Ecology of Biological Invasions of North America* (Ed. by H.A. Mooney & J.A. Drake), pp. 231–249. Springer, Berlin.

Nee, S. (1991). The relationship between abundance and body size in British birds. *Nature*, **351**, 312–313.

Nee, S., Gregory, R.D. & May, R.M. (1991). Core and satellite species: theory and artefacts. *Oikos*, **62**, 83–87.

Novotny, V. (1991). Effect of habitat persistence on the relationship between geographic distribution and local abundance. *Oikos*, **61**, 431–433.

Pagel, M.D., May, R.M. & Collie, A. (1991). Ecological aspects of the geographic distribution and diversity of mammal species. *American Naturalist*, **137**, 791–815.

Pianka, E.R. (1986). *Ecology and Natural History of Desert Lizards.* Princeton University Press, Princeton, NJ.

Pimm, S.L. (1991). *The Balance of Nature.* University of Chicago Press, Chicago.

Pulliam, H.R. (1988). Sources, sinks, and population regulation. *American Naturalist*, **132**, 652–661.

Randall, M.G.M. (1982). The dynamics of an insect population throughout its altitudinal distribution: *Coleophora alticolella* (Lepidoptera) in northern England. *Journal of Animal Ecology*, **51**, 993–1016.

Rapoport, E.H. (1982). *Areography: Geographical Strategies of Species.* Pergamon, Oxford.

Richards, O.W. (1961). The theoretical and practical study of natural insect populations. *Annual Review of Entomology*, **6**, 147–162.

Ricklefs, R.E. (1989). Speciation and diversity: the integration of local and regional processes. *Speciation and its Consequences* (Ed. by D. Otte & J.A. Endler), pp. 599–622. Sinauer, Sunderland, MA.

Ricklefs, R.E. & Cox, G.W. (1978). Stage of taxon cycle, habitat distribution, and population density in the avifauna of the West Indies. *American Naturalist*, **112**, 875–895.

Ricklefs, R.E. & Latham, R.E. (1992). Intercontinental correlation of geographical ranges suggests stasis in ecological traits of relict genera of temperate perennial herbs. *American Naturalist*, **139**, 1305–1321.

Rogers, D.J. & Randolph, S.E. (1986). Distribution and abundance of tsetse flies (*Glossina* spp.). *Journal of Animal Ecology*, **55**, 1007–1025.

Rohde, K., Heap, M. & Heap, D. (1993). Rapoport's rule does not apply to marine teleosts and cannot explain latitudinal gradients in species richness. *American Naturalist*, **142**, 1–16.

Root, T. (1988). *Atlas of Wintering North American Birds.* University of Chicago Press, Chicago.

Schall, J.J. & Pianka, E.R. (1978). Geographical trends in numbers of species. *American Naturalist*, **137**, 791–815.

Simpson, G.G. (1964). Species density of North American recent mammals. *Systematic Zoology.* **13**, 57–63.

Smith, F.D.M., May, R.M. & Harvey, P.H. (in press). Geographical ranges of Australian mammals. *Journal of Animal Ecology.*

Stevens, G.C. (1989). The latitudinal gradient in geographic range: how so many species coexist in the tropics. *American Naturalist*, **133**, 240–246.

Taylor, R.A.J. & Taylor, L.R. (1979). A behavioural model for the evolution of spatial dynamics. *Population Dynamics. 20th Symposium of the British Ecological Society* (Ed. by R.M. Anderson, B.D. Turner & R.L. Taylor), pp. 1–27. Blackwell Scientific Publications, Oxford.

Wallace, A.R. (1878). *Tropical Nature and Other Essays.* Macmillan, London.

Wiens, J.A. (1989). *The Ecology of Bird Communities: Foundations and Patterns.* Cambridge University Press, Cambridge.

Wilcove, D.S. & Terborgh, J.W. (1984). Patterns of population decline in birds. *American Birds*, **38**, 10–13.

Wilcove, D.S., McLellan, C.H. & Dobson, A.P. (1986). Habitat fragmentation in the temperate zone. *The Science of Scarcity and Diversity* (Ed. by M.E. Soulé), pp. 237–256. Sinauer, Sunderland, MA.

Wright, D.H. (1991). Correlations between incidence and abundance are expected by chance. *Journal of Biogeography*, **18**, 463–466.

4. SPATIAL DISTRIBUTION
OF MARINE ORGANISMS:
PATTERNS AND PROCESSES

MARTIN V. ANGEL

Institute of Oceanographic Sciences, Deacon Laboratory,
Wormley, Godalming, Surrey GU8 5UB, UK

SUMMARY

In the open ocean, large-scale distribution patterns in the pelagic environment are currently predominantly determined by the global thermohaline
circulation, seasonality of production cycles and interactions with mesoscale eddies. However, the imprint of past vicariance events caused by
continental drift and associated with sea-level fluctuations is still recogniz-
able. The production cycles are largely controlled by the effects of the
seasonal fluctuations in heating/cooling on the nutrient resupply. There is
a trend for species-rich communities to be associated with low productivity
regions with relatively little seasonality.

The abyssal benthic realm is similarly influenced by the seasonality of
the overlying waters, but also affected by large-scale catastrophic events
such as the long-range effects of turbidity flows and slumps caused by the
geological failure of continental slopes. The interaction between bottom
topography, the water column characteristics and biological factors can
also lead to distributional mosaics, for example the benthic fauna inhabiting the tops of some individual guyots may have been isolated over
geological time.

In coastal environments the interactions between coastal morphology,
land/ocean exchanges, meteorological and tidal conditions, create a more
complex and finer scale network of environmental boundaries. Climatic
oscillations, especially through variations in sea level, have served to
isolate populations over evolutionary time scales. Thus coastal waters
are richer in both species and ecosystem richness than their oceanic
counterparts.

Conservation protocols have to be appropriately scaled in order to
conserve either diversity (genetic, species or ecosystem) or global processes. The spatial characteristics of inshore coastal systems are small
enough for reserves to have the potential to play a useful role in conserving species and unique ecosystems. This will not be true for offshore

systems with their broader scale characteristics, nor are reserves ever likely to play a significant role in the maintenance of processes. Basin-scale management would seem to be the only feasible way of attaining conservation objectives, but the underpinning science needed remains inadequate. Pragmatically, greater difficulties may arise in surmounting the complex hurdles of common, national and international law which are being developed piecemeal without any attempts to achieve integration.

INTRODUCTION

The global oceans turn-over at time scales of less than a millennium – the times from the Atlantic, Indian and Pacific Oceans are 275, 250 and 510 years respectively (Stuiver *et al.* 1983). In the absence of any abiotic and biotic limitations to their ranges, marine organisms would rapidly become ubiquitous. So although the observed distributional patterns may have their origins in the geological past, they must be dynamically maintained through environmental interactions. Stommel (1963) first identified the linear relationship between spatial and temporal variations of physical mixing processes in the oceans. This concept was later expanded by Haury *et al.* (1978) to give a conceptual framework to explain how variations in zooplankton biomass may be related to the physical forcing and the biological responses (Fig. 4.1). It became clear that basin-scale variability, both in terms of biomass distribution and assemblage type, is predominantly determined by the thermohaline circulation patterns which are largely driven by planetary forcing (Fig. 4.2). The early zoogeographers such as Ekman (1953) recognized that not only was there a tendency for the distribution patterns to be zoned latitudinally but also that there are centres of high species diversity, which they described as being centres of radiation. However, they were unable to identify the underlying causal factors that had determined the zoogeographical patterns and that continue to maintain them. Once it had been recognized that the morphology of ocean basins has undergone radical changes during the geological past as a result of continental drift and sea-level fluctuations, careful analysis of present-day distributions began to reveal imprints of ancient ocean circulation patterns and the opening and closing of barriers between the main oceans.

The aim of this chapter is initially to review the historical factors underlying the present-day distribution patterns, and the cycles with periodicities down to decadal scales which can be expected to be having a continuing influence on biological distributions. Major changes in the

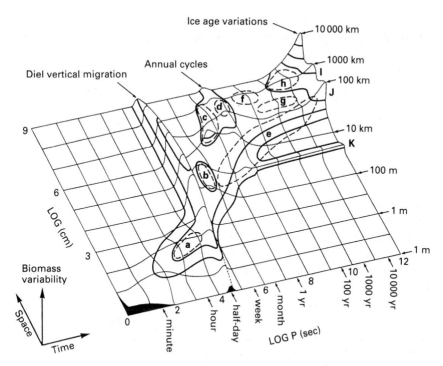

FIG. 4.1. Stommel diagram of the time/space variability of oceanic zooplankton. After Haury *et al.* (1978). a, 'micro' patches; b, swarms; c, upwelling; d, eddies and rings; e, island effects; f, ENSO effects; g, small ocean basins; h, biogeographical provinces; i, currents and oceanic fronts (length effects); j, currents (width); k, oceanic fronts (width).

structure of ocean basins and fluctuations in sea level caused both by changes in the morphology of the ocean basins and climatic cycles will be postulated as being sufficient to explain the distributional patterns of species richness. The very large scales of distributional patterns of species richness will be described, for open ocean pelagic and benthic assemblages and compared with the smaller scale shallow water patterns. The close correlation between the seasonality of production cycles and nutrient availability would seem to imply a strong causal link. Vertical patterns in the distribution of species richness in deep water will be described and shown to indicate that the link is indirect; the depths at which primary production is occurring are poorer in species than deeper depths remote from the production that supports the communities inhabiting them.

Finally, the implications to conservation policy of these observations will be discussed, and the dangers of relying too completely on ecological

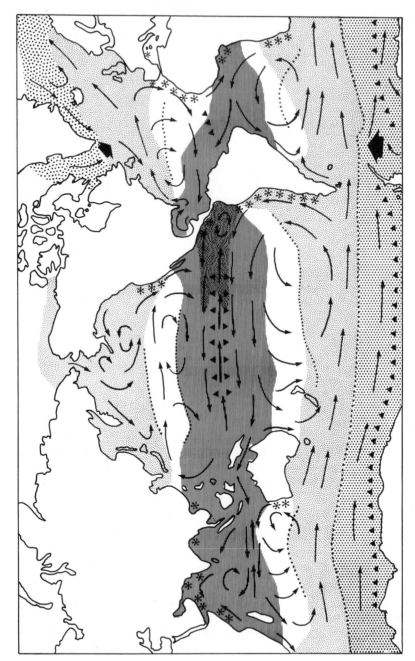

FIG. 4.2. Schematic representation of the main surface ocean current circulation with upwelling indicated by the stars, divergences by the triangles, convergences by the dots, and regions of deep-water formation by the large arrows. Superimposed are the main biogeographic zones of the pelagic realm. Based on van der Spoel and Heyman (1983).

theory based entirely on terrestrial ecosystems highlighted. In particular there is a danger in the prevailing belief that conserving regions with high diversity will also protect global processes and ensure sustainability of living resources.

CAUSAL FACTORS IN GEOLOGICAL TIME

Evolution and extinction

There has been a general rough trend for global diversity to increase linearly since the first appearance in the geological record of marine animals about 650 Ma (Sepkoski 1992). However, this trend has been punctuated with eras of diversification, stasis and extinction. There have been three main phases of diversification:

1 An explosive diversification in the early Cambrian which was followed by a stable era in the Middle and Late Cambrian.

2 A Palaeozoic phase during the Ordovician when familial and generic diversity initially increased, but was then followed by a period lasting some 200 my when diversity remained more or less constant, and finally ended in the mass extinction of the Permian.

3 A Mesozoic–Cenozoic expansion which has continued into the Neogene. The expansion of global diversity has resulted from differential expansion of groups with unequal but conservative rates of speciation.

Superimposed on this general trend of increasing diversity have been a series of mass extinction events: at the end of the Ordovician (440 Ma), in the Late Devonian (370 Ma), at the end of the Permian (250 Ma), at the end of the Triassic (220 Ma) and at the end of the Cretaceous (70 Ma). Most of these extinction events involved a loss of around 15–40% of marine species globally; but the Permian event is reported to have resulted in the extinction of around 95% of marine species (Raup 1979). Following each of these extinction events recovery to the former levels of species diversity is estimated to have taken 12–15 my through the radiation of novel taxa (Sepkoski 1992), and must have resulted in major changes in the structuring of ecosystems and maybe even their mode of functioning. Examination of fossils of individual taxonomic groups which are well preserved in deep ocean sediments, such as the benthic ostracods, reveals that periods of high extinction rates either coincide with, or immediately precede, periods when large numbers of species appear (Whatley & Coles 1991). Although it must be remembered that such sediment cores can rarely discriminate time scales $<10^4$ years.

Vicariance

Since the end of the Cretaceous, a number of major events have generated relatively sudden shifts in global and regional ecosystems which have played a substantial role in determining oceanic diversification. The residual signatures of these events are still thought to be detectable in some present-day distributions (e.g. van der Spoel *et al.* 1990). Some of these so-called vicariance events have been caused by the changes in ocean basin morphology caused by tectonic processes (continental drift and sea-floor spreading). Connections between land masses have opened and closed, altering the global patterns of ocean circulation (Haq 1984) sometimes resulting in major fluctuations in sea levels (Fig. 4.3). In the Palaeocene, 66–58 Ma and before Australasia had become separated from Antarctica and the Drake Passage between South America and the Antarctic Peninsula had opened, both high latitude surface temperatures and ocean deep waters were relatively warm (about 10°C). Ocean circulation was generally more sluggish (see Barron & Baldauf 1989). The deep waters were depleted in dissolved oxygen but enriched in nutrients, and the lysocline (the depth at which carbonates dissolve) was considerably shallower than in the modern ocean. Modern analogues are seen today in the northwestern Indian Ocean and the eastern tropical Pacific region where there are extensive oxygen minimum zones partly created by the high sedimentation of organic matter from the upwelling regions. In the Pacific the deep waters are relatively old (their source is North Atlantic deep water), so not only do they contain relatively low concentrations of dissolved oxygen, but also relatively higher concentrations of dissolved nutrients, as they have received sedimentary input for longer than comparable waters in the Atlantic (Mantyla & Reid 1983).

Despite continental drift over the last few tens of millions of years ocean circulation patterns have hardly varied because of the unchanging effects of planetary forcing. Thus in the Atlantic, migratory species like the European eel *Anguilla anguilla* have be able to adapt gradually to the increasing distances they have to migrate as the ocean basin has widened. *Anguilla* spawns in deep water in the Sargasso Sea and after hatching the larvae take about 3 years to drift across in the North Atlantic Drift to reach European rivers. The mature adults probably migrate back to the Sargasso Sea following the clockwise recirculation of the gyre. Despite the distances involved in this migration probably having more than trebled since the eels first evolved, the continuity of the large-scale gyral circulation in geological time has ensured that the eels have continued to survive the series of glaciations and their effects on the rivers of northwestern Europe.

The existence of a shallow water connection across the Isthmus of Panama has strongly influenced the distribution of shallow-living pelagic species. A connection has existed as recently as the Miocene, and estimates of when it was finally sealed range from 5–10 Ma (Smith 1985) to 3–4 Ma (Keigwin 1978; Cronin 1987, 1988). For this reason the high numbers of pelagic species which have pan-tropical distributions come as no surprise. However, it does suggest that in the absence of vicariance events, speciation in pelagic taxa is a very slow process. It is worth remembering that one of the major concerns about the proposal to build a sea-level canal was that sea-snakes would be introduced from the Pacific into the Caribbean. Shih (1979) reviewed the east/west diversity of planktonic taxa in the oceans, paying particular attention to the differences between the Atlantic and Pacific tropical and subtropical planktonic faunas. There are relatively minor morphological differences between a number of closely related taxa found in both oceans. In the genera *Rhincalanus* (Schmaus & Lehnhofer 1927) and *Sagitta* (Pierrot-Bults 1974) such differences are considered to be at the subspecific level. In contrast, the differences observed in the decapod crustacean genus *Sergestes* are considered sufficient for the populations in the two oceans to be attributed with specific rank, *S. edwardsii* and *S. brevispinatus* (Judkins 1976).

While the taxonomic significance of such differences is often determined by the whims of individual taxonomists rather than from clearly defined criteria, there are grounds for taxonomically segregating the species. For example, on the Atlantic side the decapods typically occur at mesopelagic (400–600 m) depths by day, whereas in the east tropical Pacific the strong oxygen minimum limits their bathymetric ranges to much shallower depths, possibly favouring their divergence. Trans-Pacific differences have been reported in the few other taxa studied, such as *Pontellina morii* and *P. sobrina* (Fleminger & Hulsemann 1974), and *Sagitta pacifica* and *S. bierii* (Pierrot-Bults 1974), with one of the species pair being restricted to the east tropical Pacific. However, even in some apparently cosmopolitan species, genetic isolation may be more advanced than the similarity of their morphological characters would seem to indicate. Carrillo *et al.* (1974) found they could readily breed *Acartia clausi* populations from either the Atlantic or the Pacific coasts of the USA, but were unable to interbreed the populations. Thus although the absence of clear divergence in so many of the pelagic taxa common to the Atlantic and the Pacific implies that these taxa tend to be genetically conservative, it seems clear that where there has been isolation associated with environmental change, speciation has occurred (e.g. Fleminger 1986). The conditions needed if speciation is to occur continue to be actively debated (e.g. van der Spoel & Pierrot-Bults 1979).

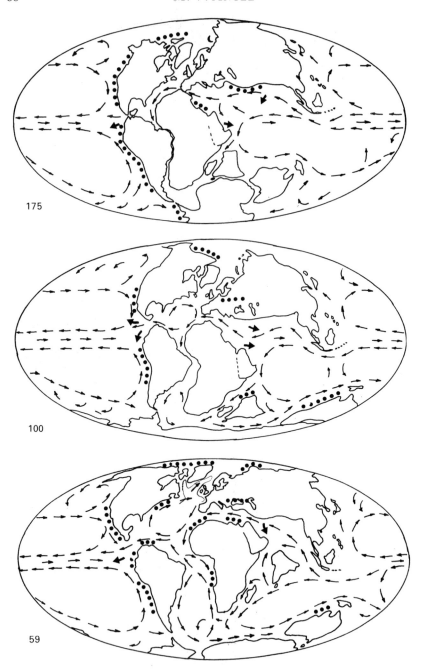

FIG. 4.3. Reconstruction of the distribution of the continental land masses and the inferred surface circulation of ocean currents over the last 175 my. The large dots indicate regions of upwelling and the large arrows regions of possible bottom water formation. Note that in

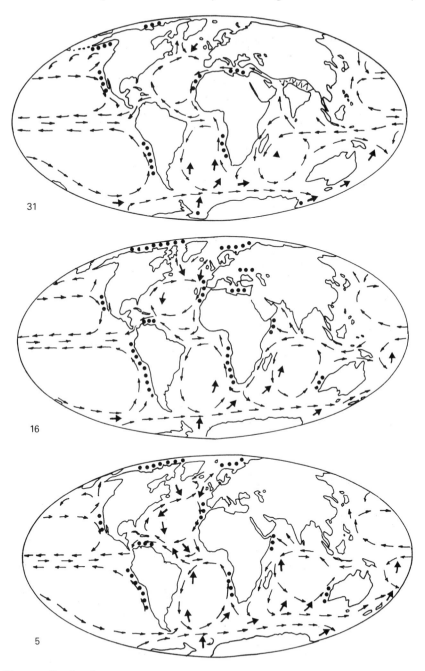

FIG. 4.3. *Continued*
many cases bottom water formation is by the sinking of warm saline water, rather than cold saline water as in the modern ocean. Based on Haq (1984).

It is generally accepted that isolation (i.e. allopatry) can and does lead to speciation, particularly in those species that are specialized in some aspect of their ecology. Whether or not sympatric speciation occurs remains controversial. Partial isolation (parapatry) linked with limited gene flow within wide-ranging species may prove to be important (Palumbi 1992). Careful analyses of many oceanic planktonic taxa have shown that high genetic diversity and morphological flexibility give rise to variations that are often described at subspecific, race or forma level (e.g. pteropods, van der Spoel 1971; chaetognaths, Bone *et al.* 1991), but generally pelagic taxa are not highly speciose. There are only about 80–100 species of planktonic Chaetognatha (Bieri 1991), around 160 species of halocyprid Ostracoda (my data), around 1900 species of marine and brackish-water Copepoda (Razouls 1981), to give a few examples.

Long-term climatic events

During the Tertiary there were at least nine cooling events, with four major events occurring 52–48, 36–35, 15–13 and most recently 2.5–2.4 Ma (Miller *et al.* 1987). Coincident with these cooling events were evolutionary turn-over events in diatom assemblages (Barron & Baldauf 1989). The establishment of the Antarctic ice cap, and hence the conditions for the evolution of the present Southern Ocean faunal assemblages, did not occur until 15–13 Ma. The northern hemisphere glaciation did not begin until the last major cooling event 2.5–2.4 Ma; this may account in part for the species diversity of the Southern Ocean being richer than in the Arctic (see also Hempel 1985).

In modern oceans, warm deep water is found in the Mediterranean (12.7°C) and the Red Sea (21.7°C); both seas lack a specific bathypelagic fauna. In the Mediterranean, the present benthic fauna is comprised mostly of pseudopopulations maintained by the influxes of pelagic larvae through the Straits of Gibraltar. During the glaciations, when the deep water must have been considerably cooler, a truly bathyal fauna existed (Bouchet & Taviani 1992). This raises the question as to whether the warm deep waters of the Palaeocene ocean also lacked a specific bathyal and bathypelagic fauna. The warm temperatures and low dissolved oxygen levels may have presented too much of a physiological demand in an environment where organic inputs are very low. Warm temperatures *per se* do not seem to be inhibitory when food supplies are abundant, as exemplified by the high localized abundances of biomass (albeit low in species richness) to be found in the immediate vicinity of hydrothermal vents (see Tunnicliffe 1991), although the elevated temperatures of the discharges are only detectable within very close range. The Red Sea is not only extremely oligotrophic but also the high temperatures throughout the water column promote the rapid microbial degradation of whatever organic material sinks out of the euphotic zone. Hence the deep-water standing crops in both the Red Sea and analogously the Mediterranean are exceptionally low (Thiel 1983; Beckmann 1984). Was this true globally in the Palaeocene ocean?

Sea-level fluctuations

Rapid changes in sea level have been produced by major tectonic events and climatic change. Repeated closure and re-opening of the Straits of

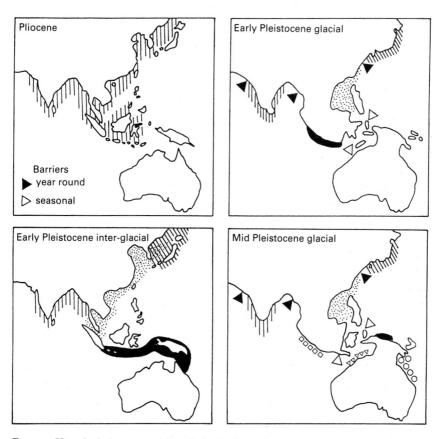

FIG. 4.4. Hypothetical sequence of speciation in the neritic copepod *Labidocera pectinata* group since the Pliocene, resulting from changes in sea level creating and removing year-round and seasonal barriers to distributions. From Fleminger (1986).

Gibraltar have resulted in global sea-level fluctuations of some 70 m within a matter of a few hundred years as the Mediterranean has dried up and refilled, leaving massive geological deposits of salt – the so-called Messinian salinity crisis (e.g. Hodell *et al.* 1986). The glacial cycles during the Quaternary have resulted in sea-level fluctuations of up to 100 m through the formation and melting of extensive ice sheets. In some regions, the interaction between the lower sea levels and the local coastal and sea-bed topography will have resulted in the isolation of local populations in deep-water basins – thus creating the conditions for allopatric and parapatric speciation. Rising sea levels will have removed the barriers to dispersion allowing the populations to recombine, which, if they had diverged sufficiently during their isolation, would have persisted as separate sibling species. Fleminger (1986) attributed the high species richness of some copepod genera (*Pontella*, *Labidocera* and *Undinula*) around the East Indies to the alternating formation and breakdown of seasonal and continental barriers during Pleistocene glacial cycles (Fig. 4.4).

Climatic influences

Climatic effects can operate independently of their influence on sea level. For example, at present in the Mediterranean there are a number of glacial relict species. These have been isolated, since the last glaciation, from their parent populations which are now to be found at latitudes >40°N (Furnestin 1979). The myctophid fish *Benthosema glaciale* (Fig. 4.5) and the euphausiid *Meganyctiphanes norvegica* have isolated populations in the Mediterranean (and in the upwelling region off the north-west coast of Africa). The coolest water temperature the Mediterranean populations experience is 12.7°C, the temperature of Mediterranean deep water, whereas the parent populations may never experience such warm water. In the North Atlantic, *Benthosema* occurs predominantly to the north of 40°N where it normally occurs by day at mesopelagic depths of 300–400 m and migrates up into the near surface waters at night usually stopping below the seasonal thermocline. It does occur south of 40°N but as a non-migrant bathypelagic species inhabiting depths of >1000 m – an example of submergence (e.g. van der Spoel & Pierrot-Bults 1979).

Analyses of sediment core samples during the CLIMAP programme (Cline & Hayes 1976) showed that 18 000 BP, at the height of the last glaciation, sea surface temperatures were substantially cooler in summertime at latitudes >40°, but at subtropical and tropical latitudes sea surface temperatures were only very slightly cooler (Fig. 4.6). The polar front

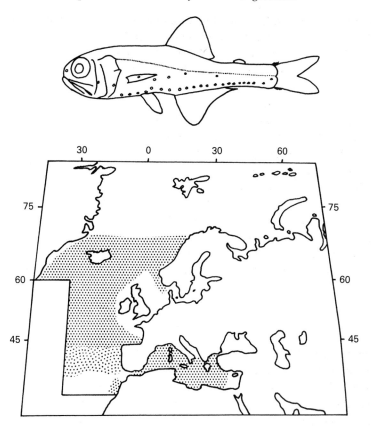

FIG. 4.5. Distribution of the myctophid fish *Benthosema glaciale* showing the distribution of its relict populations in the Mediterranean and the upwelling region off northwest Africa. The change in the hatching in the North Atlantic shows where the population submerges to become bathypelagic.

which now occurs at around 67°N to the north of Iceland, then occurred at 45°N off the European western seaboard, and free exchange was possible between the cool water species of the open Atlantic and the Mediterranean. In the intervening 10 000–20 000 generations since the isolation of these relict populations, neither *Benthosema* nor *Meganyctiphanes* have diverged morphologically; populations of some mesopelagic crustacean decapods have been separated at subspecific rank (Casanova 1977).

The factors responsible for such long-term climatic cycles are still not fully understood, but undoubtedly some astronomical cycles have

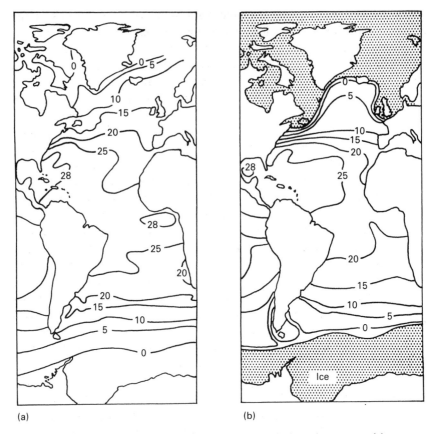

FIG. 4.6. Sea surface temperatures in the Atlantic during the boreal summer at (a) present and (b) 18 000 BP at the height of the last glaciation, showing the greater equatorwards shift in isotherms in the northern hemisphere relative to the southern. Based on Cline and Hayes (1976).

periodicities of about the right order of magnitude to explain the alternation of glacials and interglacials (e.g. Weaver 1993). The eccentricity of the Earth's orbit has a periodicity of 10^5 years, the obliquity of the Earth's axis 41 000 years, and the precession of the equinoxes 23 000 years. In the pelagic sediment record laid down during the last 1.9 my of the Quaternary, these cycles are expressed as 63 stages distinguished by variations in dominance within the fossil assemblages of foraminiferans and coccolithophores; by major geographical redistribution of some species; by fluctuations in the base of the lysocline; and in changes in ratios of stable oxygen isotopes. Correlated with some of these events is the simultaneous global extinction of nanofossil species, such as *Calcidiscus macintyrei* 1.45 Ma, *Helicosphaera selli* 1.22 Ma, and *Pseudoemiliania lacunosa* 440 ka, and the first appearance (? evolution) of *Emiliana huxleyi* 242–268 ka. Some of these stages correspond to palaeomagnetic reversals (when the polarity of the Earth's magnetic field reverses), but so far any attempts to relate reversals to substantive biological changes have been unsuccessful. One inherent difficulty is that in oxic sediments discrimination of the timing of events is restricted to a minimum of several thousand years, because of bioturbation and the low rates of sedimentation.

The record of variation is also seen in the rates of growth of corals (Montaggioni & Macintyre 1991) and changes in fossils associated with reefs (e.g. Kohn 1985; Kay 1990). There is a trend for generalist species to be more persistent and show lower rates of speciation and extinction than more specialized taxa. Any long-term cycle that results in prolonged geographical isolation is likely to have evolutionary impact on the more responsive specialist taxa than the more adaptable generalist taxa. The potential for isolation is far greater in those species with short dispersal stages, and yet species inhabiting isolated habitats, in common with terrestrial island faunas, must be subject to strong selection to reduce, or even lose, their dispersal stages. An alternative is for the life cycle to become finely tuned to a predictable characteristic of the physical environment. Even taxa with prolonged pelagic larval phases may have behaviour characteristics that serve to keep the larvae within larger scale systems so that they can return close to their starting point – the best example is probably the spiny lobsters (Phillips & McWilliam 1986). Even in some permanently pelagic taxa there are life-cycle adaptations that maintain them within specific systems. *Calanoides carinatus* is able to maintain itself within upwelling systems (Smith 1984), and there is a wide range of 'pseudo-oceanic' species that occur only in the immediate vicinity of continental slopes (Merrett 1986). But, as will be discussed below, such isolating mechanisms are more likely to operate in inshore waters than offshore.

A special case is the novel anchialine faunas discovered in cave systems in Bermuda and the Bahamas (Iliffe *et al.* 1983). These have remained isolated for about 30 my. This fauna contains several taxa whose closest relatives are now to be found only in abyssal and benthopelagic faunas. So it seems feasible that the post-Palaeocene cooling of the deep waters led to an increase in the rate of invasion of the deep ocean. But as the abyssal faunas diversified through both immigration and local radiation, the rate may have slowed as the abyssal taxa acquired the specializations needed to cope with the extreme environment of the deep ocean and became progressively more successful in keeping further invaders at bay.

Short-term climatic effects

Steele (1991) points out that in shelf seas there are striking switches in community structure that can persist for several decades (e.g. the Russell cycle, Southwood 1980). These switches have a major impact on resource exploitation (Horwood 1981; Sherman *et al.* 1981). They can also have a major political impact; in northern Europe the development of the Hanseatic League was based on the resource provided by exploitation of a particular stock of herring. The beginning of the mini-ice age led to the collapse of this Baltic herring fishery which was soon followed, in 1569, by the demise of the League, thus throwing northern Europe into a period of political instability and war (Cushing 1986). In 1626–1629 total failure of the local cod fishery in the Faeroes caused mass starvation, and again in 1675–1685; cod, which is limited by the 2°C isotherm, had disappeared from Faeroese waters. In 1695, drift ice surrounded virtually the whole of Iceland for many weeks preventing vessels from reaching port (Lamb 1977).

In the North Sea there is dependence of the size of herring stocks on the oceanic inflow, and departure from the average inflow has major impacts on the recruitment (Bailey & Steele 1991). The underlying cause may be either the changes in the advection of larvae from the spawning grounds to their nursery areas (Corten 1986) or fluctuations in primary production. In the northern North Sea, the main source of nitrate is via the inflows of North Atlantic water around the northern coast of Scotland (Dooley *et al.* 1984). These inflows seem, in turn, to be influenced by large-scale physical events, such as the 'great salinity anomaly' (Dickson *et al.* 1988), which was followed over 14 years from its genesis in the Greenland Sea, along its passage via the Labrador Current and across the North Atlantic, through the Faroe–Shetland Channel and into the West Spitsbergen Current where parts were dissipated into the Barents and Laptev Seas but some recirculated back into the Greenland Sea. Such interannual variations in the influxes of oceanic species – then described as the Lusitanian fauna – into the northern North Sea had been earlier described by Fraser (1968). The cyclic changes in the pelagic and commercial fish communities in the North Sea – the Russell cycle – also appear to be subject to climatic forcing (Southwood 1984), but it is still far from clear whether the system is also subject to top-down or bottom-up determination of the community structure. It seems hard to believe that neither the extensive exploitation of top predators in the North Sea through fishing, nor the removal of a large portion of the large mammal population in the Southern Ocean, have failed to have a substantial impact on pelagic community structure and dynamics. In the North Sea, fish stocks recovered substantially during World War II, when fishing activity was curtailed. But as yet, there has been no recovery of the large whale stocks in the Southern Ocean, despite the ban on commercial whaling over the last few years. However, there has been little effort to correlate changes in the planktonic communities with these large-scale ecosystem 'experiments'.

El Nino events

There are also shorter frequency oscillations with approximately decadal frequencies associated with sun-spot cycles. The large-scale features that generate these oscillations are often described as El Nino/Southern Ocean (ENSO) events. Every so often a major atmospheric feature, the Equatorial Tropical Convergence Zone, shifts its position longitudinally in the western Pacific. The shift in atmospheric pressure on the ocean generates a planetary wave that propagates eastwards along the equator. When this wave impinges on the west coast of the Americas, it is reflected as two new waves with different characteristics, which propagate polewards along the coastlines (Donguy 1987). There are analogous events in the Atlantic, but because the Atlantic is much narrower the time characteristics are shorter (Peterson & Stramma 1991). The waves depress the thermocline and result in an increase in sea surface temperatures. These changes to the normal ocean conditions cause widespread major perturbations to regional weather conditions in the tropics and further afield resulting, for example, in droughts in the Sahel region, low snowfall in the American Rockies, and exceptional rains in the West Indies.

Off the Peruvian and Ecuadorian coasts, there is a regular annual El Nino season, when sea surface temperatures warm and upwelling ceases, which lasts 2–3 months. But during ENSO events, the El Nino season persists anomalously for many months causing mass mortalities in pelagic fish, sea birds and marine mammals. In the Galapagos there is a prolonged wet season which actually favours the breeding of finches. Even if the time scale of such events occurs within the individual lifespans of longer-lived species, they may generate a strong selective pressure. For those species with much shorter life cycles of days or weeks, the selective pressures will be even greater.

Decadal variations in inshore waters

In shelf and slope seas, variations in riverine run-off generated by ENSO events can cause significant deviations in the nutrient supplies and suspended sediments (e.g. off California, Small 1992). Similarly, interannual variations in the wind field can alter the intensity and frequency of upwelling events (e.g. Mittelstaedt 1991; Peterson & Stramma 1991). Such variations generate substantial quantitative and qualitative oscillations in pelagic community structure, for example Tont (1987) found variations from days to years when analysing a unique time series of phytoplankton data collected off Scripps Pier. However, these were changes with time at a single geographical position and the apparent trends might not have been quite so extreme had it been possible to obtain greater latitudinal coverage.

Some (if not all) components of coastal communities are linked to disjunct geological features and their ability to disperse along the coast may be restricted. This will render them more vulnerable to large departures from the long-term mean of environmental conditions. If these perturbations, natural or anthropogenic, are either large enough or persistent enough, recruitment of keystone species may be sufficiently altered to generate substantial shifts in community structure and interactions. At worst there may be local, or even global, extinction of the most vulnerable species. However, in some ecosystems the biological structure of the community serves to buffer the community against the impact of decadal events. There are some well-documented examples in terrestrial habitats, for example in forests the presence of the trees is a major influence on soil temperature, nutrient budgeting and the water cycle (e.g. Tamm 1991). Some structured marine communities (such as kelp forests and coral reefs) may also prove to be self regulating and so be able to persist through periods of adverse climatic conditions. However, once the structure of such communities is disrupted, recovery may be slower than expected, or even prevented by their replacement by another quasi-stable community. A possible example of such climate-driven oscillations is the population oscillations of the crown-of-thorns starfish, *Acanthaster planci*. Large numbers of the starfish were first noticed in the 1970s destroying patches of reef, notably on the Great Barrier Reef. These were initially attributed solely to anthropogenic impacts, but now there is subfossil evidence which implies that human impact cannot be the sole cause of all these outbursts (Moran 1986). Another less well-known event was the basin-wide outburst of the stomatopod crustacean *Oratosquilla investigatoris* in the northwest Indian Ocean in 1965 (Losse & Merrett 1971).

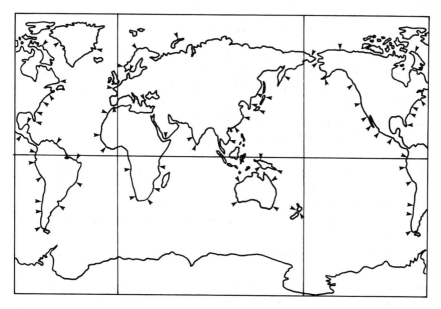

FIG. 4.7. Chart showing where there are major faunal changes in coastal communities. From van der Spoel and Heyman (1983).

PRESENT-DAY PATTERNS

Zoogeographical patterns

The large-scale patterns in the zoogeographical distribution of coastal marine species were initially described by zoogeographers such as Ekman (1953) and further developed by Briggs (1975). Regions were described throughout which the faunas were largely similar, and bounded by narrow zones of rapid faunal change (Fig. 4.7). In some instances the transitions are remarkably sharp, for example at Point Conception on the Californian coast, the faunas on either side of the headland are quite distinct. In the open ocean, data in zoogeography were slower emerging. Pickford (1946) had been the first to relate the distribution of an individual species to the distribution of a water mass. McGowan (1971) expanded the concept by demonstrating that pelagic assemblages are distributed according to the major gyral structure of the oceans. As more detailed explorations of particular groups were published (e.g. Brinton 1975, for euphausiids) it emerged that, although these patterns are clear at coarse sampling scales, they become progressively fuzzier as the coverage becomes finer. Angel and Fasham (1975) and Fasham and Foxton (1979) demonstrated that the relationship between water masses, and hence the gyral circulation, and zoogeographical distributions is maintained vertically. Once again the patterns are less clear for individual species than for assemblages. Overall syntheses of the results (van der Spoel & Pierrot-Bults 1979; van der Spoel & Heyman 1983) continued to describe the patterns empirically, but without establishing the underlying causality. Reid *et al.* (1978) demonstrated that biological characteristics are not only associated with the gyral circulation of the oceans but also with a wide range of chemical characteristics. Many of these chemical parameters have direct influence on biological processes, for example dissolved nitrate and phosphate concentrations. More recently Levitus *et al.* (1993) have published maps of the distribution of the mean annual concentrations of the main nutrients, nitrate, phosphate and silicate, in the world's oceans. These maps clearly show how some of these regional distributions of nutrients reflect the gyral current pattern in the ocean, as shown in the mean annual nitrate concentrations at a depth of 150 m (Fig. 4.8). Others illustrate global turn-over, as shown by the distribution of phosphate at 1500 m (Fig. 4.9). The impoverishment of dissolved silicate in the North Atlantic relative to the very high concentrations seen in the North Pacific and Southern Ocean has a very important bearing on diatom biology. Could this explain

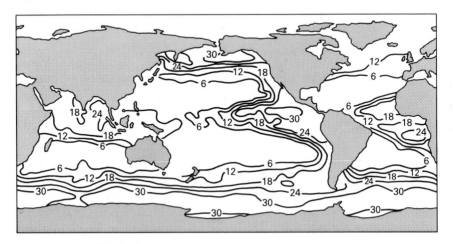

FIG. 4.8. Distribution of mean dissolved nitrate concentrations at a depth of 150 m in the world's oceans. From Levitus *et al.* (1993). Note the close similarity between the nutrient concentrations and the circulation patterns shown in Fig. 4.2.

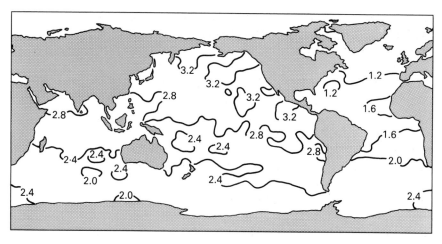

FIG. 4.9. Distribution of mean dissolved phosphate concentrations at a depth of 1500 m in the world's oceans. From Levitus *et al.* (1993). Note the relatively low concentrations in the Atlantic which spread in the deep circulation around South America into the southern Pacific and Indian Oceans.

why the coccolithophorid *Emiliana huxleyi* forms dense blooms in the North Atlantic but not in the North Pacific?

Advances in physical oceanography have provided the explanation for fine-scale sampling failing to resolve clearly these distributional patterns. The existence of persistent mesoscale eddies with diameters of 50–200 km provided the explanation for much of the patchiness of zooplankton. The internal structure and dynamics of the eddies have fine-scale effects on biological processes (e.g. Angel & Fasham 1983; Wiebe & McDougall 1986). Moreover, the eddies which often move at rates of tens of kilometres per day, transport organisms well beyond their usual ranges. However, these mesoscale eddies provide explanations for only some of the variability. Data from long time series, such as the continuous plankton recorder survey (Colebrook 1986) and studies of the Russell cycle in the English Channel (Southwood 1984) revealed long-term cyclicity in response to climatic oscillations. However, there are a few examples of species composition and relative dominance remaining highly conservative. For example, in the Bay of Biscay neither the specific composition nor the rank order of dominance changed in planktonic ostracod assemblages sampled in July 1900 and over 70 years later in April 1974 (Angel 1977). Similarly, in a decade of continuous sampling of planktonic communities in the subtropical North Pacific there have been no changes in either species composition or rank order of abundance

Euphausia pacifica

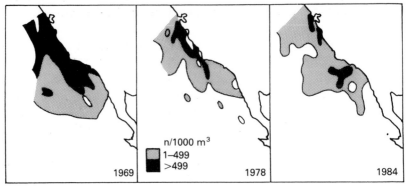

Nyctiphanes simplex

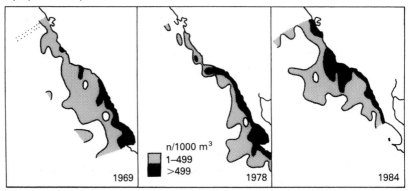

FIG. 4.10. Interannual variations in the abundances of two euphausiid species in the Californian coastal region, one a cold-water species (*Euphausia pacifica*), the other a warm-water species (*Nyctiphanes simplex*). From Brinton and Reid (1986).

(McGowan 1990), despite large perturbations by ENSO events. These events resulted in changes in standing crop but not in the species composition. During the study there were not only substantial reductions in stocks of top predators as a result of the expansion of drift netting for tuna, but also a continuing general increase in anthropogenic contaminants introduced directly via run-off or indirectly via the atmosphere (GESAMP 1991), but neither of these anthropogenic impacts have as yet caused detectable effects on the oceanic community structure (McGowan 1990). How are we to distinguish man-induced changes from natural variability? This is an important but unresolved problem, which is a consequence of the shifting nature of the boundaries of these pelagic systems, subject as they are to interannual and decadal variability, as seen in the Russell cycle in the English Channel and in the California Current (Fig. 4.10) (Brinton & Reid 1986).

The distributional boundaries to coastal assemblages appear to be less responsive to interannual variations, probably because the dominant controlling factors of the distributions are affected by much longer term climatic conditions. For example, in the coastal waters around the British Isles, productivity, turbidity and the temperature of the surface waters are strongly influenced by whether or not there is thermal stratification during the summer. The amount of energy required to stratify the water is related to the water depth and the tidal stream amplitude during spring tides. Simpson *et al.* (1977) showed that the probability of stratification occurring could be empirically estimated by the ratio h/u^{-3} (h is the water depth, u is the maximum tidal flow rate). The critical value for the ratio is 70, which coincides with where tidal fronts develop at the boundary between well-mixed waters (>70), and thermally stratified waters (<70). The factors determining the locations of these tidal fronts remain independent of climatic forcing until sea levels change.

Latitudinal gradients

In terrestrial taxa there are latitudinal gradients in species diversity, with richness increasing from poles to equator in many but not all taxa. The existence of such gradients in marine taxa has been challenged (e.g. Clarke 1992), citing, for example, marine mammals and some families of sea birds (e.g. penguins). Such departures from the 'expected' often have physiological determinants, haematypic corals are restricted to where the sea remains warm enough year-round to facilitate the production of calcium carbonate, whereas macroalgae are most diverse in the temperate waters off California, Japan, southern Australia and the coast of Brittany where the mean annual nutrient availability is greater. Similar anomalies in terrestrial taxa have not led to the rejection of the generality. In the oceans the absence of clearly identified gradients is almost certainly the result of a lack of comparable data. In the North Atlantic several pelagic groups show the expected increase in species richness towards the equator (Fig. 4.11). However, the gradient may be stepped rather than smooth, with discontinuities occurring at the major features such as the polar front and the subtropical convergences. The changes in species richness are expressed throughout the water column to depths of at least 2000 m (Fig. 4.12). Mean body size decreases towards the equator and this has allometric implications for physiological and rate processes in community

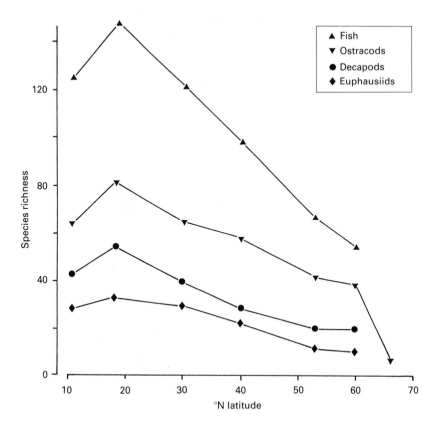

FIG. 4.11. Latitudinal changes in species richness in the top 2000 m of the water column approximately along 20°W in the northeastern Atlantic in four pelagic groups sampled with a standardized procedure. From Angel (1993).

dynamics. These stepped gradients appear to be linked with changes in the characteristics of the production cycles.

Spatial interactions with production cycles

In the open ocean the total annual quantity of primary production and its seasonality are controlled by the availability of nutrients and the light cycle. Primary production is limited to the top few tens of metres of the water column, even in the clearest of oceanic waters, because sea water both absorbs and scatters light. The amount of incident energy reaching the ocean surface is also determined by the elevation of the sun and the daylength; both decrease and become more seasonal polewards. Cloud

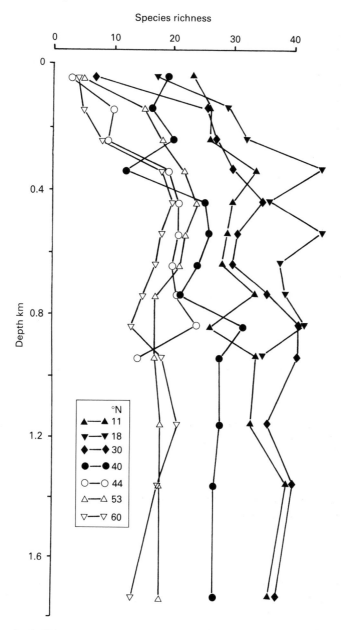

FIG. 4.12. Latitudinal changes in the bathymetric profiles of species richness of pelagic ostracods approximately along 20°W in the northeastern Atlantic sampled with a standard procedure. From Angel (1991b).

cover, which plays a significant role by intercepting and reflecting some of the radiant energy, also varies latitudinally. The surface roughness of the ocean alters its albedo, so at latitudes where mean wind speeds are stronger more radiant energy gets reflected at the surface. The transparency of the water is reduced by material in suspension, which in offshore waters may be subject to the negative feedback effect of the phytoplankton populations. Inshore, either freshwater run-off or aeolian transport may significantly reduce the transparency locally, and both are influenced by climatic oscillations and land use in the coastal zone (Small 1992).

Nutrient availability is also subject to strong climatic forcing. At the beginning of the production cycle in temperate and polar latitudes nutrient availability is dependent on the depth to which wintertime mixing of the surface waters extends and the nutrient concentrations within the deep waters ('older' deep waters, such as occur in the Pacific and Indian Ocean, are richer in nutrients than the 'younger' waters of the Atlantic, see Fig. 4.9). Production cannot get fully underway until a thermocline develops when phytoplankton is no longer being continually mixed to depths where there is insufficient light for primary production to compensate for the demands of respiration (i.e. below the compensation depth). Throughout the production season, some nutrients are being recycled within the euphotic zone but a proportion is continually being lost through sedimentation. However, since the thermal stratification limits the resupply from deep water to diffusion, which cannot keep pace with sedimentary losses, the bloom in the near surface waters eventually collapses. There is a zone of steep gradient in nutrient concentration, the nutricline, which is often, but not always, coincident with the thermocline. In exceptional circumstances, such as along the equator, the thermocline comes close to the surface raising the depth of the nutricline up into the euphotic zone so that more continuous high levels of productivity are maintained. There is also replenishment by upwelling in coastal regions, where along-shore or offshore trade winds or intermittent storm events push surface waters offshore. As a result volumes of cool nutrient-rich water well up to the surface close to the shelf-break and at the heads of canyons and flow inshore over the shelf. Generally, upwelling occurs seasonally, but even within the season upwelling is often an intermittent phenomenon driven by periods of persistent strong wind. As described above, upwelling can also be subject to strong, remote external forcing during ENSO events, as seen off Peru (Boje & Tomczak 1978) and in the Gulf of Guinea (e.g. Verstraete 1992).

At latitudes >40°, wintertime cooling of the surface waters leads to

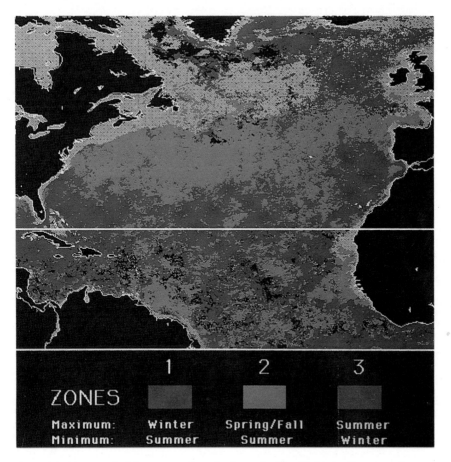

PLATE 4.1. Zones with different annual production cycles identified using five years of surface chlorophyll data observed by the Coastal Color Zone Scanner (from Campbell & Aarup, 1992).

extensive vertical mixing which not only replenishes the nutrients but also increases the annual mean availability of nutrients (Fig. 4.8). As a result the production cycle is highly seasonal, peaking in spring after the onset of stratification and sometimes with a further submaximum following the first autumnal storms. At lower latitudes thermal stratification is maintained year-round; total annual productivity is lower and spread more uniformly through the year. These zonal variations in primary production and the timing of the blooms have been thoroughly described by Campbell and Aarup (1992) based on 5-year averages of monthly means of surface chlorophyll concentrations derived from Coastal Zone Color Scanner (CZCS) images from 1979 to 1983 from the North Atlantic. The total area of the North Atlantic is $38.4 \times 10^6 \, km^2$. Some 38% of this area is tropical ocean and a further 11% could not be analysed because either data were missing or the seasonal patterns were irregular. In the North Atlantic to the north of the tropics they identified three main zones with distinct seasonal patterns (Plate 4.1, facing page 84).

Zone 1 is predominantly subtropical and covers 18% of the area. Its southern boundary follows a roughly sinusoidal track from 35°N off Cape Hatteras, to almost 30°N at 40°W, but then curving northward to about 40°N in the eastern Atlantic off the Iberian Peninsular. Typically within this zone surface chlorophyll concentrations are maximum in midwinter and minimum in later summer. The nutricline is deep, particularly in summer and there is a persistent deep chlorophyll maximum. Production remains nutrient limited throughout the year, because the wintertime resupply of nutrients is restricted. Total average production in this zone is estimated to be $84 \, g \, C/m^2/year$. New production – defined by Campbell and Aarup (1992) as the production supported by nitrate available in the wind-mixed layer at the beginning of the production cycle (note this is not the same definition for the term as originally coined by Eppley & Peterson 1979) – amounts to $18 \, g \, C/m^2/year$ equivalent to the resupply of $3.1 \, g \, N/m^2/year$.

Zone 2 is similar in area, its northern boundary lies along the northern edge of the Gulf Stream and its extension eastwards as the North Atlantic Drift reaching to about 47°N off Europe. Within it, the production cycle is characterized by phytoplankton blooms in spring and to a lesser extent in autumn. During summer conditions become oligotrophic and nutrient availability strongly limits production (resembling the conditions persisting in zone 1). The estimated total production averages $98 \, g \, C/m^2/year$, with new production contributing $24 \, g \, C/m^2/year$, equivalent to the supply of $4.2 \, g \, N/m^2/year$.

Zone 3 is smaller in area, about $2.5 \times 10^6 \, km^2$ (7% of the total). In the eastern sector it extends to the north of Iceland and then around to the south of Greenland. There the production cycle is typified by a winter minimum and a single summer maximum. Nutrient levels fluctuate but always remain high enough not to become limiting. Why the nutrient supplies in the euphotic zone are never exhausted is a subject of active debate between those who advocate limitation by trace elements such as iron (e.g. Martin *et al.* 1990; Martin 1991), and those who believe that grazing by heterotrophs is sufficient to prevent total nutrient utilization (e.g. Banse 1990; Miller *et al.* 1991). Mean estimated total production in this zone is $132 \, g \, C/m^2/year$, with new production contributing $43.4 \, g \, C/m^2/year$, equivalent to the supply of $7.2 \, g \, N/m^2/year$.

The remaining areas are equivalent in size to zone 3, and comprise both shelf seas and the complex region to the east of Newfoundland. There the production cycles showed no consistent seasonal pattern. Lewis *et al.* (1988) used dimensional analysis to demonstrate that the processes supplying nitrate to the mixed layer operate in three frequency bands. Wherever the dominant operational time band diverges from a seasonal pattern, nutrient supply, and hence productivity, becomes aseasonal. Such areas are often highly productive, and total and new production are probably similar to, or even in excess of those in zone 3.

Associated with these major shifts in general characteristics of production levels and seasonality are complex changes in biological characterization and interactions. These shifts have their origins in the primary producers. Phytoplankton communities in tropical seas are structured vertically into high- and low-light assemblages (e.g. Venrick 1982). At high latitudes where light becomes limiting during winter months the phytoplankton communities become dominated by the low-light taxa. In springtime at temperate latitudes, although nutrient concentrations are high as a result of wintertime convective mixing and light ceases to be limiting in the surface layers as the daylength increases, phytoplankton do not bloom until the surface waters have warmed sufficiently to create enough physical stability to prevent cells from being continually mixed down below the compensation depth. Once this stability is achieved, a bloom which is dominated by large diatoms rapidly develops. This diatom bloom persists for several days but then crashes spectacularly and within a day almost the total assemblage sediments out of the euphotic zone (Smetachek 1984). These crashes are often associated with the formation of large aggregates (marine snow) which, because of their size, sink rapidly into deep water (Alldredge & Silver 1988; Gardner & Walsh 1990; Lampitt *et al.* 1993). At abyssal depths, time lapse photography has revealed that large quantities of detrital material arrives on the sea floor within 4–6 weeks of the onset of the bloom (Billett *et al.* 1983; Lampitt 1985). There these large inputs of organic material stimulate seasonal activity (Gooday & Turley 1990; Pfannkuche 1993).

During the bloom a major proportion of primary production is carried out by cells >5 μm, but in post-bloom conditions after the crash, much smaller cells <5 μm make the major contribution to the primary production (Joint *et al.* 1993). Low-light species, usually dinoflagellates, become the dominant component of the phytoplankton communities and a deep chlorophyll maximum develops within the nutricline. These changes in the phytoplankton communities have a considerable impact on the animal communities, not only in the near-surface waters, but also throughout the water column down to the deep-sea floor. The large cells (>5 μm), which dominate the spring bloom, can be efficiently sieved from the water or even taken raptorially, and so are readily available to the larger macrozooplankton herbivores such as copepods and euphausiids. These species produce large, rapidly sinking faecal material. Moreover, many are diel migrants (Angel 1989) and so actively transport material down to their daytime depths thereby supplementing the sedimentary fluxes to the deep-living populations. Food chains in these high latitude systems tend to be relatively short, and the life cycles of these larger species tend to last a year or more.

In contrast, the cells <5 μm in size which dominate the post-bloom conditions at temperate latitudes and throughout much of the year at low latitudes (Fogg 1986) either enter the foodweb via mucus-web entrapment or via the microbial loop. Much smaller heterotrophic micro-organisms (ciliates and flagellates) and the mucus-web feeders (gymnosomatous pteropods, salps and foraminiferans) replace the larger macroplanktonic copepods and euphausiids as the dominant herbivores. Not only are the nano- and picoplanktonic cells too small to sediment under the influence of gravity but also the faeces of the grazers that exploit them are small and thus sink very slowly. There is more recycling because the greater warmth of the near-surface waters stimulates rapid microbial degradation of any detrital organic material. In these oligotrophic waters, the sedimentary fluxes to the deep-living populations are much lower and more evenly distributed throughout the year. The mean body size of the herbivorous population is smaller, their life cycles are shorter and food chains tend to be longer. Communities composed of taxa with short life cycles (weeks to a few months) will be responsive to short-term environmental fluctuations whereas communities composed of taxa with long cycles (several months to years) will have a much greater inertia in their responses to environmental fluctuations.

Variations in the timing of seasonal events, such as the onset of
the spring bloom in the North Atlantic, are subject to considerable
interannual variations caused by the differing weather patterns. There are
variations within a year caused by the mesoscale eddy structure generating
100 km patchiness in the onset of stratification and hence local variations
of several days in the initiation of the bloom (e.g. Savidge *et al.* 1992;
Robinson *et al.* 1993). Interannually, there are variations not only in the
seasonal timing of events, but also the quantitative expression of productivity and fluxes and in the planktonic community structure (see
the results of the Continuous Plankton Recorder Survey programme,
Colebrook 1986; Robinson *et al.* 1986). Many of the biological responses
are non-linear and so may be determined by fine-scale interactions. For
example, survival of some larval fish is determined by the availability
of food above minimum threshold concentrations at a critical phase of
development (Lasker 1975); in turn these threshold concentrations may
only persist for long enough if the frequency of storms is not too high.
There may be other 'chance' effects created, for example by key predators
attaining abundances that depress survival. This degree of variability at
fine- and meso-scales of time and space makes the unambiguous detection
of any impact resulting from anthropogenic influences a difficult task.

Spatial interactions in pelagic communities with depth

Species composition of both pelagic and benthic communities changes
with depth. The maximum richness in pelagic communities occurs at
depths of 1000–1500 m (e.g. Angel 1991a,b, 1993), well below the lower
limit for primary production. This high diversity is not spatially coincident
with where the primary production, which resources the diversity, is
occurring. Food resources reach the deep-living communities by sedimentation. Since sedimenting organic material is both intercepted and
microbially decomposed during its descent, the availability of food decreases with increasing depth. This decrease in the availability of organic
matter is reflected in the bathymetric decline in standing crop (Angel &
Baker 1982). Pelagic biomass decreases by an order of magnitude between
the surface and 1000 m and by a further order of magnitude between 1000
and 4000 m. However, metabolic rates of deeper living species are much
lower than their shallow dwelling counterparts (Childress *et al.* 1980), so
P/B ratios increase with depth.

Hydrostatic pressure alone does not appear to be the primary factor
that excludes shallow-living species from inhabiting greater depths, al-

though some very simple experiments conducted by Menzies and Wilson (1961) implied that there may be physiological barriers to invasions of water (>2700 m) by some surface-dwelling species and vice versa. Bathymetric ranges tend to be much broader in the cooler waters at high latitudes and below the warm-water sphere at low latitudes.

Vertical migrants daily experience wide ranges in bathymetric pressure, the maximum range reported is 1600–1700 m for a myctophid fish off the Azores (Angel 1989). Many pelagic species undertake ontogenetic migrations during their development which cover ranges of several hundreds of metres. The early larvae live in the near-surface waters where more food is available, and as they grow and develop they move progressively deeper. Such developmental descents are often associated with a metamorphosis to the adult form which is often morphologically adapted to environmental conditions at depth. In some taxa the ontogenetic migration is in the opposite phase. The eggs of species of *Euphausia* are heavy and sink to depths of hundreds or even thousands of metres before hatching (e.g. Hofmann *et al.* 1992). The newly hatched lecithotropic larvae begin a developmental ascent. Their upward migration is timed so that they reach the surface waters as their yolk supplies are exhausted and they have to start to feed. This is probably an adaptation to minimize predation during the early stages, but these migrations may also be selectively tuned to maintain the species within a local circulation cell. At high latitudes many species overwinter in deep water in a state of diapause; diapause is analogous to hibernation in terrestrial organisms, since feeding stops, the gut regresses and the metabolic rate slows. The abundant copepod in the temperate North Atlantic, *Calanus finmarchicus*, overwinters at depths down to 2000 m.

The role of temperature has not been fully established, and yet it must be important in view of the absence of any specifically bathypelagic fauna in isolated seas with warm deep waters – the Mediterranean (12.6°C) and the Red Sea (21.7°C). Some pelagic species which live in cold near-surface waters in polar seas, submerge at lower latitudes to bathypelagic depths where the water is as cool.

Bathymetric changes in benthic diversity

At temperate latitudes benthic biomass shows a similar bathymetric decline in standing crop to that of pelagic biomass (Lampitt *et al.* 1986), although small-scale variability can be high where localized physical interactions with the morphology of the sea floor enhance the concentrations of suspended material and so result in high biomass of suspension feeders (Rice *et al.* 1990). There are no comparable data for subtropical and tropical seas.

In many taxa species richness increases downslope to a maximum at mid-depths – 1000–2000 m for megabenthos (Angel 1991a; Fig. 4.13) and 2000–3000 m for macrobenthic infauna (Rex 1983) – below which there is a very gradual decline. Thorson (1955) postulated that pelagic larval development, which is so widespread in shallow-living taxa, is nearly or totally suspended in the deep sea. If so, gene flow would be greatly reduced, greatly increasing the trend for small-scale speciation (Palumbi 1992). However, Gage and Tyler (1991), in reviewing the existence of a characteristic life-history strategy for deep-sea organisms, conclude that since many of the larger taxa retain planktotrophic larvae, Thorson's prediction only holds true for taxa with body sizes <5 mm. Sharp increases in species richness are then to be expected in groups with smaller body sizes.

The changes observed in species composition downslope lead, by analogy with terrestrial communities, to the conclusion that the communities are zoned bathymetrically. However, in studies in which the sampling coverage has been reasonably comprehensive, species replacement with depth has proved to be continuous rather than discontinuous as would be the case if there was a zonation (Billett 1991; Merrett *et al.*

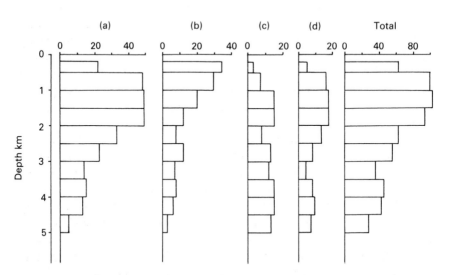

FIG. 4.13. Downslope changes in species richness of four megabenthic taxa ((a) Fish; (b) Decapoda; (c) Holothuria; (d) Asteroidea) in the Porcupine Seabight region to the southwest of Eire. Based on the Institute of Oceanographic Sciences Deacon Laboratory data. Unpublished.

1991). A thorough examination of the distributions of demersal fish throughout the North Atlantic demonstrated an absence of zonation (Haedrich & Merrett 1990). However, the results of a local study can be misleading, because the physical regime can create local conditions which lead to the formation of distributional bands of taxa. Rice *et al.* (1990) described such banding in the sponge *Pheronema carpenteri* at depths of 1000–1300 m in the Porcupine Seabight. Each band occurred close to an area where internal tides and waves interacted with steep bottom topography to enhance the local currents. The sponges, while unable to withstand the locally intensified currents, do benefit 'downstream' from the greater amounts of material in suspension. Elsewhere *Pheronema* is known to occur both shallower and deeper. Thus along continental slopes local topography (canyons, steep cliffs and slumps, e.g. Kenyon *et al.* 1978), may interact with physical processes (internal tides and waves, mesoscale eddies and shelf currents) to generate a local habitat diversity with scales as fine as 1–10 km horizontally and maybe tens of metres vertically. Extensive areas of slope remain unexplored because the underwater terrain renders them inaccessible to standard sampling procedures from surface vessels. Further regions of high abundance and biomass probably await discovery.

As with pelagic organisms, hydrostatic pressure does not always appear to be a barrier to the extension of species up- or downslope. Some abyssal bivalves have epipelagic larvae (Bouchet & Waren 1985), and larvae of bathyal species are sometimes encountered in large numbers in surface waters (e.g. Domanski 1984). Barophilic bacteria can survive being decompressed, but even quite small rises above the ambient temperature prove to be lethal. However, Young and Tyler (1993) have shown that cytoplasmic division is inhibited in larvae of the bathyal echinoid species, *Echinus affinis*, normally encountered at depths of around 2000 m, by hydrostatic pressures equivalent to depths <1000 m.

Away from the continental margins, the vast expanses of the abyssal plains are still thought to be relatively monotonous at spatial scales greater than 1 m or so. Large-scale features have been created by major geological failures of adjacent slopes resulting in slumps and extensive turbidity flows (Embley & Hallberg 1977). Such events, although very infrequent and unpredictable, may have long-term impact on the biological communities dominated by species with very slow rates of dispersal. Recovery rates are expected to be extremely slow and consequently Jumars (1981) has predicted that attempts to exploit sea-bed mineral resources in the deep ocean will have serious, long-term environmental consequences. An experimental programme, DISCOL, which has been designed to mimic the effects of sea-bed mining, is now underway and will begin to provide real data on recovery rates over the next few years (Thiel & Schreiver 1990).

As with continental slopes, there are regions of rugged terrain (mid-ocean ridges and seamounts) where the benthic assemblages are very poorly known. In the Pacific many hundreds of guyots arise from deep abyssal plains which have been isolated since their formation several million years ago. Their shallow-water and bathyal communities must consist of taxa recruited intermittently from distant guyots or continental margins and, like the islands which break the surface, probably have faunas with a high degree of endemism. Mullineaux and Butman (1990) have experimentally demonstrated a high degree of specificity in the recruitment behaviour of encrusting invertebrates on to manganese crusts, which will favour greater endemisms. How this translates into regional and global diversity is yet to be resolved.

There have been a number of attempts to estimate the total diversity of deep-sea benthos. Thorson (1955) estimated global numbers of megafaunal species to be 10^5. Much higher estimates are now being made. Grassle and Maciolek (1992) recently documented the remarkably high diversity of deep-sea benthic communities at bathyal depths (2000 m), already known to be the richest in species (Rex 1983). From their 233 30 × 30 cm box-core samples, collected off New Jersey and Delaware, they identified 798 species belonging to 171 families and 14 phyla. Extending the sampling coverage both to the north and to the south increased the numbers of species to 1597, of which 58% were new to science. Ninety per cent of the species were represented by <1% of the individuals, so most species were rare. For each kilometre the survey was extended alongslope, an additional species was recovered. Since the rate at which species are added is even greater across contours, the authors assumed that this linear rate of increase could be extrapolated to an areal increase of 1/km². Oceans deeper than 1000 m cover 3×10^8 km², so the number of macrobenthic species would be of the same order! However, because the densities of individuals are very much lower in deeper water, Grassle and Maciolek took 10^7 to be a more realistic estimate. May (1992) examines weaknesses in the extrapolation argument and suggests alternative hypotheses that significantly reduce the size of the estimate. Against the realism of Grassle and Maciolek's estimate is their assumption that regional scales of diversity are maintained downslope and into abyssal depths; an assumption not supported by data (albeit inadequate) on megabenthos. Moreover, the percentage of new taxa they found was lower than might be expected from their global estimate, even though they were studying the best-known bathyal region. However, it should be remembered that they considered only the species richness of macro-benthos; the meiobenthos is equally or even more diverse. If nothing else, their paper highlights the dilemma of trying to derive any sort of estimate on the basis of our inadequate current knowledge. Undoubtedly the benthic communities of the deep ocean are highly diverse, and may even rival some of the highest diversities to be found in terrestrial eco-systems. This is in stark contrast to the relatively low global diversity of the pelagic realm.

Is the Grassle and Maciolek estimate likely to be high or low? At regional scales the distributional patterns of megabenthic communities appear to be similar to that of the pelagic communities of overlying water column, probably as a result of the seasonality of the production cycle and the subsequent sedimentation of organic material into the deep ocean (Lampitt 1985; Berger *et al.* 1989). In the northeast Atlantic, Merrett (1987) observed that demersal fish faunas are similar over wide areas of the northeastern basin, but change suddenly to the south of 40°N, that is, at the latitude where the character of the annual production cycle changes. This relative uniformity at regional scale contrasts with the considerable heterogeneity seen at the microhabitat scale (10^{-2}–10 m). In regions of high mesoscale eddy activity in the surface waters (e.g. off the eastern seaboard of America), there are benthic storms during which high near-bottom currents erode the sediment destroying any biological structures and leaving a signature of ripple marks. Currents of 20–30 cm/s are fast enough to move and to resuspend detrital material but not to erode the sediment. So the mounds, hollows and lebenspurren generated by the bioturbating megafauna persist and tend to accumulate detritus. These detrital accumulations can be seen being slurped about by the tidal currents in video sequences assembled from time-lapse photographs (Rice, personal communication). Thus, the fine-scale variability seen between subsamples of individual box cores is often far greater than the variability between total cores. Similarly, individual cores collected by a single multicorer deployment show greater variability than occurs between the data pooled from separated deployments. There have been very few intersite comparisons completed but those that have been again suggest there is uniformity at the regional scale.

Many benthic species have a planktonic larval stage. The duration of this dispersive stage varies among different taxa from a few days to many months depending on the extent to which the larvae have to feed within the plankton. With increasing depth there is a trend for the duration of the planktonic stage to be reduced and there is a trend towards more taxa either brooding their eggs or larvae or undergoing lecithotropic development without a planktonic dispersal phase. The potential for dispersion is directly related to the duration of the planktonic stage. Scheltema (1986) described how teleplanic larvae can persist in the plankton for weeks or months by delaying their metamorphosis, and in so doing can cross oceans. The phyllosoma larvae of lobsters persist in the plankton for many months, but are not drifting passively in the sense that they have behavioural adaptations that enable them, through vertical migrations, to ride large-scale circulation features and arrive back close to where they started (Phillips & McWilliam 1986). When the large-scale circulation is temporarily perturbed by a climatic event, lobsters can appear far outside their normal range. Heydorn (1969) reported the occurrence of the South Atlantic rock lobster *Jasus tristani* at the Vema Seamount, more than 2000 km away from its normal range at Tristan da Cunha and Gough Island where it was considered to be an endemic. Within a few days of its discovery the total stock on the Seamount was fished out. There were still no lobsters to be found when the Seamount was reinvestigated in 1979, but by the end of 1980 a population had been re-established and the fishermen reported that 80 t of tails had been processed (Lutjeharms & Heydorn 1981b). Drifting buoys demonstrated a possible intermittent recruitment mechanism (Lutjeharms & Heydorn 1981a).

Widely distributed species with prolonged planktonic larval stages can be expected to have good gene flow and so low rates of speciation. Species occupying ephemeral habitats, such as the goose barnacles (*Lepas* spp.) which settle on floating debris or the borers which exploit water-logged wood (e.g. *Teredo* and *Xylophaga* spp.) as might be expected produce copious numbers of larvae which are able to persist in their dispersive phase. In contrast, species specializing in, or endemic to, small localized habitats, as with terrestrial island faunas, will be subject to strong selective pressure to shorten or even loose their dispersive phase. The specialized faunas associated with hydrothermal vents have very foreshortened dispersive phases, and yet they seem to be able to colonize new vents rapidly; individual vents are ephemeral and only persist for a few years (Tunnicliffe 1991). Palumbi (1992) has pointed out that any species with low dispersion and small population size 'tend to have significant genetic structure over small spatial scales'. He points out that the fossil record confirms that species with high larval dispersal have broader geographical ranges, greater species durations and slower rates of speciation than similar species with low dispersal (Waser & Price 1989). Palumbi (1992) argues that species with broad distributions but limited gene flow may well tend to speciate through rapid evolution of male and female gamete recognition, and that such speciation may help to explain the sympatric distributions of sibling species. Thus species that are subject to selective pressure to reduce their dispersal phase, and hence their gene flow, become more susceptible to speciation. Therefore the scaling of the environmental processes will have a direct evolutionary influence on species richness. Small sedentary benthic species in the abyssal deep ocean will tend not to have the resources for their larval stages to disperse widely within the water column. They will thus tend to speciate more freely than large and mobile species, which either can disperse pelagically, or, because of their longevity, their mobility (even if limited) may be enough to maintain sufficient gene flow to inhibit speciation.

In the open ocean, the large-scale hydrodynamics dominate the environmental processes but, as the rim of the deep ocean is approached, the shoaling sea bed with its variable geological structure begins to interact with the large-scale hydrodynamics to create a finer grained environment.

THE OCEAN RIM:
THE CONTINENTAL SLOPE AND SHELF-BREAK

Along the shelf-break, the interactions between the shallowing sea floor and mesoscale eddies, internal waves and tides help to create considerable local heterogeneity. For example, the generation of internal wave packages or solitons has been shown to have an important role in enhancing nutrient concentrations at the shelf-break (Holligan *et al.* 1985; New & Pingree 1990) and so contribute to the zone of high productivity along the shelf-break. Similar effects are seen in upwelling regions in the lee of headlands and at the heads of canyons that impinge into the continental shelf, where the topography of the sea floor creates conditions that favour upwelling (Mittelstaedt 1983). In several upwelling regions quasi-stable filaments of cool upwelled water have been described extending up to 300 km, or in one extreme case 1000 km offshore (Californian Current – Flament *et al.* 1985; southwest Africa – Lutjeharms *et al.* 1991; northwest Africa – Van Camp *et al.* 1991). The prevailing physical regimes in these regions are highly dynamic and this makes it difficult for species to stay within the system and benefit from the localized enhanced productivity. However, the general three-dimensional pattern of water movement is quite conservative with on- and offshore flows occurring at different depths.

Several species are known to have ontogenetic migrations, which are tuned to the physical processes, that are sufficiently predictable for the animals to maintain themselves within the upwelling system by moving between the off- and onshore currents. For example, in many coastal upwelling regions, there is a single dominant copepod species, *Calanoides carinatus*. During an upwelling event the population is carried offshore rapidly. There the adults and larvae sink (or swim) into deep water where the adults spawn. This is the water which at the next upwelling event emerges at the surface, already well seeded with eggs and larvae of the copepod, which are ready to dominate the exploitation of the boom in resources provided by the ensuing algal bloom (Smith 1984). Similar life-cycle adaptations have been described in the euphausiid species *Nyctiphanes capensis* and *Euphausia hanseni* in the northern Benguela upwelling system (Barange & Pillar 1992). Such behaviour will be subject to strong selection because any individuals that fail to match the timing of their migrations to the characteristics of the physical circulation will get spun out into the open ocean especially in those regions where there is high mesoscale eddy activity which can rapidly result in the import or export of passive drifters (e.g. Joyce & Wiebe 1992).

There are also 'pseudoceanic' species (Merrett 1986) which occur predominantly or even exclusively within shelf-break regimes. They too must have similar life-cycle and behavioural adaptations which allow them to ride the physical circulation to stay within the regime, despite the existence in many regions of shelf-break currents that flow persistently in one direction along the slope (e.g. Pingree & Le Cann 1989). Similarly, endemic island faunas which retain even brief planktonic dispersal stages must have life history strategies that are precisely tuned in their timing and duration to the local physical environment if they are to recruit back into the parent populations.

There are quite dramatic cross-slope changes in the pelagic community structure. The shelf-break and adjacent slope is a major transitional zone between continental shelf and oceanic regimes which have pelagic communities that are almost totally different in their specific composition and their standing crop. Along a transect from New York to Bermuda in the top 200 m, Grice and Hart (1962) observed the standing crop declined from $>1\,g\,m^3$ in neritic waters, to 0.2 g in slope waters, and further to 0.02 g in the Sargasso Sea, whereas the corresponding changes in the numbers of species were increases from 81 to 169 and 268 respectively. Where in relation to the shelf-break this change occurs has received little attention apart from a study by Hopkins *et al.* (1981) across the northern Florida shelf-break in the Gulf of Mexico (Fig. 4.14). They observed a steady increase in the species richness in euphausiids and copepods; the copepods for example increasing from less than 10 close to shore to over 50 at the shelf-break. Both mid-water fish and decapods showed sharp increases in species richness in the immediate vicinity of the shelf-break. Further offshore, the species richness of all four of the taxa studied

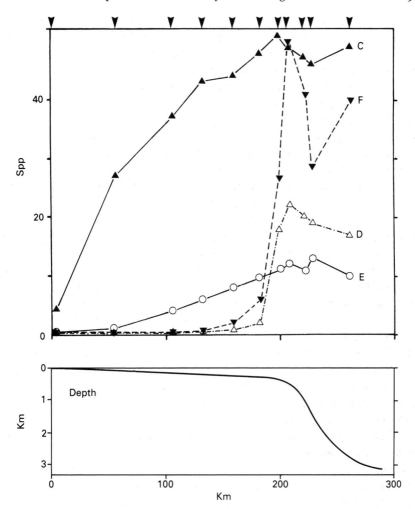

FIG. 4.14. Changes in species richness in four pelagic taxa across the shelf and continental slope of the northern coast of Florida in the Gulf of Mexico. C, copepods; D, decapod crustaceans; E, euphausiids; F, fish. Redrawn from Hopkins *et al.* (1981).

declined. The peak in species richness probably indicates that there were populations of pseudo-oceanic species, although none has been described from the region.

COASTAL DIVERSITY

In inshore and littoral environments, although there are still major faunistic boundaries (Ray 1991), which are generated by the interaction of the major oceanographic features and major coastal features (Fig. 4.7), there is general decrease in the spatial scales of environments. Moreover, coastal seas are essentially linear features bounded by land on one side and the deep ocean boundary on the other. Even though the open ocean boundary is leaky, hydrographic events impinge onto the shelf from the open ocean and there are active migrations across the shelf-break by some species, it is still a clear and well-defined boundary. Within coastal waters the scales of interaction between geological features and hydrological characteristics such as tides, local currents, prevailing winds, wave exposure and aspect become progressively smaller in time/space scales towards the land–sea interface. These interactions create very much finer environmental patchiness than seen in the open ocean. Furthermore, in some coastal ecosystems such as coral reefs, kelp beds and mangrove swamps, the biological components not only contribute to the habitat structure, and so buffer the ecosystem against short-term variations, but their influence often extends beyond their boundaries. For example, kelp beds and mangrove swamps provide nursery areas for fishes and crustaceans and are often dominant sources of detrital material to surrounding waters. The impact of experimental manipulation of such systems, for example by the removal of mangrove islands and the creation of artificial reefs, has been unexpectedly extensive.

As well as the spatial characteristics being finer in scale, temporal interactions are of higher frequency. For example, fluctuations in sea level caused by tides, weather (storm surges), climate and even isostatic geological processes, effect erosional and depositional regimes and maintain many, if not all, communities in a non-equilibrium state. The North Sea coastline has changed dramatically over the last few millennia as a result of the rise in sea level following the last glaciation, the isostatic changes in the relative levels of the land, variations in the erosional and depositional environment, changes in riverine run-off and more recently as a result of coastal engineering work and land use in the watersheds throughout the whole European region. Off California, the varved sediment deposits in the anoxic basins hold a detailed record of changes in sediment, riverine and aeolian inputs that have resulted from changes in both natural and anthropogenic processes (Small 1992). The major variations are generated highly intermittently in those years when there is very high riverine run-off. Such events are now suppressed by the damming of the rivers and the changes in land use that have followed industrialization and the expansion of irrigation.

Similarly, the building of the Aswan Dam in Egypt, which, by controlling Nile flood waters, drastically reduced the input of sediment and nutrients into the eastern Mediterranean. This catastrophically reduced the productivity of the Nile Delta region and obliterated the local coastal fishery (Dowidar 1984), so that a significant source of protein for the population was sacrificed for an increase in cotton production. In southern Japan, the clearance of the forest cover on Okinawa for agriculture resulted in extensive soil erosion and also the destruction of all the fringing coral reefs. Sheet erosion of soils from the hills surrounding the Mediterranean following forest clearance by Neolithic farmers has left many of the ports famous in Greek, Roman and biblical literature several kilometres inland. However, on some Mediterranean islands such as Crete there have been substantial isostatic changes in land levels so that archaeological remains of ports and jetties are to be found under a metre of water. The impacts of these changes on the coastal ecosystems have gone largely unrecorded.

The impact of the building of the Suez Canal is being carefully documented. Initially, the high salinities of the Bitter Lakes were a barrier to faunal exchange. However, as the lakes have become less saline the movement of species through the waterway has increased. There is flow through the canal from the Mediterranean to the Red Sea, but the movement of species has been predominantly in the opposite direction (Por 1990) from the relatively species-rich Red Sea to the species-poor Mediterranean. Distributional barriers are also being broken down by ship operations. Discharges of ballast waters taken on in Japanese waters into Australian coastal waters have already caused the introduction of a variety of exotic species, which have included some of the dinoflagellate species that are responsible for toxic red tides (Hallegraeff & Bolch 1991). Mariculture has likewise resulted in the spread of a significant number of pathogenic organisms and nuisance species (such as *Sargassum muticum*), and even marine aquaria have been blamed for the introduction of pest species. The aquarium at Monaco has been blamed for the introduction of *Caulerpa* into the Mediterranean (Meinesz & Hesse 1991). Such introductions are as potentially damaging to marine ecosystems as was the introduction of the Australian plant *Hakea* to the Cape flora in South Africa.

IMPLICATIONS FOR CONSERVATION

Conservation procedures and protocols need to be tuned to the scales in time and space at which the ecosystems function. In near-shore coastal environments these scales are relatively small and seem to be close to those prevailing in many terrestrial systems. Away from the immediately inshore waters the scales start to increase as the link between the sea-floor morphology and the hydrodynamics of the water column, especially those of the surface waters, weakens. Even in near-shore waters there are important large-scale interactions as seen in the way land-use and riverine inputs can influence communities. In the time realm, the historical context whereby the fauna and flora have become established in a particular locality is still of considerable relevance. However, the dominant scales tend to be small, so that strategic approaches evolved for terrestrial conservation, such as the use of exclusive no-take reserves, can make important contributions of ensuring that enclaves of diversity are maintained. Even so, it is still uncertain as to how effective this strategy will be in the longer term in the face of natural and man-modified climatic change. With the increasing fragmentation of natural communities, movements of species, which occurred after the last glaciation as the ice retreated and coastlines advanced inland and so ensured the continuity of species survival and system functionality, will be inhibited. Polluted estuaries and engineered coastlines may prove to be barriers to dispersal.

Such inhibiting influences are unlikely to occur in open ocean systems; indeed the opposite effect seems to be the problem. The increasing speeds of ocean-going vessels and their modes of operation, species introductions intentional and unintentional (often associated with mariculture) and the building of very large engineering projects, like the building of the Suez Canal and the proposed sea-level canal across Central America, are all contributing to the removal of barriers to dispersal. The resulting changes tend to be slow but are liable to be far reaching in terms of global diversity and also to resource management.

This raises the important and fundamental question as to whether ecosystem function and process is as, or even more, important to maintain than the integrity of the species composition of the communities within the various ecosystems. This is not to denigrate the importance of maximizing the numbers of individual taxa which can survive on Earth, but the total commitment of all resources to maintaining individual species and genetic races would seem unwise. In the oceanic context, regions of maximum diversity do not coincide with the regions where the processes fundamental to the maintenance of oceanic, and for that matter the global ecological processes, are occurring. For example, switching off the biological pump would be likely to have far more dire consequences for global climate change than the loss of tens, hundreds or even thousands of the rarer taxa on which most conservation attention seems to be focused. But perhaps the apparently high redundancy of species which seems to be widespread even in the species-poor pelagic realm is illusory, and all these rarer taxa are playing important roles within the checks and balances that maintain ecological functionality. Has the changing of very high percentages of the world's terrestrial surface into managed low-diversity systems had an impact on global processes? Has the removal of top predators from marine ecosystems by fishing and whaling influenced ocean properties? The conservation of keystone species, like the krill *Euphausia superba* in the Southern Ocean ecosystem, would seem to be given a high priority by many conservationists, but to some people it presents a potentially sustainable resource which can or even must be exploited. Can the Southern Ocean system adapt to mankind's new predation lower down the foodchain, as it appears to have done to the removal of important top predators in the past? Can oceanic resources, living and non-living (including waste management operations) be exploited more heavily without seriously impeding the processes?

The large scale of nearly all open ocean systems, both in time and space, means that conservation protocols have to be scaled similarly. The response times of the systems are also likely decadal or centurial, and once begun their inertia will make them impossible to reverse within time scales that are relevant to mankind's sociopolitical institutions. The only practical approach would seem to be to establish a global management plan for the ocean that is sufficiently flexible to accommodate our rapidly developing understanding of how the systems work. International law tends to be too inflexible in its responses by seeking to establish precisely defined boundaries, often without a proper scientific basis. For example, the outer limits of exclusive economic zones has been defined in terms of technological capability to exploit. At present there are some components of such a management in place, governing transportation of dangerous cargoes and lists of chemicals whose disposal into the oceans are banned. But the main challenge will be to control the presently free access of the open ocean for transport, recreation and resource-exploitation.

ACKNOWLEDGEMENTS

I would like to thank Dr Paul Tyler and Dr Peter Edwards for their very helpful comments on the draft of this chapter which contributed several improvements. I am also very grateful to Dr Janet Campbell of the Bigelow Laboratory for Ocean Sciences, West Boothbay Harbor for permission to reproduce Plate 4.1.

REFERENCES

Alldredge, A.L. & Silver, M.W. (1988). Characteristics, dynamics and significance of marine snow. *Progress in Oceanography*, **20**, 41–82.

Angel, M.V. (**1977**). Studies on Atlantic halocyprid ostracods: Vertical distributions of the species in the top 1000 m in the vicinity of 44°N, 13°W. *Journal of the Marine Biological Association of the United Kingdom*, **57**, 239–252.

Angel, M.V. (**1989**). Vertical profiles of pelagic communities in the vicinity of the Azores Front and their implications to deep ocean ecology. *Progress Oceanography*, **22**, 1–46.

Angel, M.V. (**1991a**). Biodiversity in the oceans. *Ocean Challenge*, **2** (Spring), 28–36.

Angel, M.V. (**1991b**). Variations in time and space: Is biogeography relevant to studies of long-time scale change? *Journal of the Marine Biological Association of the United Kingdom*, **71**, 191–206.

Angel, M.V. (**1993**). Biodiversity of the pelagic ocean. *Conservation Biology*, **7**, 760–772.

Angel, M.V. & Baker, A. de C. (**1982**). Vertical standing crop of plankton and micronekton at three stations in the North-east Atlantic, *Biological Oceanography*, **2**, 1–30.

Angel, M.V. & Fasham, M.J.R. (**1975**). Analysis of the vertical and geographic distribution of the abundant species of planktonic ostracods in the North-east Atlantic. *Journal of the Marine Biological Association of the United Kingdom*, **55**, 709–737.

Angel, M.V. & Fasham, M.J.R. (**1983**). Eddies and biological processes. *Eddies in Marine Science* (Ed. by A.R. Robinson), pp. 492–524. Springer-Verlag, Berlin.

Bailey, R.S. & Steele, J.H. (**1991**). North Sea herring fluctuations. *Climate Variability, Climate Change and Fisheries* (Ed. by M.H. Glantz), pp. 213–230. Cambridge University Press. Cambridge.

Banse, K. (**1990**). Does iron really limit phytoplankton production in the offshore subarctic Pacific? *Limnology and Oceanography*, **35**, 772–775.

Barange, M. & Pillar, S.C. (**1992**). Cross-shelf circulation, zonation and maintenance mechanisms of *Nyctiphanes capensis* and *Euphausia hanseni* (Euphausiacea) in the northern Benguela Upwelling system. *Continental Shelf Research*, **12**, 1027–1042.

Barron, J.A. & Baldauf, J.G. (**1989**). Tertiary cooling steps and paleoproductivity as reflected by diatoms and biosiliceous sediments. *Productivity of the Oceans: Present and Past* (Ed. by W.H. Berger, V.S. Smetachek & G. Wefer), pp. 341–354. John Wiley, New York.

Beckmann, W. (**1984**). Mesoplankton distribution on a transect from the Gulf of Aden to the central Red Sea during the winter monsoon. *Oceanologica Acta*, **7**, 87–102.

Berger, W.H., Smetachek, V.S. & Wefer, G. (Eds) (**1989**). *Productivity of the Ocean: Present and Past*. John Wiley, New York.

Bieri, R. (**1991**). Systematics of the Chaetognatha. *The Biology of the Chaetognaths* (Ed. by Q. Bone, H. Kapp & A.C. Pierrot-Bults), pp. 122–136. Oxford Science, Oxford.

Billett, D.S.M. (**1991**). Deep-sea holothurians. *Oceanography and Marine Biology Annual Review*, **29**, 259–317.

Billett, D.S.M., Lampitt, R.S., Rice, A.L. & Mantoura, R.F.C. (**1983**). Seasonal sedimentation of phytoplankton to the deep-sea benthos. *Nature, London*, **302**, 520–522.

Boje, R. & Tomczak, M. (Eds) (**1978**). *Upwelling Ecosystems*. Springer, Berlin and New York.

Bone, Q., Kapp, H. & Pierrot-Bults, A.C. (Eds) (**1991**). *The Biology of the Chaetognaths*. Oxford University Press, Oxford.

Bouchet, P. & Taviani, M. (**1992**). The Mediterranean deep-sea fauna: pseudopopulations of Atlantic species? *Deep-Sea Research*, **39**, 169–184.

Bouchet, P. & Waren, A. (**1985**). Revision of the Northeast Atlantic bathyal and abyssal Aclididae, Eulimidae and Epitoniidae (Mollusca, Gastropoda). *Bulletin Malacologia* (suppl. **2**), 297–576.

Briggs, J.C. (**1975**). *Marine Zoogeography*. McGraw-Hill, New York.

Brinton, E. (**1975**). Euphausiids of southeastern Asian waters. *Naga Reports*, **4**, 1–287.

Brinton, E. & Reid, J.L. (1986). On the effects of interannual variations in circulation and temperature on euphausiids of the California Current. *Unesco Technical Papers in Marine Science*, **49**, 25–34.

Campbell, J.W. & Aarup, T. (1992). New production in the North Atlantic derived from seasonal patterns of surface chlorophyll. *Deep-Sea Research*, **39**, 1669–1694.

Carrillo, E., Miller, C.B. & Wiebe, P.H. (1974). Failure of interbreeding between Atlantic and Pacific populations of the marine calanoid copepod *Acartia clausi* Giesbrecht. *Limnology and Oceanography*, **19**, 452–458.

Casanova, J.-P. (1977). *La faune pelagique profonde (zooplancton ey micronecton) de la Province Atlanto-Mediterraneenne.* PhD Thesis, L'Université de Provence (Aix-Marseille).

Childress, J.J., Taylor, S.M., Caillet, G.M. & Price, M.H. (1980). Patterns of growth, energy utilisation and reproduction in some meso- and bathypelagic fishes off Southern California. *Marine Biology*, **61**, 27–40.

Clarke, A. (1992). Is there a latitudinal diversity cline in the sea? *Trends in Ecology and Evolution*, **7**, 286–287.

Cline, R.M. & Hayes, J. (Eds) (1976). Investigation of late Quaternary paleoceanography and paleoclimatology. *Geological Society of America Memoirs*, **145**, 1–464.

Colebrook, J.M. (1986). Environmental influences on long-term variability in marine plankton. *Hydrobiologia*, **142**, 309–325.

Corten, A. (1986). On the cause of recruitment failure of herrings in the central and northern North Sea in the years 1972–78. *Journal du Conseil pour Exploration de la Mer*, **42**, 281–294.

Cronin, T.M. (1987). Evolution, biogeography, and systematics of *Puriana*. Evolution and speciation in Ostracoda lll. *Journal of Paleontology, Memoir.*, **12**, 1–71.

Cronin, T.M. (1988). Geographical isolation in marine species: Evolution and speciation in Ostracoda. *Evolutionary Biology of Ostracoda, its Fundamental and Applications.* (Ed. by T. Hanai, N. Ikeya & K. Ishizaki), pp. 871–889. Elsevier and Kodansha, New York & Tokyo.

Cushing, D.H. (1986). *Climate and Fisheries.* Academic Press, London & New York.

Dickson, R.R., Meincke, J., Malmberg, S.-A. & Lee, A.J. (1988). The 'Great Salinity Anomaly' in the northern North Atlantic 1968–1982. *Progress in Oceanography*, **20**, 103–151.

Domanski, P.A. (1984). Giant larvae: Prolonged planktonic larval phase in the asteroid *Luidia sarsi. Marine Biology*, **80**, 189–195.

Donguy, J.R. (1987). Recent advances in the knowledge of the climatic variations in the tropical Pacific Ocean. *Progress in Oceanography*, **19**, 49–85.

Dooley, H.D., Martin, J.H.A. & Ellett, D.J. (1984). Abnormal hydrographic conditions in the Northeast Atlantic during the 1970s. *Rapports et Process-Verbaux des Reunions. Conseil Internationale pour l'Exploration de la Mer*, **185**, 179–187.

Dowidar, N.M. (1984). Phytoplankton biomass and primary productivity of the south-eastern Mediterranean. *Deep-Sea Research*, **31**, 983–1000.

Ekman, S. (1953). *Zoogeography of the Sea.* Sidgwick & Jackson, London.

Embley, R.W. & Hallberg, E. (1977). Distribution and morphology of large submarine sediment slides and slumps on Atlantic continental margins. *Marine Geotechnology*, **2**, 205–228.

Eppley, R.W. & Peterson, B.J. (1979). Particulate organic matter flux and planktonic new production in the deep ocean. *Nature, London*, **282**, 677–680.

Fasham, M.J.R. & Foxton, P. (1979). Zonal distribution of pelagic Decapoda (Crustacea) in the eastern North Atlantic and its relation to the physical oceanography. *Journal of Experimental Marine Biology and Ecology*, **37**, 225–253.

Flament, P., Armi, L. & Washburn, L. (1985). The evolving structure of an upwelling filament. *Journal of Geophysical Research*, **90**, 11765–11778.

Fleminger, A. (1986). The Pleistocene equatorial barrier between the Indian and Pacific Oceans and a likely cause for Wallace's Line. *Unesco Technical Papers in Marine Science*, **49**, 84–97.

Fleminger, A. & Hulsemann, K. (1974). Systematics and distribution of the four sibling species comprising the genus *Pontellina* Dana (Copepoda, Calanoida). *Fishery Bulletin US*, **72**, 63–120.

Fogg, G.E. (1986). Picoplankton. *Proceedings of the Royal Society of London (B)*, **228**, 1–30.

Fraser, J.H. (1968). The overflow of oceanic plankton to the shelf waters of the North-east Atlantic. *Sarsia*, **34**, 313–330.

Furnestin, M.-L. (1979). Aspects of the zoogeography of the Mediterranean plankton. *Zoogeography and Diversity in Plankton* (Ed. by S. van der Spoel & A.C. Pierrot-Bults), pp. 191–153. Bunge Scientific, Utrecht.

Gage, J.D. & Tyler, P.A. (1991). *Deep-sea Biology: A Natural History of Organisms at the Deep-sea Floor.* Cambridge University Press, Cambridge.

Gardner, W.D. & Walsh, I.D. (1990). Distribution of macroaggregates and fine grained particles across a continental margin and their potential role in fluxes. *Deep-Sea Research*, **37**, 401–411.

GESAMP (1991). Global changes and the air-sea exchange of chemicals. *Reports and Studies*, **48**, 1–69.

Gooday, A.J. & Turley, C.M. (1990). Responses by benthic organisms to inputs of organic material to the ocean floor: a review. *Philosophical Transactions of the Royal Society of London, A*, **331**, 119–138.

Grassle, J.F. & Maciolek, N.J. (1992). Deep-sea richness: regional and local diversity estimates from quantitative bottom samples. *The American Naturalist*, **139**, 313–341.

Grice, G.D. & Hart, A.D. (1962). The abundance, seasonal occurrence and distribution of the epizooplankton between New York and Bermuda. *Ecological Monographs*, **32**, 287–309.

Haedrich, R.L. & Merrett, N.R. (1990). Little evidence for faunal zonation or communities in deep sea demersal fish faunas. *Progress in Oceanography*, **24**, 239–250.

Hallegraeff, G.M. & Bolch, C.J. (1991). Transport of toxic dinoflagellate cysts via ship's ballast water. *Marine Pollution Bulletin*, **22**, 27–30.

Haq, B.U. (1984). Paleoceanography: A synoptic overview of 200 million years of ocean history. *Marine Geology and Oceanography of Arabian Sea and Coastal Pakistan* (Ed. by B.U. Haq & J.D. Milliman), pp. 201–231. Van Nostrand Reinold, New York.

Haury, L.R., McGowan, J.A. & Wiebe, P.H. (1978). Patterns and processes in the time–space scales of plankton distributions. *Spatial Pattern in Plankton Communities* (Ed. by J. Steele), pp. 277–328. Plenum, New York.

Hempel, G. (1985). On the biology of polar seas, particularly the Southern Ocean. *Marine Biology of Polar Regions and Effects of Stress on Marine Organisms* (Ed. by J. Gray & M.E. Christiansen), pp. 3–33. John Wiley, Chichester.

Heydorn, A.E.F. (1969). The South Atlantic rock lobster *Jasus tristani* at Vema Seamount, Gough Island and Tristan da Cunha. *Investigation Reports of the Division of Sea Fisheries, South Africa*, **73**, 1–20.

Hodell, D.A., Elmstrom, K.M. & Kennett, J.P. (1986). Latest Miocene benthic $\delta^{18}O$ changes, global ice volume, sealevel and the 'Messinian salinity crisis'. *Nature, London*, **320**, 411–414.

Hofmann, E.L., Capella, J.E., Ross, R.M. & Quetin, L.B. (1992). Models of the early life history of *Euphausia superba* – Part 1. Time and temperature dependence during the descent-ascent cycle. *Deep-Sea Research*, **39**, 1177–1200.

Holligan, P.M., Pingree, R.D. & Mardell, G.T. (1985). Oceanic solitons, nutrient pulses and phytoplankton growth. *Nature, London*, **314**, 348–350.

Hopkins, T.L., Milliken, D.M., Bell, L.M., McMichael, E.J., Hefferman, J.J. & Cano, R.V. (1981). The landward distribution of oceanic plankton and micronekton over the west Florida continental shelf as related to their vertical distribution. *Journal of Plankton Research*, **3**, 645–659.

Horwood, J.W. (1981). Management and models of marine multispecies complexes. *Dynamics of Large Mammal Populations* (Ed. by C.W. Fowler & T.D. Smith), pp. 339–360. John Wiley, New York.

Iliffe, T.M., Hart, C.W. & Manning, R.B. (1983). Biogeography and the caves of Bermuda. *Nature, London*, **302**, 141–142.

Joint, I., Pomeroy, A., Savidge, G. & Boyd, P. (1993). Size-fractionated primary productivity in the northeast Atlantic in May–June 1989. *Deep-Sea Research II*, **40**, 423–440.

Joyce, T.M. & Wiebe, P.H. (Eds) (1992). Warm-core rings: Interdisciplinary studies of Kuroshio and Gulf Stream rings. *Deep-Sea Research*, **39**, S1–S417.

Judkins, D.C. (1976). Pelagic shrimps of the *Sergestes edwardsii* species group (Crustacea: Decapoda: Sergestidae). *Smithsonian Contributions to Zoology*, **256**, 1–34.

Jumars, P.A. (1981). Limits in predicting and detecting benthic community responses to manganese nodule mining. *Marine Mining*, **3**, 213–229.

Kay, E.A. (1990). Cypraeidae of the Indo-Pacific: Cenozoic fossil history and biogeography. *Bulletin of Marine Science*, **47**, 23–34.

Keigwin, L.D. (1978). Pliocene closing of the Isthmus of Panama based on biostratigraphic evidence from nearby Pacific Ocean and Caribbean cores. *Geology*, **6**, 630–634.

Kenyon, N.H., Belderson, R.H. & Stride, A.H. (1978). Channels, canyons and slump folds on the continental slope between S.W. Ireland and Spain. *Oceanologica Acta*, **1**, 369–380.

Kohn, A.J. (1985). Evolutionary ecology of *Conus* on Indo-Pacific coral reefs. *Proceedings of the Fifth International Coral Reef Congress, Tahiti*, **4**, 139–144.

Lamb, H.H. (1977). *Climate: Present, Past and Future. Vol. 2. Climatic History and the Future.* Methuen, London.

Lampitt, R.S. (1985). Evidence for the seasonal deposition of detritus to the deep-sea floor (Porcupine Bight, N.E. Atlantic) and its subsequent resuspension. *Deep-Sea Research*, **32A**, 885–897.

Lampitt, R.S., Billett, D.S.M. & Rice, A.L. (1986). Biomass of the invertebrate megabenthos from 500 to 4100 m in the Northeast Atlantic Ocean. *Marine Biology*, **93**, 69–81.

Lampitt, R.S., Wishner, K.F., Turley, C.M. & Angel, M.V. (1993). Marine snow studies in the Northeast Atlantic: Distribution, composition and role as a food source for migrating plankton. *Marine Biology*, **116**, 689–702.

Lasker, R. (1975). Field criteria for survival of anchovy larvae: the relation between inshore chlorophyll maximum layers and successful first feeding. *Fisheries Bulletin of the United States*, **73**, 453–462.

Levitus, S., Conkright, M.E., Reid, J.L., Najjar, R.G. & Mantyla, A. (1993). Distribution of nitrate, phosphate and silicate in the world oceans. *Progress in Oceanography*, **31**, 245–273.

Lewis, M.R., Kuring, N. & Yentsch, C.S. (1988). Global patterns of ocean transparency: Implications for the new production of the open ocean. *Journal of Geophysical Research*, **93**, 6847–6856.

Losse, G.F. & Merrett, N.R. (1971). The occurrence of *Oratosquilla investigatoris* (Crustacea: Stomatopoda) in the pelagic zone of the Gulf of Aden and the equatorial western Indian Ocean. *Marine Biology*, **10**, 244–253.

Lutjeharms, J.R.E. & Heydorn, A.E.F. (1981a). The rock lobster *Jasus tristani* on Vema Seamount: Drifting buoys suggest a possible recruiting mechanism. *Deep-Sea Research*, **28**, 631–636.

Lutjeharms, J.R.E. & Heydorn, A.E.F. (1981b). Recruitment of rock lobster on Vema Seamount from the islands of Tristan da Cunha. *Deep-Sea Research*, **28A**, 1237.

Lutjeharms, J.R.E., Shillington, F.A. & Duncombe Rae, C.M. (1991). Observations of extreme upwelling filaments in the Southeast Atlantic Ocean. *Science*, **253**, 774–776.

Mantyla, A.W. & Reid, J.W. (1983). Abyssal characteristics of the world ocean waters. *Deep-Sea Research*, **30A**, 805–833.

Martin, J.H. (1991). Iron, Liebig's law and the greenhouse. *Oceanography*, **4**, 52–55.

Martin, J.H., Broenkow, W.W., Fitzwater, S.E. & Gordon, R.M. (1990). Yes it does: a reply to the comment by Banse. *Limnology and Oceanography*, **35**, 775–777.

May, R.M. (1992). Biodiversity: bottoms up for the oceans. *Nature, London*, **357**, 278–279.

McGowan, J.A. (1971). The nature of oceanic ecosystems. *The Biology of the Oceanic Pacific* (Ed. by J.A. McGowan), pp. 7–28. Oregon State University Press, Corvallis, Oregon.

McGowan, J.A. (1990). Species dominance-diversity patterns in oceanic communities. *The Earth in Transition* (Ed. by G.M. Woodwell), pp. 395–421. Cambridge University Press, Cambridge & New York.

Meinesz, A. & Hesse, B. (1991). Introduction et invasion de l'algue tropicale *Caulerpa taxifolia* en Mediterranée nord-occidentale. *Oceanologica Acta*, **14**, 415–426.

Menzies, R.J. & Wilson, J.B. (1961). Preliminary field experiments on the relative importance of pressure and temperature on the penetration of marine invertebrates into the deep sea. *Oikos*, **12**, 302–309.

Merrett, N.R. (1986). Biogeography and the ocean rim: A poorly known zone of ichthyofaunal interaction. *Unesco Technical Papers in Marine Science*, **49**, 201–209.

Merrett, N.R. (1987). A zone of faunal change in assemblages of abyssal demersal fish in the eastern North Atlantic: A response to seasonality in production? *Biological Oceanography*, **5**, 137–151.

Merrett, N.R., Gordon, J.D.M., Stehmann, M. & Haedrich, R.L. (1991). Deep demersal fish assemblage structure in the Porcupine Seabight (eastern North Atlantic): Slope sampling by three different trawls compared. *Journal of the Marine Biological Association of the United Kingdom*, **71**, 329–358.

Miller, C.B., Frost, B.W., Booth, B., Wheeler, P.A, Landry, M.R. & Welschmeyer, N. (1991). Ecological processes in the subarctic Pacific: Iron limitation cannot be the whole story. *Oceanography*, **4**, 71–78.

Miller, K.G., Fairbanks, R.G. & Mountain, G.S. (1987). Tertiary isotope synthesis, sea level history and continental margin erosion. *Paleoceanography*, **2**, 1–19.

Mittelstaedt, E. (1983). The upwelling area off Northwest Africa – A description of phenomena related to coastal upwelling. *Progress in Oceanography*, **12**, 307–331.

Mittelstaedt, E. (1991). The ocean boundary along the Northwest African coast: Circulation and oceanographic properties at the sea surface. *Progress in Oceanography*, **26**, 307–356.

Montaggioni, L.F. & Macintyre, I.G. (Eds) (1991). Reefs as recorders of environmental changes. *Coral Reefs*, special issue, **10**, 53–125.

Moran, P.J. (1986). The *Acanthaster* phenomenon. *Oceanography and Marine Biology Annual Reviews*, **24**, 379–480.

Mullineaux, L.S. & Butman, C.A. (1990). Recruitment of encrusting benthic invertebrates in boundary-layer flows: a deep-water experiment on Cross seamount. *Limnology and Oceanography*, **35**, 409–423.

New, A.L. & Pingree, R.D. (**1990**). Evidence for internal tide mixing near the shelf break in the Bay of Biscay. *Deep-Sea Research*, **37**, 1783–1803.

Palumbi, S.R. (**1992**). Marine speciation on a small planet. *Trends in Ecology and Evolution*, **7**, 114–118.

Peterson, R.G. & Stramma, L. (**1991**). Upper-level circulation in the South Atlantic Ocean. *Progress in Oceanography*, **26**, 1–74.

Pfannkuche, O. (**1993**). Benthic response to the sedimentation of particulate organic matter at the BIOTRANS station, 47°N, 20°W. *Deep-Sea Research II*, **40**, 135–150.

Phillips, B.F. & McWilliam, P.S. (**1986**). The pelagic phase of spiny lobster development. *Canadian Journal of Fisheries and Aquatic Sciences*, **43**, 2153–2163.

Pickford, G.E. (**1946**). *Vampyroteuthis infernalis* Chun. An archaic dibranchiate cephalopod. I. Natural history and distribution. *Dana Report*, **29**, 1–40.

Pierrot-Bults, A.C. (**1974**). Taxonomy and zoogeography of certain members of the '*Sagitta serratodentata*' group (Chaetognatha). *Bijdragen tot de Dierkunde*, **44**, 215–234.

Pingree, R.D. & Le Cann, B. (**1989**). Celtic and American slope and shelf residual currents. *Progress in Oceanography*, **23**, 303–338.

Por, F.D. (**1990**). Lessepsian migration. An appraisal and new data. *Bulletin de l'Institut Oceanographique, Monaco*, special no. **7**, 1–10.

Raup, D.M. (**1979**). Size of the Permo-Triassic bottleneck and its evolutionary implications. *Science*, **206**, 217–218.

Ray, G.C. (**1991**). Coastal-zone biodiversity patterns. *BioScience*, **41**, 490–498.

Razouls, C. (**1981**). *Repertoire mondial des copepodes planktoniques marins et dans eaux saumatres: Divers systemes de classification*. Laboratoire Arago – Université Pierre et Marie Curie, Banyuls-sur-Mer, France.

Reid, J., Brinton, E., Fleminger, A., Venrick, E.L. & McGowan, J.A. (**1978**). Ocean circulation and marine life. *Advances in Oceanography* (Ed. by H. Charnock & Sir George Deacon), pp. 65–130. Plenum, New York & London.

Rex, M.A. (**1983**). Geographic patterns of species diversity in the deep-sea benthos. *Deep-Sea Biology*. 8, *The Sea* (Ed. by G.T. Rowe), pp. 453–472. Wiley-Interscience, New York & Chichester.

Rice, A.L., Thurston, M.H. & New, A.L. (**1990**). Dense aggregations of a hexantellid sponge, *Pheronema carpenteri*, in the Porcupine Seabight (northeast Atlantic Ocean), and possible causes. *Progress in Oceanography*, **24**, 179–196.

Robinson, A.R., Mcgillicuddy, D.J., Calman, J., Ducklow, H.W., Fasham, M.J.R., Hoge, F.E., Leslie, W.G., McCarthy, J.J., Podewski, S., Porter, D.L., Saure, G. & Yoder, J.A. (**1993**). Mesoscale and upper ocean variabilities during the 1989 JGOFS bloom study. *Deep-Sea Research II*, **40**, 9–35.

Robinson, G.A., Aiken, J. & Hunt, H.G. (**1986**). Synoptic surveys of the western English Channel. The relationships between plankton and hydrography. *Journal of the Marine Biological Association of the United Kingdom*, **66**, 201–218.

Savidge, G., Turner, D.R., Burkill, P.H., Watson, A.J., Angel, M.V., Pingree, R.D., Leach, H. & Richards, K.J. (**1992**). The BOFS 1990 spring bloom experiment: Temporal evolution and spatial variability of the hydrographic field. *Progress in Oceanography*, **29**, 235–281.

Scheltema, R.S. (**1986**). Epipelagic meroplankton of tropical seas: Its role for the biogeography of sublittoral invertebrate species. *Unesco Technical Papers in Marine Science*, **49**, 242–249.

Schmaus, P.H. & Lehnhofer, K. (**1927**). Copepoda 4: *Rhincalanus* Dana 1852 der Deutschen

Tiefsee-Expedition. Systematik und Verbreitung der Gattung. *Wissenschaftliche Ergebnisse der Deutschen Tiefsee-Expedition auf dem Dampfer 'Valdivia'*, **23**, 355–400.

Sepkoski, J.J. (1992). Phylogenetic and ecologic patterns in the Phanerozoic history of marine biodiversity. *Systematics, Ecology and the Biodiversity Crisis* (Ed. by N. Eldredge), pp. 77–100. Columbia University Press, New York.

Sherman, K., Jones, C., Sullivan, L., Smith, W., Berrien, P. & Ejsymont, L. (1981). Congruent shifts in sand eel abundance in western and eastern North Atlantic ecosystems. *Nature, London*, **291**, 361–366.

Shih, C.T. (1979). East–west diversity. *Zoogeography and Diversity in Plankton* (Ed. by S. van der Spoel & A.C. Pierrot-Bults), pp. 87–102. Bunge Scientific, Utrecht.

Simpson, J.H., Hughes, D.G. & Morris, N.C.G. (1977). The relation of seasonal stratification to tidal mixing on the continental shelf. *A Voyage of Discovery: George Deacon 70th Anniversary Volume* (Ed. by M.V. Angel) pp. 327–340. Pergamon Press, Oxford.

Small, L.F. (Ed.) (1992). California basin studies: Physical geological, chemical and biological attributes. *Progress in Oceanography*, **30**, 1–398.

Smetachek, V. (1984). The supply of food to the benthos. *Flows of Energy and Materials in Marine Ecosystems: Theory and Practice* (Ed. by M.J.R. Fasham), pp. 517–548. Plenum, New York & London.

Smith, D.L. (1985). Caribbean plate relative motions. *The Great American Biotic Interface* (Ed. by F.G. Stehli & S.D. Webb), pp. 17–48. Plenum, New York.

Smith, S.L. (1984). Biological indications of active upwelling in the northwestern Indian Ocean in 1964 and 1979, and a comparison of zooplankton in the southeastern Bering Sea. *Deep-Sea Research*, **31**, 951–967.

Southwood, A.J. (1980). The western English Channel – an inconstant ecosystem. *Nature, London*, **285**, 361–366.

Southwood, A.J. (1984). Fluctuations in the 'indicator' chaetognaths *Sagitta elegans* and *Sagitta setosa* in the Western Channel. *Oceanologica Acta*, **7**, 229–239.

Steele, J.H. (1991). Marine functional diversity. *BioScience*, **41**, 470–474.

Stommel, H. (1963). Varieties of oceanographic experience. *Science*, **139**, 572–576.

Stuiver, M., Quay, P.D. & Ostlund, H.G. (1983). Abyssal water carbon-14 distribution and age of the World ocean. *Science*, **220**, 849–851.

Tamm, C.O. (1991). What ecological lessons can we learn from deforestation processes in the past? *SCOPE 45: Ecosystem Experiments* (Ed. by H.A. Mooney, E. Medina, D.W. Schindler & B.H. Walker), pp. 45–58. John Wiley, Chichester & New York.

Thiel, H. (1983). Meiobenthos and nanobenthos of the deep sea. *Deep-Sea Biology*, 8. *The Sea* (Ed. by G.T. Rowe), pp. 167–230. Wiley-Interscience, New York & Chichester.

Thiel, H. & Schreiver, G. (1990). Deep-sea mining, environmental impact and the DISCOL project. *Ambio*, **19**, 245–250.

Thorson, G. (1955). Modern aspects of marine level-bottom animal communities. *Journal of Marine Research*, **14**, 387–397.

Tont, S.A. (1987). Variability of diatom species populations: From days to years. *Journal of Marine Science*, **45**, 985–1006.

Tunnicliffe, V. (1991). The biology of hydrothermal vents: Ecology and evolution. *Oceanography and Marine Biology Annual Reviews*, **29**, 319–407.

Van Camp, L., Nykjaer, L., Mittelstaedt, E. & Schlittenhardt, P. (1991). Upwelling and boundary circulation off Northwest Africa as depicted by infrared and visible satellite observations. *Progress in Oceanography*, **26**, 357–402.

van der Spoel, S. (1971). Geographical variation in *Cavolinia tridentata* (Mollusca, Pteropoda). *Bijdragen tot de Dierkunde*, **44**, 100–112.

van der Spoel, S. & Heyman, R.P. (1983). *A Comparative Atlas of Zooplankton: Biological Patterns in the Ocean*. Bunge, Utrecht.

van der Spoel, S. & Pierrot-Bults, A.C. (Eds). (1979). *Zoogeography and Diversity in Plankton*. Bunge Scientific, Utrecht.

van der Spoel, Pierrot-Bults, A.C. & Schalk, P. (1990). Probable Mesozoic vicariance in biogeography of Euphausiacea. *Bijdragen tot de Dierkunde*, **60**, 155–162.

Venrick, E.L. (1982). Phytoplankton in an oligotrophic ocean: Observation and questions. *Ecological Monographs*, **52**, 129–154.

Verstraete, J.-M. (1992). The seasonal upwellings in the Gulf of Guinea. *Progress in Oceanography*, **29**, 1–60.

Waser, N.M. & Price, M.V. (1989). Optimal outcropping in *Ipomopsis aggregata:* Seed set and offspring fitness. *Evolution*, **43**, 1097–1109.

Weaver, P.P.E. (1993). High resolution stratigraphy of Quaternary sequences. *High Resolution Stratigraphy* (Ed. by E.A. Hailwood & R.B. Kidd), pp. 137–153. Geological Society Special Publication, London.

Whatley, R.C. & Coles, G.P. (1991). Global change and the biostratigraphy of North Atlantic Cainozoic deep water Ostracoda. *Journal of Micropalaeontology*, **9**, 119–132.

Wiebe, P.H. & McDougall, T.J. (Eds). (1986). Warm core rings: Studies of their physics, chemistry and biology. *Deep-Sea Research*, **33**, 1455–1922.

Young, C.M. & Tyler, P.A. (1993). Embryos of the deep-sea echinoid *Echinus affinis* require high pressure for development. *Limnology and Oceanography*, **38**, 178–181.

5. METAPOPULATIONS AND CONSERVATION

SUSAN HARRISON

Division of Environmental Studies, U. C. Davis, Davis, CA 95616, USA

SUMMARY

Metapopulation theory is now a popular framework for understanding the threats faced by species in fragmented habitats. Spatial simulation models have recently been introduced as tools in endangered species management. Here I evaluate metapopulation theory as both a conceptual and a modelling tool in conservation.

In its strongest form, the metapopulation concept implies that species exist in a fragile balance between extinction and colonization of populations. However, on empirical grounds this appears to be a misleading model. A more realistic and less powerful definition of a metapopulation is any set of conspecific populations, possibly but not necessarily interconnected. Thus defined, metapopulation structure may have great or little relevance to a species' viability. This must be evaluated on a case-by-case basis, paying careful attention to alternative hypotheses.

Habitat fragmentation is a central concern in conservation biology, and no single generalization encompasses its possible effects. A tactic recently employed in population viability analysis for the northern spotted owl is spatially explicit computer simulation, in which real landscape geometry is integrated with demographic and dispersal data. This approach holds promise, but problems in the spotted owl case suggest a limitation: success may only be possible for extremely well-studied species.

INTRODUCTION: METAPOPULATIONS, A NEW CONSERVATION PARADIGM?

Island biogeography theory and the rise of academic conservation biology were not coincidental. In the MacArthur–Wilson model, it seemed that ecologists could make a concrete, general, and non-obvious prediction: diversity in habitat isolates is shaped by distance-dependent rates of immigration and area-dependent rates of extinction. The resulting prescriptions that reserves should be large, rounded, and close or connected to other reserves, continue to be offered as 'principles of conservation biology', despite considerable controversy (e.g. Simberloff & Abele 1982; Wilcox & Murphy 1985; Quinn & Harrison 1988; Simberloff 1988).

Island biogeography considers only aggregate diversity, not the fates of individual species. Metapopulation theory has now become the analogous paradigm for single-species populations, as evidenced in recent conservation texts (e.g. Soulé 1987; Soulé & Kohm 1989; Western & Pearl 1989; Falk & Holsinger 1991; Fiedler & Jain 1992) and in conservation strategies for the black-footed ferret, furbish lousewort, northern spotted owl, bay checkerspot butterfly, bighorn sheep, Stephens kangaroo rat, and other US threatened and endangered species (Seal *et al.* 1989; Bleich *et al.* 1989; Murphy *et al.* 1990; Thomas *et al.* 1990; Bean *et al.* 1991; Menges 1991). European landscape ecologists have likewise joined what Hanski and Gilpin (1991) term the metapopulation 'vogue' in conservation (e.g. Van Dorp & Opdam 1987; Laan & Verboom 1990; Opdam 1990; Verboom *et al.* 1991).

Like island biogeography, metapopulation theory offers the attractive prospect of a unifying principle for conservation biology. My goal here is a careful examination of the ideas and evidence underlying any such principle. I will first describe the message drawn from simple metapopulation models, and why I consider it to be false; second, present a more realistic and less unified view of metapopulations; and third, discuss the strengths and limitations of spatial simulation models as tools for analyzing the effects of fragmentation.

METAPOPULATIONS IN CONSERVATION THEORY

Metapopulation ideas and their application to conservation are well reviewed by Hanski (1989) and Hanski and Gilpin (1991), and only a brief sketch is given here. The idea of 'shifting mosaic' population dynamics originated in the empirical work of Watt (1947), Andrewartha and Birch (1954) and Ehrlich (1961) on species occupying naturally patchy and transient habitats. Levins (1970) coined the term 'metapopulation' to denote a network of extinction-prone conspecific populations. In Levins' formulation, local populations go extinct at equal and independent rates, and contribute equally to the recolonization of vacant patches of habitat. Metapopulation, or regional, persistence requires that the rate of colonization equal or exceed that of local extinction.

Variations on this single-species model are reviewed by Hanski (1991), and its extensions to competitive and predator–prey systems by Bengtsson (1991) and Taylor (1991). In nearly all such models, as in Levins' original version, regional persistence requires a positive balance between coloniz-

ation and extinction. Persistence is more likely the larger the metapopulation, in terms of number of patches, and the higher and more consistent
the rate of dispersal between patches.

Species whose habitat is naturally patchy or transient are widely held
to fit such a model (e.g. Murphy *et al.* 1990). More significantly, the
model is an increasingly popular framework for viewing the effects of
habitat fragmentation on the viability of once-continuous populations. By
creating increasingly isolated remnant populations, fragmentation is seen
as increasing rates of local extinction and decreasing those of recolonization. Thus a once-widespread species may collapse rather suddenly to
regional extinction when too few fragments of habitat remain, or when
the flow of dispersers among them is disrupted. Additional destructive
feedbacks may be generated by inbreeding (Gilpin & Soulé 1986; Gilpin
1987; see below), and by secondary 'metacommunity' effects (Nee & May
1992).

The metapopulation paradigm in conservation biology stresses the
maintenance of an extinction-colonization balance. A successful strategy
requires conserving numerous habitat patches and the potential for dispersal between them ('connectivity', in landscape ecology terms). Systems
of multiple reserves, connected by corridors, are commonly proposed as a
tactic to achieve this.

The verbal and mathematical model supporting this principle is a
simplified one, however. It omits such real-world features as variability
in the size (longevity) of habitats or local populations (Gotelli 1991;
Harrison 1991; Schoener 1991); rates of colonization that vary with
interpatch distance (Urban & Shugart 1986; McKelvey *et al.* 1992) or with
behavioural responses to conspecifics (Smith & Peacock 1990; Hansson
1991; Ray *et al.* 1991); and temporal correlation in extinction probabilities
between nearby populations (Price & Endo 1989; Harrison & Quinn
1989). It thus seems appropriate to review carefully the empirical evidence
on both natural and human-generated metapopulation dynamics.

METAPOPULATIONS IN NATURE, A META-REVIEW

A few species described in the conservation literature appear to exist as
single populations, for example, the Devil's Hole pupfish, confined to one
desert spring (Moyle & Williams 1990), and the Amargosa vole (*Microtus
californicus scirpensis*), found in a single watershed (Murphy & Freas
1988). But if it is granted that most species consist of multiple populations,
discontinuous at some scale, does it follow that most species persist in a
balance between extinction and recolonization? To answer this I briefly

summarize an earlier review of metapopulation studies (Harrison 1991), in which I found very few instances of an extinction–colonization balance. Instead, many systems bore superficial resemblances to the classic Levin-type metapopulation, but deviated from it substantially in one of several ways.

In 'mainland–island metapopulations', one or more populations persist indefinitely. Extinction and colonization occur among some populations, but are nearly irrelevant to metapopulation persistence, since this is assured by the mainland. The mainland may be a single large or high-quality habitat patch, as in the case of a metapopulation of the checker-spot butterfly that consisted (in 1987) of one population of >500 000 and 9 populations of 10–400 butterflies (Harrison *et al.* 1988; Harrison 1989). Alternatively, the same effect may be produced by substantial variation in patch or population size (or quality), as in metapopulations of five species of orb spiders on 108 Bahamanian islands (Schoener & Spiller 1987). Schoener (1991) suggests that mainland–island dynamics are the rule in many natural metapopulations, especially in long-lived species.

'Patchy populations' are found on sets of patches fragmented at a finer scale than that of the population. Species of naturally patchy habitats are typically good dispersers, and thus tend to form continuous populations over arrays of many patches. For example, three insect herbivores of ragwort (*Senecio jacobaea*) readily dispersed hundreds of metres, while ragwort patches in a 1.3 km^2 study area were seldom >50 m from other patches (Harrison & Thomas 1991). For species whose habitat has been fragmented by humans, much depends on the pattern of fragmentation. For example, although Opdam (1990) regards birds in remnant European forest patches as forming metapopulations, Haila (1986) and Rolstad (1991) counter that forest fragments are nearly always fine-grained relative to the vagility of birds, and thus produce patchy distributions of individuals rather than populations. Fine-grained fragmentation may influence viability, but to discuss it in terms of extinction and colonization seems inappropriate.

In 'non-equilibrium metapopulations', conspecific populations are virtually or completely isolated from one another. To the extent that local extinctions occur, the species is experiencing regional extinction, one population at a time. A well-known natural example is the boreal mammal populations on mountaintops in the US Southwest (Brown 1971), which have been isolated by post-Pleistocene climatic change. For many species in human-fragmented habitats, dispersal between fragments occurs so seldom (if ever) that natural recolonization is scarcely possible. Populations of red-cockaded woodpeckers in remnant conifer forests in the southeastern US (Walters 1991; Stangel *et al.* 1992), and of three endangered amphibians in old-growth forests in the Pacific Northwest (Welsh 1990) exist in such mutual isolation that, for nearly all practical purposes, each population must be regarded as an independent entity. When recolonization (either actual or potential) is sufficiently slow, the fate of a species will probably be decided by forces acting much faster than its metapopulation dynamics.

Several species in the metapopulation literature that did show a reasonable fit to the classic model shared two common features: an early-successional habitat, and an unexceptional dispersal ability. One such species is the pool frog *Rana lessonae* in temporary ponds in northern Sweden (Sjogren 1991a,b). Others are butterflies which form small colonies, are fairly sedentary, and use habitats created by small-scale disturbances such as fire or grazing (Stewart & Ricci 1988; Thomas 1991; Thomas 1994). Among successional plants, some evidence of similar dynamics exists for the endangered furbish lousewort (*Pedicularis furbishae*) on unstable riverbanks in Maine (Menges 1990, 1991), and for mosses on decaying logs in Scandinavian forests (Herben & Soderstrom 1992).

In the cases just described, changes in habitat dynamics are creating a real danger of metapopulation collapse. The pool frog is threatened by the drainage and ditching of ponds for forestry (Sjogren 1991a), and the succession-dependent butterflies by changes in land-management practices (Thomas 1991; Thomas 1994). Successful management for these species requires creating unbroken series of successional stages, in a spatial mosaic fine enough to permit constant recolonization (e.g. Warren *et al.* 1984). Large-scale strategies such as multiple reserves and corridors are less relevant.

Aside from the foregoing class of exceptions, few studies were suggestive of classic metapopulation dynamics, in either naturally or unnaturally patchy habitats. One obvious explanation is the difficulty of the field studies required. Basic tasks include identifying all the local populations and suitable habitats within a region, and estimating the rates of extinction and colonization among them. In few natural systems are populations and habitats discrete enough, or scales of spatial and time modest enough, to make these tasks tractable.

But I will venture that the more basic reason for the scarcity of 'real' metapopulations is that their essence, persistence in a balance between the extinction and recolonization of populations, is an improbable condition. To attain it, local populations must be roughly equal in size (or longevity); isolated enough to constitute separate populations; and yet interconnected enough to permit recolonization to balance extinction. Failing any of these criteria, a system falls into one or another of the above three categories of (non-) metapopulations, in each of which persistence is more dependent on within-population than metapopulation processes. This may be the case in many superficially metapopulation-like systems.

Is it coincidence that successional species provide the best approximations to classic metapopulation dynamics? I suggest it is not, for two reasons. First, a temporary habitat means that every population is regularly subject to extinction; mainlands are not possible. Second, the requirement for just enough subdivision between populations, but not too much, is less critical. As Thomas (1994) points out, habitat dynamics drive the metapopulation dynamics of successional species; disturbance and succession govern colonization and extinction. Thus, as long as the species is not such an excellent disperser as to track its shifting habitat perfectly, it will show some metapopulation features, such as absence from suitable habitats and a vulnerability to regional collapse.

Murphy et al. (1990) raise the related idea that classic metapopulation dynamics typify 'small-bodied short-lived species with high reproductive rates and high habitat specificity', and therefore form a conceptual basis for the conservation of invertebrates, annual plants and small mammals. Several caveats are in order, however. First, many such species occupy permanent habitats and show mainland–island dynamics, e.g. the bay checkerspot butterfly (Harrison et al. 1988) and Bahamanian spiders (Schoener & Spiller 1987). Second, many are extremely good dispersers and show little tendency toward population subdivision, e.g. insects on ragwort (Harrison & Thomas 1991). Third, many successional species escape in time through seed banks, longevity and facultative dormancy (Grubb 1988; Hanski 1988), rather than escaping in space through dispersal and recolonization.

Are all species truly metapopulations, if a long enough time scale is considered? It is sometimes argued that this is so, because even large populations must eventually go extinct, and long-lived species such as large mammals simply have slower metapopulation dynamics than do insects. In a limited sense this may be true. However, if we extrapolate from better-studied metapopulations at smaller scales, it seems likely that many larger-scale ones are of the mainland–island or non-equilibrium types, and also that their dynamics are often driven by long-term habitat change, as Thomas (1994) argued is true in the shorter term for butterflies. Long-term habitat change may be impossible to predict or control, and may be far from the most immediate threat to a species' survival.

REDEFINING METAPOPULATION DYNAMICS

The strongest principle that metapopulation theory can offer conservation, the idea of maintaining species in a balance between extinction and colonization, appears suspect on empirical grounds. Both natural patchiness and fragmentation are likelier to create mainlands-and-islands, patchy populations, or non-equilibrium metapopulations than sets of populations in such a delicate balance. This is not merely academic bone-picking; the metapopulation concept is being taken seriously by managers, and taken too literally could lead to the 'principles' that single, isolated populations are always doomed, or that costly strategies involving multiple connected reserves are always necessary.

It seems necessary to adopt a broader and vaguer view of metapopulations as sets of spatially distributed populations, among which dispersal and turnover are possible but do not necessarily occur. Such a definition leaves little hope for strong generalizations about the role or importance of metapopulations. A possible way forward is to ask, in each specific case, 'what is the relative importance of among-population processes, versus within-population ones, in the viability and conservation of this species?' One approach to this question is described in the next section. But bearing in mind what a difficult question it is, I first reiterate some reasons why metapopulation considerations are not always central to conservation strategies.

One is the possibility that one or a few populations are either much more persistent, or more amenable to being made so through conservation efforts, than others. Another is the possibility that populations are (unavoidably) so isolated from one another that there is little option but to manage each individually. Still another is that neighbouring populations may not be independent enough of one another for a 'metapopulation' strategy to be effective. For example, Price and Endo (1989) found that the Stephens kangaroo rat (*Dipodomys stephensi*) undergoes extreme yearly fluctuations in abundance, caused by the effect of weather on seed supplies; these are strongly correlated among sites over time. Price and Endo (1989) therefore conclude that a large reserve would be more effective than the network of small ones which the US Fish and Wildlife Service proposed on the basis of the metapopulation concept (Bean *et al.* 1991).

There must be cases in which metapopulation aspects of viability are of critical importance, however. For example, desert bighorn sheep (Bleich *et al.* 1989) and acorn woodpeckers (Stacey & Taper 1992) may exist as networks of small populations, avoiding both local and regional extinction through infrequent dispersal. The evidence is circumstantial in both cases: observations of bighorns traversing non-habitat areas, and the failure of a demographic model to produce local persistence without immigration, respectively. But this represents the sort of possibility with which large-scale conservation must contend.

SPATIAL MODELS IN CONSERVATION BIOLOGY

How can the spatial structure of populations and habitats be addressed in conservation strategies? By 'spatial structure' I now mean subdivision at any scale, along with spatial variation in habitat quality. Theory indicates that complex outcomes may arise from interactions between local population dynamics, limited dispersal, and habitat geometry (e.g. Hassell *et al.* 1991; Hastings 1993). Empirical studies indicate the importance of such phenomena as 'rescue effects' within networks of populations (Bleich *et al.* 1989; Stacey & Taper 1992); fine-grained habitat fragmentation (Urban & Shugart 1986; Opdam 1990); edge effects (Haila 1986; Temple 1986), and source-sink dynamics (Howe *et al.* 1991; McKelvey *et al.* 1992). But despite the growing recognition of its importance, spatial structure is not yet incorporated in most population viability analyses, studies that attempt to determine the conservation requirements of specific species or populations (Soulé 1987; Boyce 1992).

New techniques for spatial simulation modelling promise to change this situation (Burgman *et al.* 1993). Spatial models typically represent populations as occupying habitat 'cells', each of which is a pair territory or other spatial subunit. Dispersal is restricted to neighbouring cells. Both within-cell dynamics and dispersal patterns can be made as complex and realistic as data allow. Habitat geometry and quality can be tailored to match real landscapes. Like more abstract metapopulation models, these show the potential for habitat fragmentation to cause population collapse. However, there is considerably greater capacity for demographic realism. The best existing example, a recent model of the northern spotted owl (Lamberson *et al.* 1992; McKelvey *et al.* 1992), serves to illustrate well, both the strengths and limitations of this new technique.

The spotted owl (*Strix occidentalis caurina*) is an archetypal victim of fragmentation. Its once-continuous habitat, ancient coniferous forest in the northwestern US, now exists only in small remnants totalling 20% of its former extent. Owls require territories of 1000 ha or more, and do not readily cross young forest. Preservation of a viable owl population is required both by the Endangered Species Act and by the laws governing the US Forest Service, owner of 70% of the owl's remaining habitat.

Under considerable pressure from timber and environmental interests to neither under- nor over-estimate the owl's requirements, the US Government assigned a commission of wildlife biologists to determine how much forest, in what spatial configuration, must be set aside for the owl.

The Interagency Scientific Commission (ISC) (Thomas *et al.* 1990) assembled available biological and demographic information on the owl, and assessed possible reserve designs using spatial simulation models (described in detail by Lamberson *et al.* 1992; McKelvey *et al.* 1992). In the models, each cell represented an owl pair territory, within which pair dynamics were modelled using field-measured rates of pair formation, fecundity and survival. Dispersal was estimated from studies of radio-tagged juvenile owls. Building on earlier work by Lande (1987, 1988a) and Doak (1989), the models linked owl demography to forest fragmentation through juvenile dispersal success. To join the breeding population, a newly fledged owl must find a vacant suitable territory, and its chances of success depend on the proportion of the landscape that is forested.

Like earlier spatial models of the owl population (Lande 1987, 1988a; Doak 1989), these models demonstrated the existence of a critical habitat threshold. When less than about 20% of the landscape was forested, recruitment of juveniles to the breeding population was too low to balance mortality of adult territory holders, and the population collapsed. When the model was elaborated to require juveniles to find mates as well as empty territories, the result was a second threshold, an owl density below which mate-finding failure led to extinction.

The 'territory-cluster model' examined the effects of patch configuration for a fixed total area of forest. Territories were arranged in clusters representing fragments of forest. Juveniles were assumed to disperse more readily within than between clusters. This model showed that the owl population seldom persisted in a universe of five-territory clusters, but persisted much better in clusters of 10. Gains in persistence were much smaller as cluster size increased from 10 to 20.

The ISC Strategy called for 7.7 million acres of old-growth forest to be preserved in a system of patches ('habitat conservation areas' or HCAs) each large enough for ⩾20 territories and ⩽12 miles from another HCA. Between HCAs, the forest was to be harvested lightly in a manner consistent with owl dispersal. Results from the models were presented in support of this design. The ISC strategy was hailed by some conservation biologists as the most advanced and scientific population viability analysis ever performed (Murphy & Noon 1992).

However, in ensuing legal battles, critics identified several weak points in the models (Harrison *et al.* 1993). One problem concerned the search behaviour of dispersing juvenile owls. In the cluster model, juveniles first search thoroughly within their home cluster, and if unsuccessful, move away in linear paths until reaching a new cluster. The ISC report notes that the results depend fairly strongly on this very efficient search behaviour. Little information exists on how real juvenile movements respond to variable landscapes. A second problem was habitat geometry; the cluster model portrayed an abstract universe of round patches and no edges. More advanced versions of the model have shown that both irregular forest patches and 'sink' effects, i.e. juveniles settling in the unsuitable inter-patch matrix, may greatly affect the viability of the owl population (McKelvey *et al.* 1992).

The most serious issue was the equilibrium nature of the model. Designated HCAs presently contain only 30% (on average) old-growth forest, the rest being logged forest which will regrow in 50–100 years' time. The ISC strategy would meanwhile allow cutting of 500 000 old-growth acres outside of HCAs. In the worst case, according to the ISC, the owl population could decline by 50–60% during this period of net habitat loss. What neither mentioned nor analyzed was the possibility that the owl could become extinct during this period. Recent information suggests that near-term extinction is a serious risk; analysis of 4–7 years' data from 2000+ marked owls shows the population declining by 7.5–10% per year throughout its range (Anderson & Burnham 1992).

A final conclusion to emerge from courtroom scrutiny of the owl models is that monitoring is not an adequate safeguard. The sparsity of data, the longevity of adult owls, and the possible existence of a pool of 'floaters' (owls lacking territories), combine to produce a long time lag between the point when a minimum threshold of habitat or population density is reached and the time when the resulting population crash is evident. It may be true of rare species in general that when declines become statistically detectable, the fate of the population is already sealed (Taylor & Gerrodette 1993).

Ruling in environmentalist lawsuits in 1991 and 1992, a US district judge barred the Forest Service from selling timber under the above-described plan, ruling that it carried unacknowledged risks to the owl. The message of this story is that a usefully realistic spatial model demands an enormous amount of information. Even for as well-studied an endangered species as the spotted owl, conclusions remain open to doubt because of assumptions that must be made about key aspects of demography and dispersal.

US conservation legislation places strong emphasis on population viability. For example, the Forest Service is required to maintain viable populations of all native vertebrates, and to designate 'management indicator species' whose well-being ensures that of the entire forest eco-system. There is thus a ready market for improved techniques, including spatial simulation models.

For example, a recent Draft Management Plan for the Flathead National Forest in Montana included spatial models for the pine marten, barred owl and pileated woodpecker, the forest's three management indicator species. These analyses overlooked all impacts of logging other than on the dispersal success of juveniles of the three species. Juveniles were assumed to move in straight paths between forest patches, and to wait at occupied patches until the patch owner died. Dispersal success was a linear function of distance, with parameters estimated from the following sources: nine banded pileated woodpeckers, the assumption that barred owls disperse identically to spotted owls, and an unspecified number of marked pine martens in a 1959 study. This model was used to evaluate three management options, and purportedly showed the adequacy of Alternative 2 (setting aside 10% of loggable old-growth habitat).

Unfortunately, this is not an isolated example of the misuse of models in population viability analysis (Doak & Mills, in press). The idea that estimates of extinction probabilities for rare and little-studies species can serve as guidelines for managing ecosystems seems ill-advised. One sensible alternative may be to compare the abundances of species of concern in forests of varying degrees of fragmentation (e.g. Rosenberg & Raphael 1986).

A GENETIC CODA

Genetic concerns are part of the reason for the interest in metapopula-tions in conservation. Existing theory suggests that metapopulation struc-ture may have profound effects on total genetic variation (see for example, Lacy 1987; Lande & Barrowclough 1987; Wade & McCauley 1988; McCauley 1991). The subdivision of a species into partly isolated sub-populations can either increase or decrease total genetic variation, because while each subpopulation loses variation due to drift, the differentiation among subpopulations increases. However, local extinctions cause the genetic variation both within and between subpopulations to be ultimately lost. The effective size of a metapopulation with turnover may be several orders of magnitude smaller than its census size (Gilpin 1991).

Theory further suggests that the consequences of turnover will vary among species. Species that naturally exist as metapopulations should be low in heterozygosity, but also relatively immune to inbreeding depression because they lack genetic load. Recently fragmented populations, in contrast, are expected to undergo a substantial loss of heterozygosity, accompanied by inbreeding depression. Inbreeding may accelerate local extinction, in turn increasing the loss of genetic variation, a feedback process that Gilpin and Soulé (1986) term an extinction 'vortex'.

One attempt to apply such theory in viability analyses is a proposed strategy for reintroducing the black-footed ferret, *Mustela nigripes* (Foose 1989). Now nearly extinct in the wild, the ferret may once have formed metapopulations in which local populations inhabited prairie-dog colonies (Brussard & Gilpin 1989). Despite very low heterozygosity, and episodes of canine distemper, the captive population has done surprisingly well and may soon be ready for reintroduction (May 1989). Foose's (1992) plan would 'use fragmentation to advantage' by distributing individuals among colonies with the goal of maximizing total genetic variation.

Empirical evidence on the importance of metapopulation genetics is extremely scarce. One of the few studies to address the genetic consequences of turnover in the field is Sjogren's (1991a,b) work on *Rana lessonae*. Frogs in the northern Swedish metapopulation averaged <1% heterozygosity at 31 loci, while more stable populations in Central Europe averaged 5% at 28 loci. Yet the Swedish frogs showed no reduced fertility or other signs of inbreeding depression. As expected, turnover reduces heterozygosity, yet produces little effect on fitness in this species.

There is less evidence for the converse expectation that genetic problems will be serious in recently fragmented species. For example, the red-cockaded woodpecker exists in populations of about 5–1000 that have been isolated for about 100 years. Heterozygosity at 16 loci in 26 populations was quite high within populations (average heterozygosity = 7.8%) regardless of their sizes, and among-population variation was also high (FST = 14%) (Stangel *et al.* 1992). There were no signs of either reduced variability or reduced fitness as a consequence of fragmentation.

Lande (1988b) contends that genetic problems are unlikely be major threats to the viability of wild populations. Populations small enough to inbreed severely are either naturally small, thus immune to inbreeding depression, or recently reduced, in which case they nearly always face more immediate threats of extinction. None the less, genetic principles remain the basis for most viability analyses, such as recent plans for the red-cockaded woodpecker (Escano 1991) and golden-cheeked warbler (USFWS 1992). This serves to illustrate the power of rules and formulas, even weakly supported ones, in conservation biology.

CONCLUSIONS

Many if not most threatened species manifest some of the hallmarks of metapopulation dynamics, such as patchy distributions, local population turnover, and low heterozygosity. On close examination, however, few well-studied species appear to fit the classic metapopulation model or to exhibit the dynamic fragility it implies. The metapopulation concept is a useful tool for thought, but should not be applied too broadly or too literally. Most importantly, it should not be the basis for concluding that single populations are always doomed, or that multiple interconnected reserves are always a necessary strategy.

In assessing the importance of metapopulation proccesses in con-servation, there is little choice but to consider each case individually, with the various alternatives firmly in mind. Useful questions to ask include: Are data available on the distribution of populations and habitats, on local demography, and on the rates and patterns of regional dispersal? Do some, or all, local populations seem likely to become extinct within a short time in the absence of immigration? Are populations possibly connected enough (actually or potentially) that dispersal could lead to rescue or recolonization? If the answer is yes to all these questions, spatial simulation models may be a valuable tool for synthesizing and interpreting the available data, and for 'testing' alternative conservation strategies.

The value of metapopulation theory in conservation biology, I con-clude, is very analogous to that of island biogeography theory: it suggests hypotheses to test and important data to gather, but can produce neither powerful generalizations nor ready-to-use formulas. As Doak and Mills (in press) suggest, it may be both unrealistic and dangerous to promote general 'principles of conservation biology', as is sometimes done on the grounds that non-academics must be presented with simple rules. The alternative is to accept that conservation biology is an essentially empirical science, in which theory provides guidance, but in which each case re-quires its own appropriate hypotheses; and that in the practical arena, we may do better to explain than to hide the complexity and uncertainties involved.

ACKNOWLEDGEMENTS

I thank Robert M. May and Chris D. Thomas for comments.

REFERENCES

Anderson, D.R. & Burnham, K.P. (1992). Demographic analysis of northern sportted owl populations. Recovery Plan for the Northern Spotted Owl, Appendix C. US Fish and Wildlife Service, Portland, Oregon, USA.

Andrewartha, H.G. & Birch, L.C. (1954). *The Distribution and Abundance of Animals*. University of Chicago Press, Chicago.

Bean, M.J., Fitzgerald, S.G. & O'Connell, M.A. (1991). Reconciling conflicts under the Endangered Species Act. WWF, Washington.

Bengtsson, J. (1991). Interspecific competition in metapopulations. *Metapopulation Dynamics: Empirical and Theoretical Investigations*. (Ed. by M.E. Gilpin & I. Hanski), pp. 219–237. Academic Press, London.

Bleich, V.C., Wehausen, J.D. & Holl, S.A. (1989). Desert-dwelling mountain sheep: conservation implications of a naturally fragmented distribution. *Conservation Biology*, 4, 383–390.

Boyce, M.S. (1992). Population viability analysis. *Annual Review of Ecology and Systematics*, 23, 481–506.

Brown, J.H. (1971). Mountaintop mammals: nonequilibrium insular biogeography. *American Naturalist*, 105, 467–478.

Brussard, P.F. & Gilpin, M.E. (1989). Demographic and genetic problems of small populations. *Conservation Biology and the Black-Footed Ferret*. (Ed. by U.S. Seal, E.T. Thorne, M.A. Bogan, & S.H. Anderson), pp. 37–48. Yale University Press, New Haven.

Burgman, M.A., Ferson, S. & Akcakaya, H.R. (1993). *Risk Assessment in Conservation Biology*. Chapman & Hall, New York.

Doak, D. (1989). Spotted owls and old growth logging in the Pacific Northwest. *Conservation Biology*, 3, 389–396.

Doak, D. & Mills, S. (1993). A useful role for theory in conservation. *Ecology*, (in press).

Ehrlich, P.R. (1961). Intrinsic barriers to dispersal in the checkerspot butterfly *Euphydryas editha*. *Science* 134, 108–109.

Escano, R.E.F. (1991). Technical review draft: Long-term red-cockaded woodpecker management strategy. USDA Forest Service, Southern Region, Atlanta, GA.

Falk, D.A. & Holsinger, K.E. (1991). *Genetics and Conservation of Rare Plants*. Oxford University Press, Oxford.

Fiedler, P.L. & Jain, S.K. (1992). *Conservation Biology: the Theory and Practice of Nature Conservation, Preservation and Management*. Chapman & Hall, New York.

Foose, T.L. (1989). Species survival plans: the role of captive propagation in conservation strategies. *Conservation Biology and the Black-footed Ferret* (Ed. by U.S. Seal, E.T. Thorne, M.A. Bogan & S.H. Anderson), pp. 210–222. Yale University Press, New Haven.

Gilpin, M.E. (1987). Spatial structure and population viability. *Viable Populations for Conservation* (Ed. by M.E. Soule), pp. 125–139. Cambridge University Press, Cambridge.

Gilpin, M.E. (1990). Extinction of finite metapopulations in correlated environments. *Living in a Patchy Environment* (Ed. by B. Shorrocks & I.R. Swingland), pp. 177–186. Oxford Science Publications, Oxford.

Gilpin, M.E. (1991). The genetic effective size of a metapopulation. *Metapopulation Dynamics: Empirical and Theoretical Investigations* (Ed. by M.E. Gilpin & I. Hanski), pp. 165–175. Academic Press, London.

Gilpin, M. & I. Hanski. (1991). *Metapopulation Dynamics: Empirical and Theoretical Investigations.* Academic Press, London.

Gilpin, M.E. & Soulé, M.E. (1986). Minimum viable populations: processes of species extinction. *Conservation Biology: The Science of Scarcity and Diversity* (Ed. by M.E. Soulé), pp. 19–34. Sinauer, Sunderland, MA.

Gotelli, N.J. (1991). Metapopulation models: the rescue effect, the propagule rain, and the core–satellite hypothesis. *American Naturalist,* **138**, 768–776.

Grubb, P.J. (1988). The uncoupling of disturbance and recruitment, two kinds of seed bank, and persistence of plant populations at the local and regional scales. *Annales Zoologici Fennici,* **25**, 23–36.

Haila, Y. (1986). Northern European land birds in forest fragments: evidence for area effects? *Wildlife 2000: Modeling Habitat Relationships of Terrestrial Vertebrates* (Ed. by J. Verner, M.L. Morrison & C.J. Ralph), pp. 315–320. University of Wisconsin Press, Madison, Wisconsin.

Hanski, I. (1988). Four kinds of extra long diapause in insects: a review of theory and observations. *Annales Zoologici Fennici,* **25**, 37–54.

Hanski, I. (1989). Does it help to have more of the same? *Trends in Ecology and Evolution,* **4**, 113–114.

Hanski, I. (1991). Single-species metapopulation dynamics: concepts, models and observations. *Metapopulation Dynamics: Empirical and Theoretical Investigations* (Ed. by M.E. Gilpin & I. Hanski), pp. 17–38. Academic Press, London.

Hanski, I. & Gilpin, M.E. (1991). Metapopulation dynamics: brief history and conceptual domain. *Metapopulation Dynamics: Empirical and Theoretical Investigations* (Ed. by M.E. Gilpin & I. Hanski), pp. 3–16. Academic Press, London.

Hansson, L. (1991). Dispersal and connectivity in metapopulations. *Metapopulation Dynamics: Empirical and Theoretical Investigations* (Ed. by M.E. Gilpin & I. Hanski) pp. 89–103. Academic Press, London.

Harrison, S. (1989). Long-distance dispersal and colonization in the Bay checkerspot butterfly. *Ecology,* **70**, 1236–1243.

Harrison, S. (1991). Local extinction in a metapopulation context: an empirical evaluation. *Metapopulation Dynamics: Empirical and Theoretical Investigations* (Ed. by M.E. Gilpin & I. Hanski), pp. 73–88. Academic Press, London.

Harrison, S. & Quinn, J.F. (1989). Correlated environments and the persistence of metapopulations. *Oikos,* **56**, 293–298.

Harrison, S. & Thomas, C.D. (1991). Patchiness and dispersal in the insect community on ragwort (*Senecio jacobaea*). *Oikos,* **62**, 5–12.

Harrison, S., Murphy, D.D. & Ehrlich, P.R. (1988). Distribution of the bay checkerspot butterfly, *Euphydryas editha bayensis*: evidence for a metapopulation model. *American Naturalist,* **132**, 360–382.

Harrison, S., Stahl, A.M. & Doak, D. (1993). Spotted owl update: US judge rejects Forest Service plan. *Conservation Biology,* **7**, 1–4.

Hassell, M.P., Comins, H.N. & May, R.M. (1991). Spatial structure and chaos in insect population dynamics. *Nature,* **353**, 255–258.

Hastings, A.M. (1993). Conservation and spatial structure: theoretical approaches. *Lecture Notes in Biomathematics* (Ed. by S.A. Levin). Vol. 100.

Herben, T. & Soderstrom, L. (1992). Which habitat parameters are most important for the persistence of a bryophyte species on patchy, temporary substrates? *Biological Conservation,* **59**, 121–126.

Howe, R.W., Davis, G.J. & Mosca, V. (1991). The demographic significance of 'sink' populations. *Biological Conservation*, **57**, 239–255.

Laan, R. & Verboom, B. (1990). Effect of pool size and isolation on amphibian communities. *Biological Conservation*, **54**, 251–262.

Lacy, R.C. (1987). Loss of genetic diversity from managed populations: interacting effects of drift, mutation, immigration, selection and population subdivision. *Conservation Biology*, **1**, 143–158.

Lamberson, R.H., McKelvey, K., Noon, B.R. & Voss, C. (1992). A dynamic analysis of northern spotted owl viability in a fragmented forest landscape. *Conservation Biology*, **6**, 505–512.

Lande, R. (1987). Extinction thresholds in demographic models of territorial populations. *American Naturalist*, **130**, 624–635.

Lande, R. (1988a). Demographic models of the northern spotted owl (*strix occidentalis caurina*). *Oecologia*, **75**, 601–607.

Lande, R. (1988b). Genetics and demography in biological conservation. *Science*, **241**, 1455–60.

Lande, R. & Barrowclough, G.F. (1987). Effective population size, genetic variation, and their use in population management. *Viable Populations For Conservation* (Ed. by M.E. Soulé), pp. 87–124. Cambridge University Press, Cambridge.

Levins, R. (1970). Extinction. *Lectures on Mathematics in the Life Sciences*, **2**, 75–107.

May, R.M. (1989). Black-footed ferret update. *Nature*, **339**, 104.

McCauley, D.E. (1991). Genetic consequences of local population extinction and recolonization. *Trends in Ecology and Evolution*, **6**, 5–8.

McKelvey, K., Noon, B.R. & Lamberson, R.H. (1992). Conservation planning for species occupying fragmented landscapes: the case of the northern spotted owl. *Biotic Interactions and Global Change* (Ed. by P.M. Kareiva, J.G. Kingsolver & R.B. Huey), pp. 424–450. Sinauer, Sunderland, MA.

Menges, E.S. (1990). Population viability analysis for an endangered plant. *Conservation Biology*, **4**, 52–62.

Menges, E.S. (1991). The application of minimum viable population theory to rare plants. *Genetics and Conservation of Rare Plants* (Ed. by D.A. Falk & K.E. Holsinger), pp. 45–61. Oxford University Press, Oxford.

Moyle, P.B. & Williams, J.E. (1990). Biodiversity loss in the temperate zone: decline of the native fish fauna of California. *Conservation Biology*, **4**, 275–284.

Murphy, D.D. & Freas, K. (1988). Habitat-based conservation: the case of the Amargosa vole. *Endangered Species Update* **5(6)**, 6.

Murphy, D.D. & Noon, B.R. (1992). Integrating scientific methods with habitat conservation planning: reserve design for northern spotted owls. *Ecological Applications*, **2**, 3–17.

Murphy, D.D., Freas, K.E. & Weiss, S.B. (1990). An environment – metapopulation approach to the conservation of an endangered invertebrate. *Conservation Biology*, **4**, 41–51.

Nee, S. & May, R.M. (1992). Dynamics of metapopulations: habitat destruction and competitive coexistence. *Journal of Animal Ecology*, **61**, 37–40.

Opdam, P. (1990). Metapopulation theory and habitat fragmentation: a review of holarctic breeding bird studies. *Landscape Ecology*, **5**, 93–106.

Price, M.V. & Endo, P.R. (1989). Estimating the distribution and abundance of a cryptic species, *Dipodomys stephensi* (Rodentia, Heteromyidae), and implications for management. *Conservation Biology*, **3**, 293–301.

Quinn, J.F. & Harrison, S. (1988). Effects of habitat fragmentation and isolation on species richness: evidence from biogeographic patterns. *Oecologia*, **75**, 132–140.

Ray, C., Gilpin, M. & Smith, A.T. (1991). The effect of conspecific attraction on metapopulation dynamics. *Metapopulation Dynamics: Empirical and Theoretical Investigations* (Ed. by M.E. Gilpin & I. Hanski), pp. 123–134. Academic Press, London.

Rolstad, J. (1991). Consequences of forest fragmentation for the dynamics of bird populations: conceptual issues and the evidence. *Metapopulation Dynamics: Empirical and Theoretical Investigations* (Ed. by M.E. Gilpin & I. Hanski), pp. 149–163. Academic Press, London.

Rosenberg, K.V. & Raphael, M.G. (1986). Effects of forest fragmentation on vertebrates in Donglas-fir forests. *Wildlife 2000: Modeling Habitat Relationships of Terrestrial Vertebrates* (Ed. by J. Verner, M.L. Morrison & C.J. Ralph), pp. 263–272. University of Wisconsin Press, Madison, Wisconsin.

Schoener, T.W. (1991). Extinction and the nature of the metapopulation. *Acta Oecologica*, 12, 53–75.

Schoener, T.W. & Spiller, D.A. (1987). High population persistence in a system with high turnover. *Nature*, 330, 474–477.

Seal, U.S., Thorne, E.T., Bogan, M.A. & Anderson, S.H. (1989). *Conservation Biology and the Black-footed Ferret*. Yale University Press, New Haven.

Simberloff, D. (1988). The contribution of population and community biology to conservation science. *Annual Review of Ecology and Systematics*, 19, 473–511.

Simberloff, D.S. & Abele, L.G. (1982). Refuge design and island biogeographic theory: effects of fragmentation. *American Naturalist*, 120, 41–50.

Sjogren, P. (1991a). Extinction and isolation gradients in metapopulations: the case of the pool frog (*Rana lessonae*). *Metapopulation Dynamics: Empirical and Theoretical Investigations* (Ed. by M.E. Gilpin & I. Hanski), pp. 135–147. Academic Press, London.

Sjogren, P. (1991b). Genetic variation in relation to demography of peripheral pool frog populations (*Rana lessonae*). *Evolutionary Ecology*, 5, 248–271.

Smith, A. & Peacock, M.M. (1990). Conspecific attraction and determination of metapopulation colonization rates. *Conservation Biology*, 3, 320–323.

Soulé, M.E. (Ed.) (1987). *Viable Populations For Conservation*. Cambridge University Press, Cambridge.

Soulé, M.E. & Kohm, K.A. (1989). *Research Priorities For Conservation Biology*. Island Press, Covelo, CA.

Stacey, P.B. & Taper, M. (1992). Environmental variation and the persistence of small populations. *Applications*, 2, 18–29.

Stangel, P.W., Lennartz, M.R. & Smith, M.H. (1992). Genetic variation and population structure of red-cockaded woodpeckers. *Conservation Biology*, 6, 283–290.

Stewart, M.M. & Ricci, C. (1988). Dearth of the blues. *Natural History*, 5, 64–70.

Taylor, A. (1991). Studying metapopulation effects in predator-prey systems. *Metapopulation Dynamics: Empirical and Theoretical Investigations* (Ed. by M.E. Gilpin & I. Hanski), pp. 305–323. Academic Press, London.

Taylor, B. & Gerrodette, T. (1993). The uses of statistical power in conservation biology: the vaquita and northern spotted owl. *Conservation Biology*, 7, 489–500.

Temple, S.A. (1986). Predicting impacts of forest fragmentation on forest birds: a comparison of two models. *Wildlife 2000: Modeling Habitat Relationships of Terrestrial Vertebrates* (Ed. by J. Verner, M.L. Morrison & C.J. Ralph), pp. 301–304. University of Wisconsin Press, Madison, Wisconsin.

Thomas, C.D. (1994). The ecology and conservation of butterfly metapopulations in fragmented landscapes. *Ecology and Conservation of Butterflies* (Ed. by A.S. Pullin), Chapman & Hall, London.

Thomas, J.A. (1991). Rare species conservation: case studies of European butterflies. *Symposia of the British Ecological Society*, 31, 149–197.

Thomas, J.W., Forsman, E.D., Lint, J.B., Meslow, E.C., Noon, B.R. & Verner, J. (1990). A conservation strategy for the northern spotted owl. *Report of the Interagency Scientific Committee to Address the Conservation of the Northern Spotted Owl.* p. 427. USDA Fish and Wildlife Service.

Urban, D.L. & Shugart H.H. (1986). Avian demography in mosaic landscapes: modeling paradigm and preliminary results. *Wildlife 2000: Modeling Habitat Relationships of Terrestrial Vertebrates* (Ed. by J. Verner, M.L. Morrison & C.J. Ralph), pp. 273–280. University of Wisconsin Press, Madison, Wisconsin.

US Fish and Wildlife Service (USFWS) (1992). Draft golden-cheeked warbler recovery plan. USDI Fish and Wildlife Service, Austin, TX.

US Forest Service (1992). Draft environmental impact statement: Flathead Forest land and resource management plan, Amendment 16, Old-Growth MIS Standards. USDA Forest Service, Kalispell, MT.

Van Dorp, D. & Opdam, P. (1987). Effects of patch size, isolation and regional abundance on forest bird communities. *Landscape Ecology*, **1**, 59–73.

Verboom, J., Lankester, K. & Metz, J.A.J. (1991). Linking local and regional dynamics in stochastic metapopulation models. *Metapopulation Dynamics: Empirical and Theoretical Investigations* (Ed. by M.E. Gilpin & I. Hanski), pp. 39–55. Academic Press, London.

Wade, M.J. & McCauley, D.E. (1988). Extinction and colonization: their effects on the genetic differentiation of local populations. *Evolution*, **42**, 995–1005.

Walters, J.R. (1991). Application of ecological principles to the management of endangered species: the case of the red-cockaded woodpecker. *Annual Review of Ecology and Systematics*, **22**, 505–523.

Warren, M.S., Thomas, C.D. & Thomas, J.A. (1984). The status of the heath fritillary butterfly, *Mellicta athalia* Rott., in Britain. *Biological Conservation,* **29**, 287–305.

Watt, A.S. (1947). Pattern and process in the plant community. *Journal of Ecology*, **35**, 1–22.

Welsh, H. (1990). Relictual amphibians and old-growth forests. *Conservation Biology*, **3**, 309–319.

Western, D. & Pearl, M. (1989). *Conservation Biology in the 21st Century.* Oxford University Press, Oxford.

Wilcox, B.A. & Murphy, D.D. (1985). Conservation strategies: the effects of fragmentation on extinction. *American Naturalist*, **125**, 879–887.

6. CONSERVING INSECT HABITATS IN HEATHLAND BIOTOPES: A QUESTION OF SCALE

N. R. WEBB AND J. A. THOMAS

Furzebrook Research Station, NERC Institute of Terrestrial Ecology, Wareham, Dorset BH20 5AS, UK

SUMMARY

The widespread decline in the populations of many European species has been attributed loosely to habitat loss. This loss has occurred both through fundamental destruction of biotopes and as a result of changes in the management of surviving biotopes. The combined effects of these processes has altered the distributions and availability of patches of habitat for different species within the landscape: habitat patches have changed in frequency, in size, in their degree of isolation and in their continuity in time – both within and between patches of the biotope.

We concentrate on the example of the heathlands of southern England and consider the habitat requirements for three species of insect. We distinguish clearly between the heathland patches (the biotope patches) and the habitat patches of the insects. Changes in both space and time for the biotope itself (patches of heathland within the landscape), and of habitat patches of these characteristic species within patches of the heathland biotope are described. This indicates the problem confronting these species as they track the occurrence of their habitats in space and time over the landscape. To be effective, conservation must maintain within the landscape the dynamics of this system of habitat and biotope patches. Only when this is achieved will the characteristic species survive.

INTRODUCTION

The aim of this chapter is to take a fresh look at how a region, containing the island remnants of a once extensive biotope, can support the diversity of insect species for which it was famed. Ecologists have traditionally approached this problem by first describing the rate and extent of fragmentation of semi-natural vegetation into islands, before considering how the changes in patch sizes, shapes, edge lengths and isolation affect species turnover, dispersal, and biodiversity. Their conclusions form the rationale for much conservation planning (May 1975; Margules *et al.* 1982; Reed 1983; Usher 1986). Given that it is seldom possible to conserve more than a small proportion of the biotope patches, it has been important to address the following questions. Is greater species richness achieved by establishing a few large or more small nature reserves (the SLOSS argument) (e.g. Gilpin & Diamond 1980; Wilcox & Murphy 1985; Quinn & Harrison 1988; Spellerberg 1991)? What shapes are most appropriate, and should they be widely dispersed or aggregated in the landscape? Is biodiversity increased by linking reserves with corridors of seminatural vegetation?

In general, conservation practice – particularly reserve ownership, acquisition and management – has been at the scale of the biotope patch, usually an entire wood or heathland or grassland. These are areas which humans perceive as units. By coincidence, this scale may be appropriate for vertebrates and some plants, but autecological research on insects (e.g. butterflies; Thomas 1991) shows that many species utilize small, specific parts of their biotopes. Populations may be confined to a topographical feature, to a narrow and ephemeral seral stage, or to a combination of such factors, leaving the rest of the biotope as alien to them, as the reclaimed land surrounding biotopes is to the conservationist. Thus at any point in time, a single island of biotope could contain several, one, or no patches of the habitat of one species, and different amounts, differently distributed, of the habitat of some other species. It follows that the same set of biotopes can present different opportunities or problems of isolation when viewed from the perspective of different insect species.

Although insect populations can generally be supported by smaller areas than can vertebrates, the distribution and continuity of insect habitat patches may instead be limiting (Thomas 1984; Warren 1992). For example, a wide range of population structures exists even within so small a group as the British butterflies. Mobile species such as *Cynthia cardui* L., which fly freely across Europe, can be contrasted with the 78% of British butterfly species which live in closed populations (Thomas 1984). For these it may take several years and generations before new habitat patches within 200–500 m of a population are colonized, and gaps of 1–10 km may represent almost insuperable barriers (Thomas 1991; Thomas *et al.* 1992). For such species, several discrete populations can exist within a single nature reserve (or biotope patch), each with its own dynamics and isolated from its neighbours by different extents.

In this chapter, we examine from a dynamic viewpoint one of the best studied and most intensively conserved areas of semi-natural habitat in Britain: the heathlands of southeast Dorset. We take three insects species – an ant, a grasshopper and a butterfly – whose habitats have been precisely defined, and calculate how the abundance and continuity of their specific habitat patches varied across this landscape over a 9-year period. We consider whether the availability of habitat for each species was satisfactory, how far it fell short of the potential available on individual heaths and in the landscape as a whole, and how the shortfall could be reduced. In the case of the butterfly, whose powers of dispersal were known, we also estimate the minimum number of metapopulations that may exist in the landscape under current management. We indicate how this could be changed into fewer larger metapopulations by concentrating conservation management on a few key areas.

PREVIOUS STUDIES OF DORSET HEATHLAND

The biotope

The Dorset heathlands have developed on poor, acidic soils derived from Tertiary deposits. Sparse oak forest covered much of this area until about 4500 years ago when forest clearances created the conditions for heathland development (Webb & Haskins 1980). This landscape was a patchwork of different vegetation types, the pattern of which was determined by variations in topography and soils. These were environmental resource patches (*sensu* Forman & Godron 1986). Disturbance, as a result of human activities, led to the spread of dwarf shrub vegetation which created a series of disturbance patches overlying the pattern of topography and soils. This position existed roughly at the late Iron Age and Roman times (*c.* 2000 BP) when the extent of these heathlands was greatest (Webb & Haskins 1980). Subsequently, as a result of changed or intensified land use on the ground surrounding these large heathland patches or continents (Fig. 6.1), a series of remnant patches of heathland was created. These remnant patches are secondary in that they are remnants of the disturbance patches and not remnants of the mosaic of primary vegetation. Over the past 150 years, the total area of heathland has declined by about 85% from 300 km² to under 60 km², and the number of patches increased considerably. Today, heathland species are confronted with a mosaic of small patches lying in a matrix of farmland, forest and urban areas (Fig. 6.1) (Moore 1962; Webb & Haskins 1980;

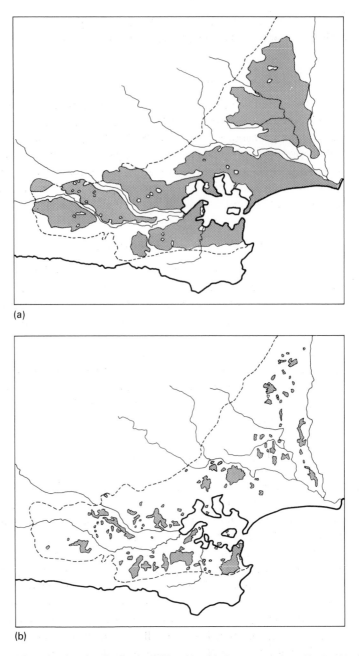

(a)

(b)

FIG. 6.1. Maps showing the distribution of heathland in Dorset, southern England in (a) 1759 and in (b) 1978. In the course of two centuries, the formerly large continuous areas have been reduced in extent and considerably fragmented. After Webb and Haskins (1980).

Webb 1990). The management and use of these heathlands have followed well-established practices over many centuries (Webb 1986).

The number and size of biotope patches

To relate biodiversity to patch size and isolation it is necessary to define heathland patches, and for the Dorset heathlands a variety of approaches has been used (e.g. Moore 1962; Webb & Haskins 1980). We have adopted the approach of Chapman *et al.* (1989) who used vegetation cover to define 141 heathland fragments (biotope patches). From the human perspective this exercise was successful as it produced a pattern similar to that on a map or aerial photograph.

The biodiversity of heathland patches

Early studies on these heaths reported positive relationships between species richness to heath area and degree of isolation (Moore 1962). Later, Webb and Hopkins (1984) confirmed these patterns, but for invertebrates found the relationship between species richness and area to be weak. However, Hopkins and Webb (1984) showed that heathland spiders tended to be more abundant on large heaths than small and that those species contributing most to that effect tended to be restricted to the larger heathlands by poorer powers of dispersal. The weakness of these relationships was attributed to edge effects, which were greatest on small fragments (Webb *et al.* 1984).

Two problems of the early studies

By comparing species richness with biotope size or isolation, these early analyses did not include an important dynamic of these fragments: the fact that to remain as heathland they need to be managed. Such management assumes there will be local extinctions of species, with recolonization from neighbouring populations or from any section of a population surviving the disturbance. In effect, there has been an assumption of patch dynamics in the conservationists' concept of rotational habitat management and in this respect conservation practice went a step further than the biogeographic theory. On occasions, the management of suites of reserves has also been proposed, but usually on the grounds of increasing species diversity within the total area protected (Usher 1986), rather than as an integrated approach to maintain continuity of a particular habitat type across a landscape, which would result in fewer larger metapopulations of characteristic species. If the components of a suite of reserves were poorly selected, the outcome of metapopulation dynamics may not be easily predicted (Nee & May 1992).

THE OCCURRENCE OF INSECT HABITATS WITHIN BIOTOPES

The approach which examined the numbers of species in relation to the size and degree of isolation of heathland fragments was clearly one relevant to conservation science at that time. It is now equally clear that results would be inconclusive because of the disparity of scale at which the heaths were examined compared to that at which the insects were examined. While ownership may continue at the scale of heathland fragments, the conservation of invertebrates will need to be on a finer scale within the boundaries of the fragments.

We therefore approached this problem from an insect's viewpoint. By combining the results of autecological studies with parameters measured during the original heathland surveys, we have been able to generate and map the distribution of the habitat patches available to a few insect species in Dorset in 1978–87, to examine interspecific differences in these and see how they differ from the biotope patches perceived by humans. The exercise was possible only for species whose habitats have been precisely defined, but can be done across the whole landscape. This is because a database was established containing digital information from 3110 grid squares, each 200 m × 200 m, covering the entire Dorset heathlands (Webb & Haskins 1980; Webb 1990). For each grid square, 184 attributes were recognized, representing vegetation composition and structure (100 attributes), topography and physical characteristics (36 attributes) and land use (48 attributes). For clarity the worked examples will be illustrated by a subset of the squares although the statistics have been generated for all 3110 examples (Fig. 6.2). It should be noted that the subset of heaths presented for illustration is somewhat atypical, since almost all are managed nature reserves.

If, for example, we define the requirements of a hypothetical thermophilous species as being pioneer heath (the seral stage with the warmest microclimate; Thomas 1993) with a southerly aspect, we find that although individually both attributes are in themselves common, their co-occurrence is much rarer. When the dimension of time is added, indicating the persistence or continuity of habitat, then the number of locations at which this occurs is extremely few (Fig. 6.3).

For the comparatively sedentary insects to which this current exercise

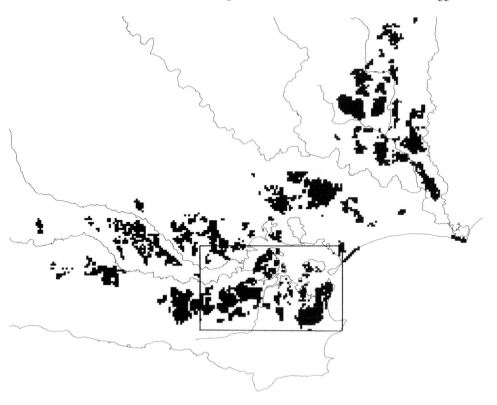

FIG. 6.2. A map of the distribution of heathland in Dorset, southern England, between 1978 and 1987. The box marks an area which is used in the illustrations in this chapter; however, statistics have been calculated for the entire patchwork of the heathlands.

is limited, our new unit of scale is therefore the habitat patch: we can measure the number, size, distribution and continuity of islands of their habitat patches in the Dorset landscape, and also examine whether the number and sizes of these habitat islands within individual heaths (biotope patch) is related to the size of the biotope patch.

THE HABITATS OF THREE INSECTS ON DORSET HEATHLAND

The three insects selected as examples are *Plebejus argus* (Lepidoptera: Lycaenidae), *Chorthippus vagans* (Orthoptera: Acrididae) and *Myrmica sabuleti* (Hymenoptera: Formicidae). Each species has a wide range of foodplants which occur in every one of the 3110 grid squares comprising

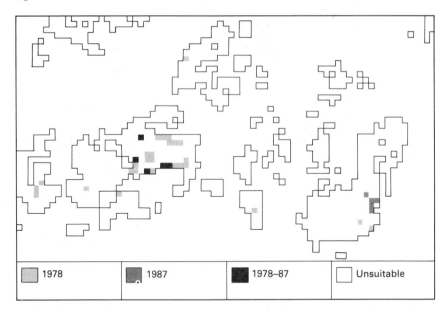

FIG. 6.3. The distribution of an arbitrarily defined habitat consisting of pioneer south-facing heath. The light squares indicate the distribution in 1978, the dark in 1987, and the black squares where the habitat has persisted over the 9 years. This indicates how, despite the large area of heathland spread over 35 patches, this simply defined habitat is rare, scattered and is discontinuous in time.

the heathland biotope, and each is sufficiently sedentary for us to assume that the boundary of each habitat patch forms the boundary (if the patch is occupied) of a separate population of the insect. The extent to which populations may be linked into metapopulations is examined later.

The habitat of *P. argus* in southern England was defined from Read (1985), supplemented by Thomas (1985a,b); for *C. vagans* (Orthoptera: Acrididae) from Ragge (1965) and Marshall and Haes (1988); and for *M. sabuleti* (Hymenoptera: Formicidae) from Thomas (1993 and unpublished). Together, they represent a series from a species that inhabits a fairly broad niche (*P. argus*) to one that has exacting habitat requirements (*M. sabuleti*). Thus, *P. argus* is defined as requiring dry or humid heath, of any aspect, that is in the pioneer or building phase of a succession (= approximately the first 10 years after a fire or cut); *M. sabuleti* is defined as requiring dry heath with a warm south-facing aspect, containing >50% grass species, and which is in the pioneer phase (approximately the first 5 years) after a disturbance. The habitat of *M. sabuleti* is similar to the theoretical example shown in Fig. 6.4, and

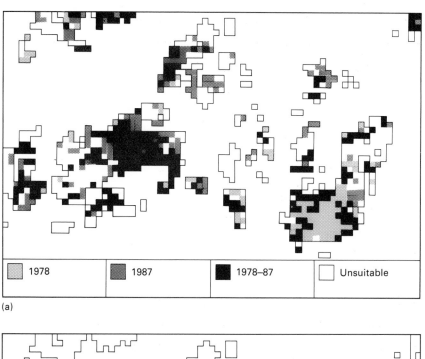

(a)

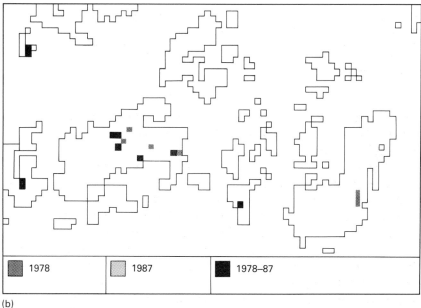

(b)

FIG. 6.4. (a) A map showing the distribution of habitat for the silver-studded blue (*Plebejus argus*) on a subset of the heathland biotope patches in Dorset. Even for this widely distributed species on these heathlands, only about 40% of the biotope provides suitable habitat. (b) The ant *Myrmica sabuleti* is a species with more exacting habitat requirements and the map shows how little habitat is available within the patches of heathland biotope.

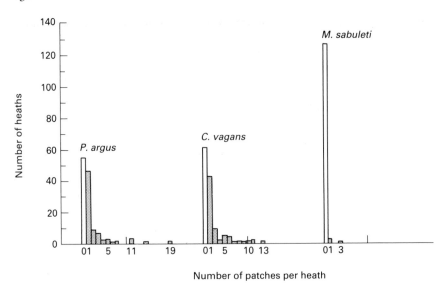

Fig. 6.5. The number of patches of habitat per heath (biotope patch) for silver-studded blue (*Plebejus argus*), heath grasshopper (*Chorthippus vagans*) and the ant *Myrmica sabuleti* in 1978. Note that the majority of heaths lack patches of the habitat.

is probably representative of the habitats of other ground-dwelling, thermophilous heathland invertebrates, a group that contains an important proportion of the rarities for which the Dorset heathlands are famed.

The habitat of *P. argus* was comparatively common, but patchily distributed, during the survey period (Fig. 6.4a). In 1978, there were 172 habitat patches distributed over 137 heaths, yet about 40% of the heaths did not contain any patch of habitat for this species at all (Fig. 6.5). On the other hand, one heath contained no fewer than 19 separate patches of *P. argus* habitat in 1987. The mean size of each habitat patch was large for *P. argus* (a population can be supported by less than 1 ha of habitat), and about half the size of the average heath (Table 6.1). The total area of habitat available for this species in the landscape increased by 18% over the 9 years, to exist as slightly more, slightly larger habitat patches. Perhaps most important of all, there was continuity of habitat somewhere within the same island of habitat at 88% of these locations between 1978 and 1987 (Table 6.1).

The Dorset heathland landscape also contained a large number of patches of the habitat of *C. vagans*, which again increased in number and

TABLE 6.1. Number, size and continuity of habitat patches of two species on Dorset heaths

	Number of patches			Mean size of patches (ha)		
	1978	1987	1978 & 1987	1978	1987	1978 & 1987
P. argus	172	190	152	28	30	10
C. vagans	84	117	66	7	9	6
M. sabuleti	42	6	0	10	6	–
Heathland biotope	137	131	–	57	61	–

TABLE 6.2. Isolation: distances between the nearest neighbouring habitat patches

	Mean distance (km)			Maximum distance (km)		
	1978	1987	1978–87	1978	1987	1978–87
P. argus	0.40	0.33	0.08	5.6	8.7	17.9
C. vagans	0.76	0.53	0.13	5.0	2.0	2.1
M. sabuleti	0.96	2.83	9.52	3.8	8.7	18.0
Heathland biotope	1.3	1.2	–	18.8	18.8	–

mean size between 1978 and 1987, with some continuity of habitat in 79% of them (Table 6.2). However, the situation was very different for *M. sabuleti* (Fig. 6.4b). In exactly the same heathland biotope, there was no more than 410 ha of its habitat in 1978, and this declined to just 36 ha by 1987. There was little or no continuity of habitat over this period, and even in 1978, the whole landscape contained just 42 patches of this important type of habitat (Table 6.1). These were distributed between 16% of the heaths in 1978, but occurred on only 3% of heaths 9 years later (Fig. 6.5). The rapid change in the availability of habitat for this thermophilous insect reflects successional trends in the heathland biotope, and is discussed in a later section.

For species that depend on ephemeral successional stages, the mean and maximum distances between habitat patches are of prime importance (Table 6.2). For *Plebejus argus*, the mean distance between patches of habitat was similar in both years. More significantly, if every original suitable patch had been occupied, this species had only to move a mean distance of 80 m to track the occurrence of new habitat between 1978 and 1987. Even this modest distance may represent a considerable barrier to this insect. The maximum natural single-step colonization distances that have been observed or deduced for *P. argus* in other landscapes in which new habitat islands were generated are 0.60–1.00 km (Thomas & Harrison 1992; Thomas *et al.* 1992). These, however, are the maximum distances recorded as being bridged: the distance across which there is a, say, 50% chance that a new patch will be colonized (during the *c.* 10-year period that they remain suitable for this species) is likely to lie somewhere between this extreme and the distance up to which adults are regularly recorded as dispersing from populations during mark–recapture experiments: this latter distance is no more than 25–50 m (Thomas 1983; Read 1985; Ravenscroft 1992). This inference is consistent with data collected by R.J. Rose (personal communication), who found that only 31% of the apparently suitable habitat patches for *P. argus* on a sample of Dorset heathland were occupied by the butterfly in 1992. We deduce that the remaining two-thirds of all patches were too isolated to have been colonized, rather than that it is a substantial error in our definition of *P. argus* habitat. This definition has been experimentally confirmed by the success of introductions of this butterfly into isolated islands of unoccupied habitat elsewhere (Oates & Warren 1990; Ravenscroft 1992; Thomas *et al.* 1992). It is also reassuring for this entire exercise that Rose recorded *P. argus* in only a few of the grid squares that we designated as containing none of this species' habitat in 1991, and that in every one of these cases, the 'unsuitable' square of heathland was immediately adjacent to a square containing 'suitable' habitat and a population of this butterfly.

Not every insect is as sedentary as *P. argus*. However, it is disturbing that the mean distance between neighbouring patches of the more exacting thermophilous species, *M. sabuleti*, was three times greater in 1987 than in 1978, and was more than double the distance that humans perceive as separating the biotope islands of heathland (Table 6.2). More disturbing still is our estimation that *M. sabuleti* needed to cross an average distance of >9 km if populations were to track the regeneration of the few new patches of its habitat in 1978–87 (Table 6.2).

Relatively few heathland fragments contain any patch of the least or most specialized of our three examples, *M. sabuleti* (Fig. 6.5). It is also interesting to note that even where more than one habitat patch occurs on the same heath, the nearest neighbour is quite frequently on a different heath. In the case of *P. argus*, 14 out of 27 instances, and for *M. sabuleti*, 4 out of 11 instances in 1978. This has implications for conservation, as the nearest neighbour patch of habitat may be on a heath which is under the control of someone else – as it is with the high proportion of habitat patches that occur as a single patch per heath.

Conservation: utilization of potential habitat in biotopes

The above examples refer to the situation for three species in the years 1978 and 1987. Despite the comparative abundance of lowland heath that survives in Dorset, and the fact that much of it is managed for nature conservation, this biotope appears to have been a very unsatisfactory source of habitat for the specialist, *M. sabuleti*, and a less than satisfactory source for the more generalist species, *P. argus*. In the first case the few habitat patches became fewer and smaller over the 9 years, and such new habitat as was generated was, on average, 9–10 km away from the nearest old patch. In the second case, there was an apparently satisfactory area of habitat, but new patches appear to have been created too far from old ones for more than about a third of them to be colonized.

We can now explore whether this situation could be improved by altering the management of the landscape, or whether the soils and topography of the biotope make most of it intrinsically unsuitable for these species. In other words, how fully did the occurrence of their habitats fulfil the potential for these biotope islands? In a sense, fulfil-ment of potential is a measure of conservation success, although it should be noted that not all potential habitat patches should be filled at any one time. Continuity is almost certainly more important, and the availability of habitat for one species has to be integrated with the needs of other species.

In the case of *P. argus* (Fig. 6.6a), we estimate that 44% of all potential breeding grounds was in a suitable condition (i.e. was in the pioneer and building stages of the succession) in 1978, and 48% was suitable in 1987. Given the requirement of many other species for the mature and degenerate phases of the heathland succession, this appears to be a satisfactory state of affairs. The only improvement open to conservationists would be to ensure that new successions are started rather closer to pioneer or building heath than occurs at present; say, within 25–40 m of a patch rather than the current average of 80 m away.

The situation for *M. sabuleti* (Fig. 6.6b) was far less satisfactory. Only 29% of the warm southerly-facing slopes that can potentially support this insect were in a suitable successional stage to do so in 1978, and this had fallen to 2% in 1987 (Table 6.3); it should also be noted that the potential habitat of this species was rare on these heathlands to start with. This is disturbing because we believe that many of the other rare species for which these heathlands are noted also need warm microclimates, and clearly this special type of habitat was almost lost from the Dorset

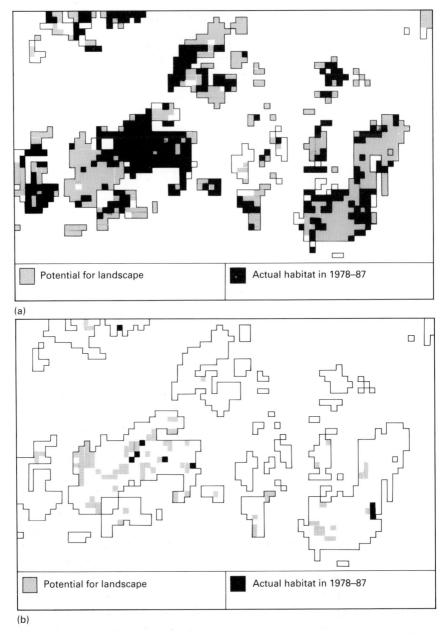

| Potential for landscape | Actual habitat in 1978–87 |

(a)

| Potential for landscape | Actual habitat in 1978–87 |

(b)

FIG. 6.6. (a) A map showing the potential that a portion of the Dorset heathland landscape has to contain habitat for the silver-studded blue (*Plebejus argus*) (light shading). The dark shading indicates the area of habitat existing in the period 1978–87 which amounts to about 40–50% of the heathland biotope. This contrasts with the ant *Myrmica sabuleti* (b) in which not only is the potential habitat much less, but that occurring was 29% in 1978 and only 2% in 1987.

TABLE 6.3. Potential and actual amounts of two habitats available on Dorset heaths in 1987

	Number of patches	Mean size (ha)	Total size (ha)	Mean distance (km)
P. argus				
Potential	142	20.4	2897	1.5
Actual	190	7.4	1406	1.7
M. sabuleti				
Potential	141	3.4	47.9	2.5
Actual	6	1.5	9	14.1

landscape. This unsatisfactory situation is much more likely to occur with species that inhabit narrow early-successional niches. On these heaths, conservationists have made the mistake of allowing a uniform cover of dwarf shrub vegetation to develop.

Templet theory and dispersal

Templet theory (Southwood 1977) would lead us to predict that the many heathland species with narrow niches would have considerable powers of dispersal, thus allowing them to track the actual occurrence of their ephemeral habitats in these biotopes between 1978 and 1987. In fact, the few species that have been studied, in this and other biotopes, tend to be much less mobile than expected (see Thomas 1991 for examples among butterflies).

Thomas (1991) and Erhardt and Thomas (1991) have suggested that this failure to match Southwood's predictions may be because we are studying local races of species that have evolved to track the dynamics of their habitat patches under the traditional methods of heathland management that prevailed over several millennia until the turn of this century. Not only were the biotope islands larger and nearer together, but numerous patches were regularly managed by burning, cutting or grazing which would have delayed or arrested succession. This would have created far more continuity in both time and space for species that possess early successional habitats. If we are correct, conservationists are having to deal with invertebrate species whose population dynamics were moulded over perhaps 5000 generations by a recently obsolete form of land management.

Metapopulations

Having considered species patches and the occurrence of individual populations within and between biotopes, we move up the hierarchy and consider habitat availability from the viewpoint of metapopulations.

This exercise is possible only for *P. argus*, because, at best, only anecdotal accounts exist on the dispersal of the other heathland insects whose habitats are sufficiently well known for us to map. However, for the purposes of this exercise, we can assume that new patches within 1 km of an existing colony have a chance of colonization during the period they remain suitable, but those further away are extremely unlikely to be colonized (here, we say never) (Thomas & Harrison 1992). This assumption enables us to map the chains of habitat patches that link one population with another. It should be noted that these patches are not corridors in the conventional sense, where individual insects may rest while travelling between existing areas of habitat. Instead they conform to the stepping-stone model of island occupancy (Gilpin & Hanski 1991), being areas where new component populations of the metapopulation are established and from which further populations will be established.

Figure 6.7a shows the potential distribution of *P. argus* metapopulations in Dorset, assuming that conservationists maintain all potential breeding patches at a suitable successional stage (i.e. all dry or humid heath is maintained in a pioneer or building phase, which would involve all areas being cut or burnt at least once every 10 years). Nine of the 13 potential metapopulations are small and isolated, and seven of them consist of a single component population. Four are very large, with up to 37 populations in them.

This, however, is only the potential for this landscape. If key areas were left to grow (in this example) into mature or degenerate heathland, they can produce gaps large enough to divide one metapopulation into two or more. For the period 1978–87, we estimate that the distribution of *P. argus* habitat was such that there were 19 possible metapopulations, of which 13 were very small (Fig. 6.7b).

This analysis suggests a modification of our previous advice to conservationists (Webb 1990; Webb & Vermaat 1990), which was to ignore small heaths because strong edge effects created a high species-richness composed largely of species that were unrepresentative of heathland. Reasoning from the metapopulation point of view, strategically placed small heaths would provide vital links, if correctly managed. Isolated small heaths could still be ignored.

It also follows that, although we concluded that the proportion of

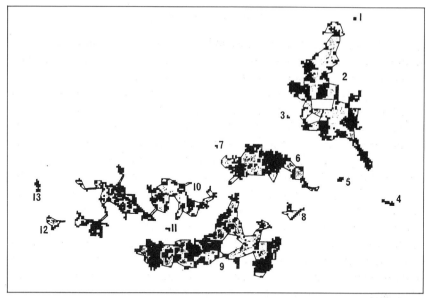

(a)

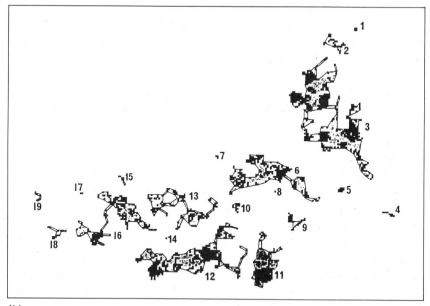

(b)

FIG. 6.7. (a) A map showing the potential metapopulations of silver-studded blue (*Plebejus argus*) over the entire Dorset heathlands. This contrast with (b) the potential metapopulations calculated for the period 1978–87.

heathland that was in a suitable successional stage for *P. argus* was satisfactory under the piecemeal management that occurred across this landscape in 1978–87, conservationists should take particular care to maintain greater continuity of habitat on the small islands that link other populations, and also on the nearest edges of larger heaths, to prevent the occurrence of gaps too large for this species to cross. This suggests that there should be greater co-operation among the aims of heathland managers than currently exist.

Although it is improbable that the situation in nature is anything like as clearcut as our Figs. 6.7a and b imply, this exercise is perhaps useful in identifying areas of science where information is poor or absent. For example, we assumed linear distance to be the only barrier to dispersal, but the nature of the intervening land is also likely to be important, whether it be woodland, grassland and so on. We already possess data on the different types of land that separate heathland biotope patches in Dorset, but unfortunately the effects of these barriers to different species are unknown and need studying; we would expect to find considerable interspecific differences. It should eventually be possible to generate a permeability factor between two populations of any species.

DISCUSSION

This example of an autecological approach to island biogeography and conservation is based on very few species, and even for these there are gaps in information, notably on the ability of their populations to cross unsuitable areas of heathland and other land, and to track the temporary occurrences of their habitat patches. Clearly, any conclusions must be regarded as provisional pending further research. Yet we believe the initial exercise to be worthwhile because it suggests an extra dimension to the traditional approach to biogeography; there are already clear implications for conservation.

The main result was to demonstrate that the apparently stable archipelago of island biotopes was anything but stable when seen through the eyes of three insect species. Moreover for each insect species, these same islands of biotope contained very different amounts of habitat, differently distributed and with different dynamics.

In most previous studies of island biogeography, each biotope island has been considered as a stable unit, and analyses have been made of the biodiversity of these islands assuming a range of fixed land sizes, each a fixed distance away from one another. It was then presumed that this allowed a different range or number of species to establish on each island at any one time due to regular, but differing, rates of extinction and colonization among species. This study shows that there may be an extra dynamic variable in the system – the occurrence (or not) of individual species' patches of habitat – which might bear little relation to the occurrence of biotope islands. In our example, even for the species with the broadest niche, *P. argus*, there was no patch of habitat available, and by definition no possibility of a population existing, on about 40% of the biotope islands in the landscape in each study year. On the other hand, one heath, which would be classed as a single large island by the traditional biogeographer and conservationist, contained 29 separate patches of *P. argus* habitat in 1987, each well beyond the trivial flight range of this species and each, from the butterfly's viewpoint, representing a small separate habitable island. With *M. sabuleti*, 84% of heaths contained no patch of habitat in 1978 and 97% of heaths was uninhabitable in 1987, making its islands in this landscape one to two orders of magnitude smaller, more isolated and more dynamic than they appear to the scientist who confines his study to the whole biotope.

The importance of this extra parameter of dynamic habitat patches within biotopes may be greater in the landscape we studied than in some other systems, for ours is a secondary biotope created several millennia ago by man and still subject to successional change. Most other terrestrial European biotopes are similar in this respect (see Thomas 1991), but the 'primary habitats' that prevail in the tropics, and in many tropical islands, may accord more with the traditional parameters used by biogeographers.

Most further scientific implications of considering habitat rather than biotope islands will be discussed in a different paper (Thomas *et al.*, unpublished). Here, we will simply examine whether the actual and potential availability of habitat was related to biotope size for two species, before considering some implications for conservation.

For *P. argus* and *M. sabuleti*, there was a significant correlation between the number of habitat islands per heath and biotope size, accounting for 90% and 43% respectively of the observed variation in 1978 (Fig. 6.8a). But no relationship existed for *M. sabuleti* in 1987 due to the virtual disappearance of its habitat. The total area of habitat was also correlated with biotope size for both species, accounting for 87% and 26% of recorded variation in 1978 (Fig. 6.8b). For the conservationist wishing to assess the ability of biotope size to predict the potential of an individual island to provide the habitat of these two species, this second relationship is closer (Fig. 6.8c), and accounts for 99% and 49% of the observed variation. This near perfect correlation is not surprising for *P. argus*, since almost all Dorset heathland biotope becomes suitable for breeding if it is maintained in the first *c.* 10 years of vegetational succession. The locally warm habitat of *M. sabuleti* depends in part on a

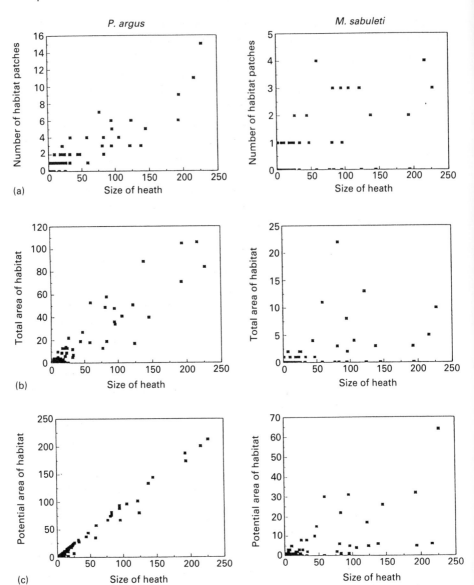

FIG. 6.8. Relationships between the size of biotope islands and the habitat patches of *Plebejus argus* and *Myrmica sabuleti* contained within them. (a) Number of habitat patches per biotope. (b) Area of habitat per biotope. (c) Potential area of habitat per biotope.

patchily distributed topographical feature and, although there is a fair correlation between the occurrence of this and biotope size, two of the four largest heaths contained very little potential habitat (Fig. 6.8c). Even in 1978, when an exceptionally high proportion of Dorset heathland was in the pioneer phase of regeneration, two of the four largest heaths in the landscape contained none of its habitat at all (Fig. 6.8b).

In Britain and Europe, the conservationist generally has a fair idea of the initial richness of island biotopes, and island biogeography is more important in helping to suggest what size and configuration of reserves are most likely to maintain the existing biodiversity. There is no need to select Dorset heaths on the basis, say, of their topographic diversity, on the assumption that these will support the greatest diversity of insect habitats when it is already known which of these particular heaths actually supports the greatest range of insects. However, all else being equal, traditional biogeographic studies of this landscape have indicated that conservationists should concentrate their resources on the largest (Webb & Hopkins 1984) and least isolated heaths (Moore 1962). Our analysis partly supports this view, but with two modifications. First, some small heaths may be the most important parts of the entire biotope, because they form links between otherwise separate metapopulations.

Second, the ability of the entire landscape to support species can be transformed by the way in which each biotope island is managed. Although Dorset heathland is one of the most intensively conserved biotopes of Europe, the frequency and distance apart at which even the unexacting habitat for *P. argus* was generated during 1978–87 was inadequate for about two-thirds of the habitat patches available in the landscape to be used. And the representation of the much rarer, but more highly valued, locally warm habitat of *M. sabuleti* was completely inadequate. For both species, the availability of habitat patches in the landscape could be transformed: in the first case by making burns or clearings smaller but closer to one another; in the second case by maintaining a much higher proportion of the biotope, and especially the areas that have southerly aspects, in early successional stages. Particular emphasis should be placed on integrating the management of adjoining heaths, and, as we have already seen, the management of strategically placed small heaths, and the edges of large ones that are opposite these, merit the most careful management of all.

Conservationists may justifiably say that we have considered the needs of just three insects, and have found them to be very different. How can they hope to cater for all the individual species for which Dorset heathlands are famed? As yet, the habitats of very few species have been defined, but we believe the problem will prove to be much less complex than it at first appears. With further research, we expect it will be possible to divide all the characteristic species into a comparatively small number of groups, each requiring a similar habitat type, and hence occupying similar areas that need similar management. In this chapter, we have already written of *M. sabuleti* as representing other ground-dwelling thermophilous invertebrates. At the time of writing, we suspect that practical conservationists will consider the needs of no more than about 10 different groups.

ACKNOWLEDGEMENTS

We are extremely grateful to R.J. Rose and R.T. Clarke for making many of the analyses described in this chapter.

REFERENCES

Chapman, S.B., Clarke, R.T. & Webb, N.R. (1989). The survey and assessment of heathland in Dorset, England for conservation. *Biological Conservation*, 47, 137–152.

Erhardt, A. & Thomas, J.A. (1991). Lepidoptera as indicators of change in the semi-natural grasslands and lowland and upland Europe. *Conservation of Insects and their Habitats* (Ed. by N.M. Collins & J.A. Thomas), pp. 213–236. Academic Press, London.

Forman, R.T.T. & Godron, M. (1986). *Landscape Ecology*. Wiley, New York.

Gilpin, M.E. & Diamond, J.M. (1980). Subdivision of nature reserves and the maintenance of diversity. *Nature*, 285, 567–568.

Gilpin, M.E. & Hanski, I. (1991). *Metapopulation Dynamics: Empirical and Theoretical Investigations*. Academic Press, London.

Hopkins, P.J. & Webb, N.R. (1984). The composition of the beetle and spider faunas on fragmented heathlands. *Journal of Applied Ecology*, 21, 935–946.

Margules, C.R., Higgs, A.J. & Rafe, R.W. (1982). Modern biogeographic theory: are there lessons for nature reserve design? *Biological Conservation*, 24, 115–128.

Marshall, J.A. & Haes, E.C.M. (1988). *Grasshoppers and Allied Insects of Great Britain and Ireland*. Harley, Colchester.

May, R.M. (1975). Patterns of species abundance and diversity. *Ecology and Evolution of Communities* (Ed. by M.L. Cody & J.M. Diamond), pp. 81–120. Belknap Press, Harvard.

Moore, N.W. (1962). The heaths of Dorset and their conservation. *Journal of Ecology*, 50, 369–391.

Nee, S. & May, R.M. (1992). Dynamics of metapopulations: habitat destruction and competitive coexistence. *Journal of Animal Ecology*, 61, 37–40.

Oates, M. & Warren, M.S. (1990). *A Review of Butterfly Introductions in Britain and Ireland*. Worldwide Fund for Nature, Godalming.

Quinn, J.F. & Harrison, S.P. (1988). Effects of habitat fragmentation and isolation on species richness: evidence from biogeographical patterns. *Oecologia*, 75, 132–140.

Ragge, D.R. (1965). *Grasshoppers, Crickets and Cockroaches of the British Isles*. Warne, London.

Ravenscroft, N.O.M. (1992). The declining status of the butterfly *Plebejus argus* L. on the Sandlings of East Anglia. *Entomologist*, **111**, 88–94.

Read, M. (1985). *The silver-studded blue conservation report.* MSc Thesis, University of London.

Reed, T.M. (1983). The role of species–area relationships in reserve choice: a British example, *Biological Conservation*, **25**, 263–271.

Southwood, T.R.E. (1977). Habitat, the templet for ecological strategies. *Journal of Animal Ecology*, **46**, 337–365.

Spellerberg, I.F. (1991). Biogeographical basis for conservation. *The Scientific Management of Temperate Communities for Conservation* (Ed. by I.F. Spellerberg, F.B. Goldsmith & M.G. Morris), pp. 293–322. Blackwell Scientific Publications, Oxford.

Thomas, C.D. (1983). The Ecology and Status of Plebejus Argus in North West Britain. MSc Thesis, University of Wales.

Thomas, C.D. (1985a). Specializations and polyphagy of *Plebejus argus* (Lepidoptera: Lycaenidae) in North Wales. *Ecological Entomology*, **10**, 325–340.

Thomas, C.D. (1985b). The status and conservation of the butterfly *Plebejus argus* (Lepidoptera: Lycaenidae, in north-west Britain. *Biological Conservation*, **33**, 29–51.

Thomas, C.D. & Harrison, S. (1992). The spatial dynamics of a patchily distributed butterfly species. *Journal of Animal Ecology*, **61**, 437–446.

Thomas, C.D., Thomas, J.A. & Warren, M.S. (1992). Distributions of occupied and vacant butterfly habitats in fragmented landscapes. *Oecologia*, **92**, 563–567.

Thomas, J.A. (1984). The conservation of butterflies in temperate countries: past efforts and lessons for the future. *The Biology of Butterflies* (Ed. by R.I. Vane-Wright & P.R. Ackery), pp. 333–353. Academic Press, London.

Thomas, J.A. (1991). Rare species conservation: case studies of European butterflies. *The Scientific Management of Temperate Communities for Conservation* (Ed. by I.F. Spellerberg, F.B. Goldsmith & M.G. Morris), pp. 149–197. Blackwell Scientific Publications, Oxford.

Thomas, J.A. (1993). Holocene climate change and warm man-made refugia may explain why a sixth of British butterflies possess unnatural early-successional habitats. *Ecography* (in press).

Usher, M.B. (1986). *Wildlife Conservation Evaluation.* Chapman & Hall, London.

Warren, M.S. (1992). Butterfly populations. *The Ecology of Butterflies in Britain* (Ed. by R.L.H. Dennis), pp. 73–92. Oxford University Press, Oxford.

Webb, N.R. (1986). *Heathlands.* Collins, London.

Webb, N.R. (1989). Studies on the invertebrate fauna of fragmented heathland in Dorset, UK, and the implications for conservation. *Biological Conservation*, **47**, 153–165.

Webb, N.R. (1990). Changes on the heathlands of Dorset. England, between 1978 and 1987. *Biological Conservation*, **51**, 273–286.

Webb, N.R. & Haskins, L.E. (1980). An ecological survey of heathlands in the Poole Basin, Dorset, England in 1978. *Biological Conservation*, **17**, 281–296.

Webb, N.R. & Hopkins, P.J. (1984). Invertebrate diversity on fragmented *Calluna* heathlands. *Journal of Applied Ecology*, **21**, 921–933.

Webb, N.R. & Vermaat, A.H. (1990). Changes in vegetational diversity on remnant heathland fragments. *Biological Conservation*, **53**, 253–264.

Webb, N.R., Clarke, R.T. & Nicholas, J.T. (1984). Invertebrate diversity on fragmented *Calluna* heathland: effects of surrounding vegetation. *Journal of Biogeography*, **11**, 41–46.

Wilcox, B.A. & Murphy, D.D. (1985). Conservation strategy: the effects of fragmentation on extinction. *American Naturalist*, **125**, 879–887.

7. DECLINING FARMLAND BIRD SPECIES: MODELLING GEOGRAPHICAL PATTERNS OF ABUNDANCE IN BRITAIN

SIMON GATES, DAVID WINGFIELD GIBBONS,
PETER C. LACK AND ROBERT J. FULLER
*British Trust for Ornithology, The Nunnery, Nunnery Place, Thetford,
Norfolk IP24 2PU, UK*

SUMMARY

Several farmland bird species are undergoing a decline in population and contraction in range in Britain. Reasons for these changes are poorly understood, as is the potential impact of future environmental changes.

The geographical variation in abundance in Britain of eight declining farmland bird species, as shown by the 1988–91 breeding bird atlas, was modelled in relation to climate, topography and land use (particularly agriculture), using multivariate logistic regression modelling. Sample units were 10-km squares of the National Grid.

On average, across species, about 45% of the variation in breeding bird abundance was explained by the models and, for most species, the modelled pattern was very similar to the real pattern shown from fieldwork. For some species the models included variables which may have been causal, while for others they were probably simply correlates of these.

The predicted effects of possible future agricultural change (converting 15% or 30% of cereals to grassland) were small when compared to those of climate change (increase in mean temperature of 1°C or 3°C), although the latter did not consider changes in land use and other climate variables consequent upon a rise in temperature.

However, the models were unsuccessful in predicting long-term declines, even though all eight species are known to have declined in abundance by at least 50% during 1968–91. Possible reasons for this are discussed.

INTRODUCTION

Results from the British Trust for Ornithology (BTO)'s long-running Common Birds Census have shown that breeding populations of several species of farmland birds have declined in Britain (Marchant *et al.* 1990), and that these declines have been most marked since the mid-1970s (Marchant & Gregory 1994). Population declines of some of these species have also been noted elsewhere in Europe (e.g. Hustings 1988, 1992; Robertson & Berg 1992).

O'Connor and Shrubb (1986) established several links between declining bird populations and changing farming practices, although proving causal relationships has, with only a few exceptions (e.g. Potts 1986, Stowe *et al.* 1993), been difficult. O'Connor and Shrubb (1986) suggested that the switch from spring to autumn sowing of cereals, accompanied by the loss of winter stubble fields, a move away from crop rotations, intensification and an increase in pesticide usage might be important factors. Together with destruction and deterioration of non-crop habitat, such as drainage and loss of hedgerow, these changes are likely to have led to a reduction in the food supply of farmland birds (Fuller *et al.* 1991). Furthermore, Marchant and Gregory (1994) have shown that declines in seed-eating passerines have been greatest on arable farms.

In some cases the population declines have led to range contraction and local extinction (Gibbons *et al.* 1993), but very little is known about the factors underlying distributions and abundance of birds. In this chapter we model patterns of geographical variation in abundance of eight declining farmland bird species using data from the 1988–91 breeding bird atlas organized by the BTO, Scottish Ornithologists' Club and Irish Wildbird Conservancy (Gibbons *et al.* 1993), and data on climate, topography and agriculture. We test whether or not the models can help in an understanding of the causes of the declines by predicting, retrospectively, their abundance in 1969 and comparing these values with those for 1988–91. In addition we use the models to predict the effects of possible future environmental change, both climatic and agricultural.

We chose to analyse the geographical pattern of abundance of each species rather than simply its distribution (presence/absence). This was because the environmental factors which set the limits to the range of a species may be different from those which determine abundance within its range, and analysis of presence/absence data can detect only the first of these. Use of abundance data is therefore more sensitive.

METHODS

Bird abundance data

The eight farmland species chosen for study were: grey partridge, *Perdix perdix* (L.); lapwing, *Vanellus vanellus* (L.); turtle dove, *Streptopelia turtur* (L.); skylark, *Alauda arvensis* L.; tree sparrow, *Passer montanus* (L.); linnet, *Carduelis cannabina* (L.); reed bunting, *Emberiza schoeniculus* (L.); and corn bunting, *Miliaria calandra* (L.): Populations of each of these species declined by 50% or more (range 54.3–85.1%) on farmland during 1968–91 (Marchant *et al.* 1990; BTO unpublished data).

An index of relative abundance of each species in each 10-km National Grid square was calculated from data collected during atlas fieldwork (Gibbons 1991; Gibbons *et al.* 1993). This consisted of the number of 2-km squares ('tetrads') in each 10-km square where a species was recorded, out of the total number that were visited (up to a maximum of 25). It was not reduced to a proportion for analysis, because the logistic regression method used derived information from both numerator and denominator; a 10-km square in which a species was recorded in 10 tetrads out of 20 that were visited was given more weight than one where it was recorded in five out of 10.

Environmental data

Environmental data were obtained from three sources:

1   Areas of crops and numbers of livestock from the 1988 June agricultural census of the Ministry of Agriculture, Fisheries and Food (MAFF) and the Department of Agriculture and Fisheries for Scotland, converted to 10-km grid squares by the Edinburgh University Data Library.

2   Monthly average temperatures for each 10-km square, not corrected to sea level, provided by J.R.G. Turner and J. Lennon (Leeds University Genetics Department).

3   Other environmental variables taken from the Institute of Terrestrial Ecology (ITE) Land Characteristics Data Bank (Ball *et al.* 1983).

All variables given as an area in the original data were converted to a proportion of the land in each 10-km square to make them consistent with the bird data.

These data sets contained over 200 variables, most of which were either unlikely to be causally related to abundance of any bird (e.g. number of farm workers, greenhouse area and foreshore length) or were subsets of other variables (e.g. 22 different kinds of vegetables are recorded by the agricultural census). We therefore selected a set of 48 that we considered could potentially have a role in determining bird distributions for inclusion in analyses. A full list is given in the Appendix.

Bird abundance and full environmental data were available for 2684 10-km squares in Britain. The Scilly Isles, Lundy, and the Isle of Man were excluded from the analyses because agricultural data were not collected for these islands. Similarly, Shetland and most of Orkney were excluded because no temperature data were available. Although bird data were available for Ireland, environmental data were not, so Ireland was also excluded from the analyses.

Analyses

Multiple logistic regression was used to relate the abundance index of each species in each 10-km square to the environmental data, using SAS logistic procedure (SAS Institute 1990), which fits models using maximum likelihood methods. Various ordination techniques were considered as a means of reducing the number of environmental variables, but were rejected because the variables produced are often difficult to interpret.

There are no firmly established methods of selecting a 'best' multiple regression model from such a large set of variables (see e.g. James & McCulloch 1990). We therefore produced several alternative models for each species using a subset of the 10-km squares, and then tested them by predicting abundance in randomly selected subsets of the remainder of the data. The analytical method is detailed below.

We attempted to take account of the fact that some environmental variables were non-linearly related to the abundance indices. This is often ignored in studies relating animals and plants to environmental variables, but it seems intuitively likely that for some environmental factors there will be an optimum at which abundance is highest (Westman 1980), and for others the relationship with abundance will increase to an asymptote. To do this, first we fitted the square of each variable, as well as the variable itself, into a quadratic regression model; this fitted a Gaussian curve to the data. If the addition of the quadratic term provided a statistically significant improvement ($P < 0.001$) over the linear model, the quadratic term was retained and used in all subsequent analyses. Second, we took log and square root transformations of each explanatory variable and entered these into univariate models. Once again, if the transformation yielded a significant improvement over non-transformation, the transformed variable was used in subsequent analyses. In many cases transformation linearized a relationship, making inclusion of the quadratic term unnecessary, and in other cases it converted an asymmetric peaked relationship into a symmetrical one, to which a Gaussian response curve could then be fitted.

Spatial autocorrelation is a problem in the analysis of geographically referenced data (Cliff & Ord 1973). We have not explicitly dealt with this here, but we have attempted to reduce its effects by using only a quarter of the data, sampled from a regular grid, to produce the models. Thus no neighbouring 10-km squares were included in any one analysis. Randomly chosen subsets of the remaining three quarters of the data were then used as independent data sets for testing the models.

Analytical procedure

For each species, univariate regressions of the abundance index on each of the environmental variables were calculated, using a transformation of the environmental variable and including the X^2 terms if these improved the fit (see above). Each of the eight variables explaining the largest amount of deviance was then used as the basis for a model. For each of these eight, we needed to select a number of other variables to make a set from which a multiple regression model would be produced. We arbitrarily decided to produce the final models from sets of 20 variables; including more than this would have led to a great increase in computer processing time, probably with little increase in the amount of deviance explained. We generated these sets of 20 variables by fitting each of the eight variables that had been chosen to be the basis of a model into a two variable model with every one of the other 47 variables, including X^2 terms if necessary. The 19 variables which explained the most additional deviance were then included in the set from which a model was produced. Where easting and northing were included in this set, an alternative set excluding them was also produced. This was done because easting and northing are themselves unlikely to exert a controlling influence on abundance, and are likely to be acting as a substitute for other variables. This gave us a maximum of 16 sets of variables for each species, from each of which a model was produced. In some cases fewer than 16 models were produced, either because the same set of variables was selected twice, or because easting or northing were not included in all of the sets.

All 20 variables and their quadratic terms were entered into each model, and a backward selection method was used to remove variables that contributed little explanatory power. The variable that gave the smallest reduction in deviance explained on its removal was deleted from the model, and this was repeated until no variable could be removed without a reduction in deviance equivalent to $P < 0.001$. A maximum of 16 models (eight including easting and northing, and eight excluding these variables) was produced for each species and the frequency with which each variable occurred was recorded. Some subsets of variables were strongly intercorrelated, and hence it is not surprising that the same ones turned up in some of the models, but if a variable was generally important we would expect it to be included in the models whatever the starting set.

To test these models, each was fitted to 50 randomly selected subsets (of about the same size as used to build the models) drawn from the 75% of the data that was not used to build the models, and the amount of deviance explained by each model for each subset was recorded. For each of the 50 random subsets of data, the models were ranked according to the amount of deviance explained, and the mean amount of deviance explained by each model across all subsets was calculated. The 'best' model was taken to be that which was most consistently ranked the highest, provided that there was no other model which contained fewer parameters and which explained the same (or not significantly less) average amount of deviance.

The best model for each species excluding easting and northing was used predictively in two ways. First, we predicted the abundance in each 10-km square in 1969. For this the agricultural census data for 1988 were replaced by those for 1969, although a few of the variables in the two censuses did not correspond exactly (see Appendix). All other variables remained unchanged. The mean predicted abundance across all squares in 1969 was calculated, and this was compared with that for 1988.

Second, as an example of how these models could be used, we have predicted the pattern of abundance of each species under two scenarios of climatic change and two of land-use change: increases in breeding season and winter temperatures of 1°C and 3°C, and conversion of 15% and 30% of cereal area to grassland. The climate changes are those expected by 2025 and the end of the 21st century, respectively, with current rates of emission of greenhouse gases (Houghton *et al.* 1990), and the changes in land use are similar to those currently being introduced to reduce agricultural surpluses (MAFF 1992). For the temperature rises we have taken no account of any changes in land use or rainfall that might occur and which are necessary if fully realistic predictions are to be made. The mean abundance index across all squares was calculated for each species under each scenario and predicted distribution maps for each were plotted.

RESULTS

Results from the univariate regressions, the data transformations, and the two variable models are not presented, as these were used simply for choosing the variables to fit into the backward selection models.

Results of the backward selection models including easting and northing are presented in Table 7.1, and those excluding easting and northing in Table 7.2. The number of occasions on which each variable occurred in the models (a maximum of eight models including, and a maximum of eight models excluding, easting and northing) is shown for each species. Corn bunting is not included in Table 7.2 because easting and northing

TABLE 7.1. Frequency of occurrence of variables in the models when easting and northing were included. The total number of models for each species is given. For brevity species codes are used in the table: P = grey partridge, L = lapwing, TD = turtle dove, S = skylark, TS = tree sparrow, LI = linnet, RB = reed bunting, CB = corn bunting. The shape of the univariate relationship between each species abundance and each variable is classified as follows: linear relationships, + = positive, − = negative; curvilinear relationships, +v = positive concave, +x = positive convex, −v = negative concave, −x = negative convex, +o = mid-variable peak, −o = mid-variable trough

Species	P	L	TD	S	TS	LI	RB	CB
No. of models	6	8	5	7	6	8	8	7
1 Area of farmland			1+		1+		4+o	1+
2 Young grassland		2+x		1+o		5+x	1+o	
3 Other grassland	3+x	2+x		1+o		1+x	6+o	7+o
4 Total grassland	5+x	5+x		7+o	1+x	3+x	3+x	7+o
5 Wheat		2+o	3+v	7+v				3+
6 Winter barley		1+		1+		2+	5+o	
7 Spring barley	6+v		2+o	5+o		1+x	1+x	
8 Oats							1+o	
9 Total cereal	4+	2+		2+				2+
10 Vegetables						1+		
11 Orchards								
12 Soft fruit		1+v		1+				
13 Total horticulture								
14 Beet				1+o		3+o	1+	
15 Oilseed rape				1+		6+v		7+
16 Total non-cereal crops	2+		2+	2+	5+	1+x	1+	
17 Fallow				1+	4+o	3+x	5+	
18 Total crops and fallow	3+v			6+v		6+		
19 Grass/agricultural land	4+o	1+o	2−	5+o	1−		1+o	7−x
20 Dominance index		3+x				4+x		1+o
21 Cattle		1+x	1+o		1+x		1+x	4+o
22 Pigs						8+x	6+x	2+
23 Sheep		5+o		4−x				
24 Easting	6+o		5+	3−o				
25 Northing	6+o	8+o	2−x	2+o	6+o	6+o	6+o	
26 Urban				7+o	3+o			
27 Agricultural land class 1		1−o						7+
28 Agricultural land class 2								4+v
29 Agricultural land class 3							6+o	
30 Agricultural land class 4		1+					3+	
31 Agricultural land class 5		1+o		2−	6−	8−x	3−	

Continued on p. 160

TABLE 7.1. *Continued*

Species No. of models	P 6	L 8	TD 5	S 7	TS 6	LI 8	RB 8	CB 7
32 Agricultural land class 6			1+0	1−x				
33 Mean altitude		6−o				2−	8−	4−
34 Height difference	3−x	6−	1−	8−		3−x	8−	
35 River							3+0	
36 Lakes						7−	5−x	
37 Woodland		6−		8−			4−	2−
38 Brown earths		6−x		7+0	4−		6−	7−x
39 Rendzinas or calc. soils	1+			3+				
40 Gley soils	2+	2+					2+	
41 Humus etc.								
42 Peaty podzols		1−x						
43 Peaty gleys		6+	1−	7+				
44 Deep peaty soils and peats		1+0				1−	1−	4−
45 Immature or skeletal soil				1−				
46 Rainfall	2−	8−x				4−	1−	4−
47 Breeding season temperature	5+x	4+0	3+x			4+x		1+
48 Winter temperature	2+0	7+0			6+0	6+x	1+x	7+0

TABLE 7.2. Frequency of occurrence of variables in the models when easting and northing were excluded. Other definitions as for Table 7.1. The shape of each relationship is only shown for species–variable combinations not included in Table 7.1

Species No. of models	P 7	L 6	TD 7	S 6	TS 8	LI 6	RB 6
1 Area of farmland							3
2 Young grassland		1	3+0			4	2
3 Other grassland		2	2+0	1	2+x	3	3
4 Total grassland	7	3		5		1	4
5 Wheat		1	4	6	2+		1+
6 Winter barley	5+	1		3	4+	2	3
7 Spring barley	5	1+	3	4		1	2
8 Oats							1
9 Total cereal	5	1		2			
10 Vegetables						1+0	1
11 Orchards		1−					
12 Soft fruit		2		2			
13 Total horticulture						1+0	
14 Beet				2		3	
15 Oilseed rape	4+					5	

Continued

TABLE 7.2. *Continued*

Species	P	L	TD	S	TS	LI	RB
No. of models	7	6	7	6	8	6	6
16 Total non-cereal crops	2		3	2	8	3	3
17 Fallow	1+0		4+x	1	3	2	
18 Total crops and fallow	2		2+	5	1+	3	
19 Grass/agricultural land	6	1	6	1	1		
20 Dominance index	1+0	4			1+x	5	
21 Cattle			6		1		1
22 Pigs					1+	6	4
23 Sheep		5		3			
24 Easting	−	−	−	−	−	−	−
25 Northing	−	−	−	−	−	−	−
26 Urban			4+0	6	4	1+	
27 Agricultural land class 1		2					
28 Agricultural land class 2						1+x	
29 Agricultural land class 3						5	
30 Agricultural land class 4							1
31 Agricultural land class 5		3			6		2
32 Agricultural land class 6		1−				5+	
33 Mean altitude		4				4	6
34 Height difference	6	3		6		2	6
35 River							3
36 Lakes						5	5
37 Woodland	4−	6		6	7−		5
38 Brown earths	3−	4		5	4		5
39 Rendzinas or calc. soils	1			1			
40 Gley soils	3	1			2+		1
41 Humus etc.							
42 Peaty podzols	1−	3					
43 Peaty gleys		5		5			
44 Deep peaty soils and peats				1−0		1	1
45 Immature or skeletal soil				3			
46 Rainfall	7	6			2−	4	1
47 Breeding season temperature	7	5	7		7+x	4	1+
48 Winter temperature	2	6	7+0		8	3	1

were not included in any of the starting sets of 20 variables entered into the backward selection models. The shape of the relationship between the abundance of each species and each environmental variable is also noted. These are only approximate and were determined by plotting, for each species, the predicted line of logit (abundance) against each variable (transformed, or with a X^2 term added, as necessary) to show whether the relationship was positive, negative, or curvilinear. In many cases the number of models produced was less than eight, because even though different sets of variables were entered into the eight starting models, some converged to retain exactly the same variables.

For each species particular variables occurred repeatedly in the models. Easting and/or northing were important variables for all species except skylark and corn bunting. Interpretation of the effects of geographical position is difficult (see above), and a comparison of Tables 7.1 and 7.2 shows that the effects of other variables were often masked by its inclusion. When easting and northing were excluded the following variables were important (i.e. were included in all, or all bar one, of the models) for each species. For each variable, the sign of the relationship, + = positive, − = negative and ○ = mid-variable peak, is given:

Grey partridge: Total grassland (+), breeding season temperature (+), grass as a proportion of agricultural land (○), height difference (−), rainfall (−).

Lapwing: Peaty gleys (+), sheep (○), breeding season (○) and winter temperatures (○), woodland (−), rainfall (−).

Turtle dove: Breeding season temperature (+), cattle (○), winter temperature (○), grass as proportion of agricultural land (−).

Skylark: Wheat (+), total crops and fallow (+), peaty gleys (+), total grassland (○), urban (○), brown earths (○), height difference (−), woodland (−).

Tree sparrow: Total non-cereal crops (+), breeding season (+) and winter temperatures (○), woodland (−).

Linnet: Oilseed rape (+), dominance index (+), pigs (+), agricultural land class 6 (+), lakes (−).

Reed bunting: Agricultural land class 3 (○), mean altitude (−), height difference (−), lakes (−), woodland (−), brown earths (−).

Corn bunting: Oilseed rape (+), agricultural land class 1 (+), other grassland (○), total grassland (○), winter temperature (○), grass as a proportion of agricultural land (−), brown earths (−).

Table 7.3 shows the mean amount of deviance explained by the models for each species (with easting and northing included and excluded). On average, across species, about 45% of the deviance was explained, but there was a wide range. This was probably due partly to how widespread was each species, as more widespread species live in a greater variety of habitats and are thus more difficult to model. It was probably also an effect of the census methodology used to estimate abundance, which detected less variation in abundance of very common species (see Gibbons 1991; Gibbons *et al.* 1993).

Table 7.4 shows the results of the validation of the models against 50 randomly selected subsets. The table shows the 'best' model for each

TABLE 7.3. The percentage of deviance explained by the models for each species

	Including easting and northing			Excluding easting and northing		
Species	No. of models	Mean	Range	No. of models	Mean	Range
Grey partridge	6	41.3	40.2–41.8	7	38.5	37.0–39.9
Lapwing	8	46.8	45.2–48.3	6	45.7	43.3–47.7
Turtle dove	5	75.2	74.6–76.0	7	71.0	70.2–72.0
Skylark	8	31.8	28.1–33.8	6	32.1	28.4–33.6
Tree sparrow	6	45.5	44.3–47.7	8	44.3	43.1–45.3
Linnet	8	49.2	46.7–51.7	6	49.3	47.1–50.7
Reed bunting	8	32.4	30.3–35.1	6	31.9	29.5–33.7
Corn bunting	7	52.9	51.7–55.7	–	–	–

TABLE 7.4. The 'best' models for each species

Including easting and northing		Excluding easting and northing	
Grey partridge			
Constant	−3.6086	Constant	6.5843
Log^2(other grassland)	0.1771	Log^2(total grassland)	−0.1683
Log^2(total grassland)	−0.2083	Square root(winter barley)	5.1878
Log(spring barley)	0.8096	Log(spring barley)	0.8135
Log^2(spring barley)	0.0635	Log^2(spring barley)	0.0690
Easting	0.2651	Square root(total cereals)	−6.7664
(Easting)^2	−0.00278	Log(oilseed rape)	0.2309
Northing	0.0680	Log(grass/agricultural land)	−1.1287
(Northing)^2	−0.00081	Woodland	−1.8915
Log(gley soils)	0.0396	Rendzinas	0.8554
Rainfall	−0.00104	Rainfall	−0.00187
(Breeding season temperature)^2	−0.0194	(Breeding season temperature)^2	−0.0169
(Winter temperature)^2	0.0367		
Lapwing			
Constant	0.7744	Constant	−1.1868
Log(wheat)	−0.0784	Log(total grassland)	0.2235
Log(dominance index)	0.4451	Log(winter barley)	−0.1853
Sheep	0.1913	Log(dominance index)	0.3278
(Sheep)^2	−0.0303	(Sheep)^2	−0.0131
Northing	0.0544	Log(agricultural land class 1)	0.3614
(Northing)^2	−0.00032	Log^2(agricultural land class 1)	0.0309
Log(agricultural land class 4)	0.0471	Log(mean altitude)	−0.6810
(Agricultural land class 5)^2	0.7397	Woodland	−1.9771
Log(height difference)	−0.2786	Brown earths	1.7994
Woodland	−1.2017	(Brown earths)^2	−2.0451
Brown earths	1.6524	Peaty gleys	0.8627
(Brown earths)^2	−2.0487	(Rainfall)^2	-6.4×10^{-7}

Continued on p. 164

TABLE 7.4. *Continued*

Including easting and northing		Excluding easting and northing	

Lapwing (contd.)

Peaty gleys	0.8549	Breeding season temperature	1.7157
(Rainfall)^2	-6.44×10^{-7}	(Breeding season temperature)^2	-0.0945
Winter temperature	-0.2282	Winter temperature	-0.5447

Turtle dove

Constant	-17.34	Constant	-21.19
Log(wheat)	0.6485	Spring barley	-6.971
Log^2(wheat)	0.05768	Log^2(fallow)	-0.0202
Log(total non-cereals)	0.4460	Grass/agricultural land	-3.762
Log(easting)	4.822	Log(cattle)	-0.8121
(Northing)^2	-0.0005665	Log^2(cattle)	-0.2087
		Log(urban)	-0.7919
		Log^2(urban)	-0.0741
		Breeding season temperature	1.711
		(Winter temperature)^2	-0.08393

Skylark

Constant	2.9794	Constant	1.4118
Log(young grassland)	-0.1096	Log(total grassland)	-0.5502
Log^2(total grassland)	-0.0389	Log^2(total grassland)	-0.0690
(Spring barley)^2	-17.7305	(Wheat)^2	18.5682
Log(grass/agricultural land)	-1.1033	Spring barley	5.2509
Log^2(grass/agricultural land)	-0.1159	Square root(soft fruit)	-6.1383
Sheep	0.2031	Total non-cereals	9.0817
(Sheep)^2	-0.0268	(Total crops and fallow)^2	-6.8015
Log(easting)	-0.5192	Sheep	0.2256
(Northing)^2	0.000083	(Sheep)^2	-0.0275
Urban	-1.2945	Urban	-1.5208
Log(height difference)	-0.2842	Log(height difference)	-0.3305
Woodland	-1.9666	Woodland	-2.7230
Brown earths	1.1208	Brown earths	1.0574
(Brown earths)^2	-1.5549	(Brown earths)^2	-1.3862
Peaty gleys	1.1758	Rendzinas	0.6467
		Peaty gleys	1.1426
		Square root(skeletal soils)	-0.6963

Tree sparrow

Constant	-16.4647	Constant	-38.6005
Log(total non-cereals)	0.4622	Log(winter barley)	-0.2040
Grass/agricultural land	1.0683	Log(total non-cereals)	0.4633
Log(northing)	6.6727	Agricultural land class 5	-1.8394
Log^2(northing)	-0.8944	Woodland	-2.5326
Agricultural land class 5	-1.7783	Log(gley soils)	0.0455
Brown earths	-0.6161	Breeding season temperature	6.0670
Winter temperature	2.3010	(Breeding season temperature)^2	-0.2587
(Winter temperature)^2	-0.3491	Winter temperature	1.9481

Linnet

Constant	4.5525	Constant	0.8951
Log(young grassland)	-0.2436	Log(young grassland)	-0.3336
Log(total grassland)	0.2822	Log(total grassland)	0.4039
Log(oilseed rape)	0.7638	Log(winter barley)	-0.1158

Continued

TABLE 7.4. *Continued*

Including easting and northing		Excluding easting and northing	
Linnet (contd.)		Log(oilseed rape)	0.8204
Log(oilseed rape)	0.0534	Log^2(oilseed rape)	0.0567
Log(total crops and fallow)	0.1435	Log^2(total non-cereals)	−0.0253
Log^2(pigs)	−0.0101	Log^2(pigs)	−0.0115
Northing	0.0354	Log(urban)	0.0580
(Northing)^2	−0.00038	Agricultural land class 5	3.6566
Agricultural land class 5	3.8384	(Agricultural land class 5)^2	−4.0734
(Agricultural land class 5)^2	−4.6203	Height difference	−0.00126
Lakes	−13.387	Lakes	−14.2905
Rainfall	−0.00081	Square root(peaty soils)	−0.6043
Breeding season temperature	−0.3734	Rainfall	−0.00062
Winter temperature	0.5919	Breeding season temperature	0.6461
		(Breeding season temperature)^2	−0.0460
		Winter temperature	0.0628
		(Winter temperature)^2	−0.3377
Reed bunting			
Constant	3.0236	Constant	−0.9603
Log(total grassland)	0.3491	Log(young grassland)	−1.2608
Log(vegetables)	−0.0939	Log^2(young grassland)	−0.1010
Log(fallow)	0.1063	Log(other grassland)	−3.5705
Log(cattle)	−0.3675	Log^2(other grassland)	−0.4587
Log^2(cattle)	−0.0528	Log(total grassland)	3.9573
Northing	0.0318	Log^2(total grassland)	0.4771
(Northing)^2	−0.00026	Log(vegetables)	−0.0849
(Agricultural land class 3)^2	−1.1825	(Agricultural land class 3)^2	−1.0897
Agricultural land class 5	−0.7691	Agricultural land class 5	−0.6748
Log(mean altitude)	−0.4770	Log(mean altitude)	−0.4365
Log(height difference)	−0.3432	Log(height difference)	−0.3108
Square root(lakes)	2.1433	Square root(lakes)	2.3996
Brown earths	−0.4182	Log(woodland)	−0.0507
Log(peaty soils)	0.0523	Square root(gley soils)	0.4366
		Log(peaty soils)	0.0467
Corn bunting			
Constant	−4.2987		
Log(other grassland)	−1.3525		
Log(total grassland)	2.0878		
Log(total cereal)	−0.3754		
Log(oilseed rape)	0.2645		
(Grass/agricultural land)^2	−5.5190		
Square root (agricultural land class 1)	1.6405		
Log(agricultural land class 2)	0.3908		
Log^2(agricultural land class 2)	0.0373		
Woodland	−4.6646		
(Brown earths)^2	−1.5125		
Log(deep peaty soils)	0.1363		
Winter temperature	2.8560		
(Winter temperature)^2	−0.2727		

species; this was the highest ranking model for each, again separately including and excluding easting and northing. In general, the 'best' model included those variables which occurred most frequently in the individual species models as shown in Tables 7.1 and 7.2.

As an illustration of the power of these models to predict the distributions of these species, Figs 7.1, 7.2 and 7.3 show for the corn bunting, turtle dove and tree sparrow, respectively, (a) the 'real' geographical pattern of abundance as determined from fieldwork (modified from Gibbons *et al.* 1993); and (b) the geographical pattern of abundance from the 'best' model for that species. In the latter case the 'best' model excluding easting and northing has been used. For all three species there are marked similarities between the real and modelled data.

Similar maps for the other five species have been produced, but are not presented because of lack of space. However, the information contained in them is summarized in Table 7.5. This table shows, for each species, the mean index of abundance (across all 10-km squares in Britain) of each species under a number of different scenarios. For each species there was a close agreement between the 'real' mean (calculated from fieldwork) and that obtained from the 'best' model (in each case, excluding easting and northing). The 'real' and 'best' model maps for the species not presented also showed similar spatial patterns.

Table 7.5 also shows some predictions for the effects of climate and agricultural change on each species. The predicted effects of the switch from cereals to grass were generally small when compared to those of temperature rise, although species responded in different ways. Particularly marked were the predicted effects of temperature rise on grey partridge, lapwing, tree sparrow (all showing predicted population declines) and turtle dove (a population increase). Fig. 7.2(c) shows the

---

FIG. 7.1. (*Opposite.*) (a) Geographical variation in relative breeding density of the corn bunting in Britain and Ireland from 1988 to 1991 fieldwork (modified from Gibbons *et al.* 1993). Increasing dot sizes refer to increasing levels of abundance. The data are frequency of occurrence in 2-km squares (tetrads) within 10-km squares. See Gibbons (1991) and Gibbons *et al.* (1993) for a full explanation. The values for increasing dot sizes are: 0.0400–0.1000, 0.1001–0.1905, 0.1906–0.3600, 0.3601–1.000. These class limits are based on quartiles, and there are thus the same number of dots in each class. (b) As for (a) except for Britain only, and the data are predicted from the corn bunting 'best' model rather than from fieldwork. The values for increasing dot sizes are: 0.0400–0.0828, 0.0829–0.1445, 0.1446–0.2254, 0.2255–0.8465. Values of <0.04 were sometimes predicted from the models, but as the fieldwork would not have picked up species below this level of abundance, these values have been excluded from the map to make it more comparable with (a). Open circles are 10-km squares with missing environmental data.

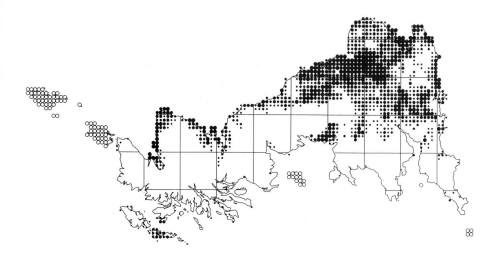

(b)

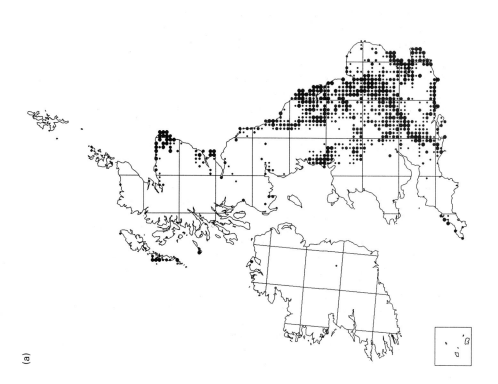

(a)

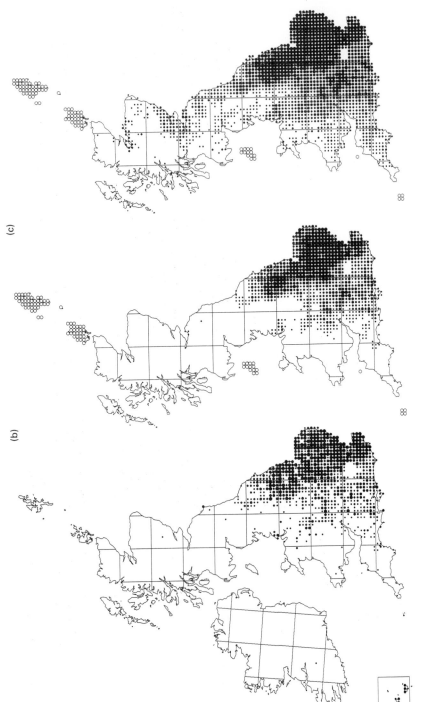

Fig. 7.2. (a) As for Fig. 7.1(a) but for turtle dove. The values for increasing dot sizes are: 0.0400–0.1250, 0.1251–0.2821, 0.2822–0.5556, 0.5557–1.000. (b) As for Fig. 7.1(b) for turtle dove. In this case the 'best' model excluding easting and northing was used. The values for increasing dot sizes are: 0.0400–0.1225, 0.1226–0.2467, 0.2468–0.4671, 0.4672–0.8138. (c) The predicted geographical pattern of abundance of the turtle dove from the 'best' model, but following a 3°C rise in temperature. The values for increasing dot sizes are: 0.0400–0.1135, 0.1136–0.4418, 0.4419–0.8189, 0.8190–0.9808. Other details as for Fig. 7.1(a) and (b).

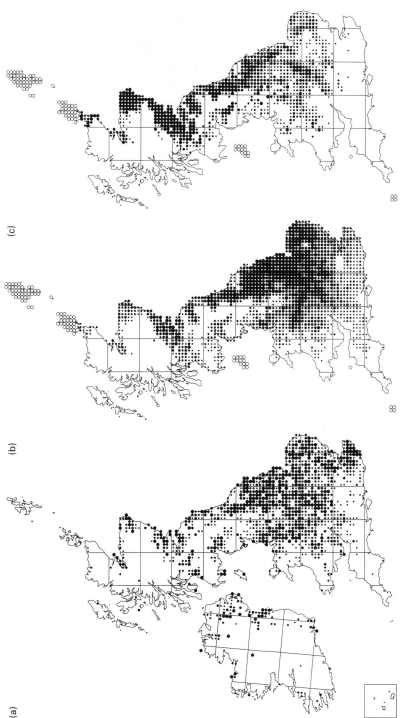

(a)

(b)

(c)

FIG. 7.3. (a) As for Fig. 7.1(a) but for tree sparrow. The values for increasing dot sizes are: 0.0400–0.1000, 0.1001–0.1667, 0.1668–0.3043, 0.3044–1.000. (b) As for Fig. 7.1(b) but for tree sparrow. The 'best' model excluding easting and northing was used. The values for increasing dot sizes are: 0.0400–0.0880, 0.0881–0.1437, 0.1438–0.2160, 0.2161–0.4345. (c) The predicted geographical pattern of abundance of the tree sparrow from the 'best' model, but following a 1°C rise in temperature. The values for increasing dot sizes are: 0.0400–0.0562, 0.0563–0.0750, 0.0751–0.1121, 0.1122–0.2570. Other details as for Fig. 7.1(a) and (b).

TABLE 7.5. Mean indices of abundance of each species across all 10-km squares from 1988 to 1991 fieldwork (= real), from the best model (excluding easting and northing), from various agricultural and climate scenarios, and from fitting the 1969 agricultural statistics into the best models

| | | | Predictions from: | | | | |
| | | | Temperature rise | | Cereals to grass | | |
	Real	Best model	+1°C	+3°C	15%	30%	1969
Grey partridge	0.118	0.122	0.087	0.038	0.121	0.122	0.051
Lapwing	0.331	0.333	0.169	0.022	0.342	0.351	0.729
Turtle dove	0.108	0.103	0.170	0.288	0.091	0.079	0.047
Skylark	0.626	0.606	–	–	0.597	0.587	0.585
Tree sparrow	0.093	0.092	0.040	0.003	0.094	0.097	0.201
Linnet	0.387	0.376	0.382	0.412	0.392	0.403	0.440
Reed bunting	0.192	0.193	–	–	0.195	0.196	0.207
Corn bunting	0.073	0.073	0.108	0.086	0.069	0.063	0.044

predicted pattern of abundance of the turtle dove following a 3°C rise in temperature, and Fig. 7.3(c) that of the tree sparrow following a 1°C rise.

Table 7.5 also shows that the 'best' models were unable to predict consistently the decline in abundance of all eight species when using the 1969 agricultural statistics. For only four species (lapwing, tree sparrow, linnet and reed bunting) did the models predict higher abundances in 1969 than in 1988.

DISCUSSION

Ecological interpretation of the models

The models we have produced explain a reasonable amount of the variation in abundance (28–76%) and appear to be able to predict with a good degree of accuracy the present patterns of abundance of the eight species. However, their interpretation in terms of the ecology of the species is not simple. The large number of variables suggests that none of them has an overriding effect, but that a large number contribute a relatively small amount of explanatory power. Also some of them are not connected in any obvious way with the ecology of the species in question. This raises the question of whether the species' abundance may be largely determined by variables that were not measured, and the variables in our data set appear in models only because they are correlated with these unmeasured factors. Two major factors which have been suggested as responsible for the decline of farmland birds, and which could not be explicitly put into our models because of lack of data at a suitable geographic scale, are pesticide inputs and cereal sowing dates (O'Connor & Shrubb 1986). The pesticide-treated area of cereals has increased markedly over the past two decades (e.g. Davis *et al.* 1990, 1991), and there is strong evidence that some are detrimental to birds. For example, in the case of the grey partridge, pesticides reduce the supply of invertebrate food for the young (Potts 1986). Changes in sowing dates have not been linked to changes in the survival or breeding success of any bird species, but the lapwing (Shrubb & Lack 1991) and skylark (Schläpfer 1988), have been shown to prefer spring-sown crops and avoid autumn-sown ones.

Analysis based on correlation rather than experiment cannot establish causality; we cannot conclude that certain variables are responsible for the observed pattern of abundance, but only suggest which ones may play a part. However, production of several independent models for each species allowed us to assess which variables were consistently good predictors of abundance, and it is likely that those variables that occurred most frequently in the models for each species were those that played the most important role in determining abundance (Tables 7.1 and 7.2). If we had produced only one model for each species, there would have been no reliable way of assessing the relative importance of different variables (James & McCulloch 1990).

The models are likely to contain variables which are both causative and correlative, and it is not easy to tell them apart. The important causative factors are those that influence the survival and breeding success of the birds, and in some cases it is possible to suggest ways in which the variables in the models may do this. However, only a few of these are associations which have been established by previous studies. For example, both linnet and corn bunting have been shown to favour oilseed rape fields (Lack 1992), although in this earlier study the reed bunting showed the strongest preference for this crop but it did not appear consistently in the models for this species. For other variables, it is possible to think up an explanation of how they could affect the species' ecology, though they have not been suggested as determinants of breeding success or survival by previous studies. For example, rainfall is negatively associated with the pattern of abundance of both of the large species, lapwing and grey partridge, in which the young leave the nest soon after hatching. It is plausible to suggest that the absence of these species from some areas may be due to the influence of high rainfall on chick survival. Other variables which occur consistently in the models are harder to explain, and we must assume that they are merely correlates of other, unmeasured, variables. For example, there is no obvious way in which density of pigs or the area of lakes could affect linnet abundance.

Using the models for prediction

Our predictions of the possible effects of future environmental change, though simplistic, produced the interesting result that temperature change had a much larger effect than land-use change. There are two important caveats to the use of these models for prediction. First, changes in land use and other climatic variables, such as rainfall, with increasing temperature were not taken into account, but these must be included if fully realistic predictions are to be made. Second, we are assuming that the variables in our models are those that really do underlie the pattern of abundance, or, if they are only correlates of the underlying variables, that the correlations will not change. Nevertheless, the predicted abundance of these species seems to be more sensitive to changes in temperature than to changes in land use. This is an intriguing suggestion, because land use has in the past been assumed to be more important than climate in determining bird distributions in Britain (e.g. O'Connor & Shrubb 1986). There are some contradictions here; the turtle dove is a declining species in a supposedly warming climate, yet our models suggest that this species should increase in numbers and range under such conditions. However, this species is a long-distance migrant and conditions in its wintering grounds could determine its abundance in Britain.

Breeding season and winter temperature appeared in the models for five species. Relationships with winter temperature were often peaked, and, if this is a causal relationship, it implies that the population could change in either direction with an increase or decrease in temperature. If a temperature change results in more of the country becoming optimal, the range will expand, but if a temperature rise moves most of the country away from the optimum, the range will contract. Similar arguments apply to other variables, such as grassland, that have peaked relationships.

The predictions of mean abundance for 1969 from the 'best' models of 1988 are higher for only four out of the eight species, although all are known to have declined by at least 50%. Thus the models did not accurately predict the populations in 1969. There are several possible reasons for this. First, as noted above, the models may not have contained the right variables, but only correlates of them, and there is no guarantee that any correlations between variables will remain the same over time.

For example, the distributions of wheat and oilseed rape were very similar in 1988, but quite dissimilar in 1969 as oilseed rape was planted much less extensively then. Second, only for agricultural variables do we have measures of differences between 1969 and 1988. We have no data on changes in any climatic variables or, for example, the area of woodland or urban land, all of which may have changed. Third, the recording of the agricultural census data changed slightly, but significantly, between 1969 and 1988, and so the two data sets do not contain exactly the same variables. For example, 'young grassland' is up to 5 years old in 1988, but up to 7 years old in 1969. All of these may have introduced errors into our estimates of bird abundance for 1969.

Interactions and geographical position

In our analyses we have not taken into account any interactions between the explanatory variables, although they too may play a role in determining abundance. This was because there is an enormous number of potential interactions, and we have no idea which of them may have a causal role. There is no method, apart from knowledge of the ecology of the species, of deciding which interactions may be of importance, and for most of the species considered here, such data do not exist. Nevertheless, it is conceivable that the effect on abundance of a land cover variable could depend on the level of, for example, rainfall. This is effectively saying that the factors that determine abundance may vary between different regions, and, in particular, the factors that determine abundance in the centre of the range may be different from those operating at the periphery. The best way to understand this may be to carry out analyses based on smaller regions. Such analyses may suggest different variables as determinants of abundance than analyses at the national scale, because there is less variation in factors such as rainfall and temperature, and hence they are unlikely to explain much of the variation in abundance within regions.

We carried out analyses both including and excluding easting and northing. For some species (e.g. turtle dove) they appeared to have some explanatory power, as they were among the best univariate predictors, and models including them both explained more of the deviance and contained fewer parameters. Moreover, in the rankings of the models, those containing easting and northing were consistently ranked higher than those not containing these variables. However, in other cases (e.g. skylark, reed bunting) there was very little difference between models including and excluding these variables, either in amount of deviance explained or rank order; and for one species, corn bunting, easting and northing never appeared in the models at all.

There are two reasons why geographical position could itself show a strong correlation with abundance, rather than acting as a substitute for other, unmeasured variables. First, for migratory species, eastern Britain is closer to continental Europe, from where the birds arrive in the spring, and this may explain greater abundance of migratory birds in the east. Of the species included here, this could apply only to the turtle dove. Second, a species' range may be restricted not by habitat or climate but by some past event, such as persecution, which has excluded them from part of their potential range that they have not been able subsequently to recolonize. Neither of these reasons seems very likely to apply to the species studied here, and it is more likely that easting and northing are acting as substitutes for unmeasured variables. Thus, though they may be useful for describing the current pattern of abundance, it is probably unwise to use them for predictions. We therefore used models excluding them to predict 1969 and future abundances, despite the fact that some of these models explained less of the variance.

This work has shown that the pattern of abundance of farmland birds can be predicted from environmental variables and, in some cases, we can suggest factors that may play a causal role. The declines in population and range contractions of these birds may be due to changes in these factors. However, biotic as well as environmental factors could play a role; predation, competition or disease could lead to population changes, but nothing is known about these processes on a national scale. To discover if any of the variables we have highlighted may determine the pattern of abundance, intensive fieldwork studying the effects of them on breeding success and survival of the birds is necessary.

ACKNOWLEDGEMENTS

This work was funded by the Natural Environment Research Council and the Royal Society for the Protection of Birds. We thank members of the British Trust for Ornithology, the Scottish Ornithologists' Club and the Irish Wildbird Conservancy for undertaking the enormous amount of fieldwork required for the 1988–91 breeding bird atlas. We are grateful to John Turner and Jack Lennon (Leeds University Genetics Department), Alison Bayley and Peter Burnhill (University of Edinburgh Data Library), and the Institute of Terrestrial Ecology for providing the environmental data in a suitable format for analysis, and to David Hill and Mark Avery for valuable discussion throughout the duration of this work.

APPENDIX
ENVIRONMENTAL VARIABLES USED IN THE MODELS

Variable	Units	Source
1 Area of farmland	propn	Ag.stats
2 Young grassland (<5 years old)*	propn	Ag.stats
3 Other grassland (>5 years old; excluding rough grazing)*	propn	Ag.stats
4 Total grassland (= 2 + 3)	propn	Ag.stats
5 Wheat (*Triticum* spp.)	propn	Ag.stats
6 Autumn-sown ('winter') barley (*Hordeum* spp.)*	propn	Ag.stats
7 Spring-sown ('spring') barley (*Hordeum* spp.)	propn	Ag.stats
8 Oats (*Avena* spp.)	propn	Ag.stats
9 Total cereal crops (= 5 + 6 + 7 + 8 + other minor cereals)	propn	Ag.stats
10 Vegetables	propn	Ag.stats
11 Orchards	propn	Ag.stats
12 Soft fruit	propn	Ag.stats
13 Total horticultural crops (= 10 + 11 + 12 + other minor horticultural crops)	propn	Ag.stats
14 Sugar and fodder beet (*Beta vulgaris*)*	propn	Ag.stats
15 Oilseed rape (*Brassica napus oleifera*)*	propn	Ag.stats
16 Total non-cereal crops (= 13 + 14 + 15 + other minor non-cereal crops)	propn	Ag.stats
17 Fallow	propn	Ag.stats
18 Total crops and fallow (= 9 + 16 + 17)	propn	Ag.stats
19 Grass as a proportion of agricultural land (= 4/[4 + 18])	propn	Ag.stats
20 Dominance index (max of 2, 3, 9, 10, 11, 12 and 13)	propn	Ag.stats
21 Cattle	no./ha	Ag.stats
22 Pigs	no./ha	Ag.stats
23 Sheep	no./ha	Ag.stats
24 Easting of National Grid		ITE
25 Northing of National Grid		ITE
26 Urban	propn	ITE
27 Agricultural land class 1	propn	ITE
28 Agricultural land class 2	propn	ITE
29 Agricultural land class 3	propn	ITE
30 Agricultural land class 4	propn	ITE
31 Agricultural land class 5	propn	ITE
32 Agricultural land class 6	propn	ITE
33 Mean altitude	m	ITE
34 Difference between highest and lowest point	m	ITE
35 River frequency score		ITE
36 Lakes	propn	ITE
37 Woodland	propn	ITE
38 Brown earth variants	propn	ITE
39 Rendzinas or calcareous soils	propn	ITE
40 Gley soils	propn	ITE

Continued on p. 176

Appendix (Continued)

Variable	Units	Source
41　Humus or iron podzols or brown podzolic soils	propn	ITE
42　Peaty podzols	propn	ITE
43　Peaty gleys	propn	ITE
44　Deep peaty soils and peats	propn	ITE
45　Immature or skeletal soil	propn	ITE
46　Mean annual rainfall	mm	ITE
47　Mean breeding season temperature (April–July)	°C	Leeds
48　Mean winter temperature (December–February)	°C	Leeds

* Variables differed slightly between 1969 and 1988. In 1969 young grassland was <7 years old, other grassland was >7 years old; autumn-sown and spring-sown barley were not considered separately, but autumn-sown barley was almost non-existent at that time and has thus been set to zero; fodder beet was not included with sugar beet; and oilseed rape was not considered separately from other forms of rape.

Abbreviations

Propn = extent of variable expressed as a proportion of land in the 10-km square.
Ag.stats = data from 1988 June agricultural census organised by Ministry of Agriculture, Fisheries and Food in England and Wales, and the Department of Agriculture and Fisheries for Scotland. Similar data for 1969 were used for model testing. These data were converted to the 10-km square grid by the University of Edinburgh Data Library.
ITE = data from the Institute of Terrestrial Ecology's Land Characteristic Data Bank (Ball *et al.* 1983).
Leeds = temperature data from the Meteorological Office, converted to the 10-km square grid by the University of Leeds Genetics Department. The data are true temperatures, not corrected to sea level.

Notes

Agricultural land classes vary from 1 (high quality arable land) to 5 (low quality); class 6 is non-agricultural land.
The agricultural statistics date from the same period as the bird abundance data, 1988.
The ITE Land Characteristics data are from maps dating from the early 1980s. Many of the variables, for example soils and altitude, will show no differences between this period and 1988–91, when the bird data were collected. However, a few, such as rainfall which was the average of 1941–70, and urban areas and woodland which were both taken from Ordnance Survey maps of the time, may have changed slightly.
The temperature data were 30-year averages of 1941–70, and they may have changed slightly by 1988.
Neither the ITE data nor the temperature data were ideal in the sense of being for earlier periods than the bird data, but, as they are the best data available for these variables, we have used them in our analyses, making the assumption that they have not changed.

REFERENCES

Ball, D.F., Radford, G.L. & Williams, W.M. (1983). A land characteristic data bank for Great Britain. ITE occasional paper of ITE project 534.

Cliff, A.D. & Ord, J.K. (1973). *Spatial Autocorrelation.* Pion, London.

Davis, R.P., Garthwaite, D.G. & Thomas, M.R. (1990). Pesticide usage survey report 78: arable farm crops 1988. MAFF Publications, London.

Davis, R.P., Garthwaite, D.G. & Thomas, M.R. (1991). Pesticide usage survey report 85: arable crops 1990. MAFF Publications, London.

Fuller, R.J., Hill, D.A. & Tucker, G.M. (1991). Feeding the birds down on the farm: perspectives from Britain. *Ambio*, **20**, 232–237.

Gibbons, D.W. (1991). The new atlas of breeding birds in Britain and Ireland: an overview. *Sitta*, **5**, 11–18.

Gibbons, D.W., Reid, J.B. & Chapman, R.A. (1993). *The New Atlas of Breeding Birds in Britain and Ireland: 1988–1991.* T. & A.D. Poyser, London.

Houghton, J.T., Jenkins, G.J. & Ephraums, J.J. (1990). *Climate Change: the IPCC Scientific Assessment.* Cambridge University Press, Cambridge.

Hustings, F. (1988). European monitoring studies of breeding birds. SOVON, Beek, The Netherlands.

Hustings, F. (1992). European monitoring studies of breeding birds: an update. *Bird Census News*, **5**, 1–56.

James, F.C. & McCulloch, C.E. (1990). Multivariate analysis in ecology and systematics: panacea or pandora's box? *Annual Review of Ecology and Systematics*, **21**, 129–166.

Lack, P.C. (1992). *Birds on Lowland Farms.* Her Majesty's Stationery Office, London.

MAFF (1992). Arable area payments: explanatory booklet. MAFF Publications, London.

Marchant, J.H. & Gregory, R.D. (1993). Recent population changes among seed-eating passerines in the United Kingdom. *Proceedings of the 12th International Conference of IBCC and EOAC.* (Ed. by W. Hagemeijer & T. Verstrad).

Marchant, J.H., Hudson, R., Carter, S.P. & Whittington, P.A. (1990). *Population Trends in British Breeding Birds.* British Trust for Ornithology, Tring.

O'Connor, R.J. & Shrubb, M. (1986). *Farming and Birds.* Cambridge University Press, Cambridge.

Potts, G.R. (1986). *The Partridge. Pesticides, Predation and Conservation.* Blackwell Scientific Publications, Oxford.

Robertson, J. & Berg, A. (1992). Status and population changes of farmland birds in southern Sweden. *Ornis Svecica*, **2**, 119–130.

SAS Institute Inc. (1990). *SAS/STAT User's Guide, Version 6, Fourth Edition, Vol. 2.* SAS Institute Inc., Cary, NC.

Schläpfer, A. (1988). Populationsökologie der Feldlerche *Alauda arvensis* in der intensiv genutzten Agrarlandschaft. *Der Ornithologische Beobachter*, **85**, 309–371.

Shrubb, M. & Lack, P.C. (1991). The numbers and distribution of lapwings *V. vanellus* nesting in England and Wales in 1987. *Bird Study*, **38**, 20–37.

Stowe, T.J., Newton, A.V., Green, R.E. & Mayes, E. (1993). The decline of the corncrake *Crex crex* in Britain and Ireland in relation to habitat. *Journal of Applied Ecology*, **30**, 53–62.

Westman, W.E. (1980). Gaussian analysis: identifying environmental factors influencing bell-shaped species distributions. *Ecology*, **61**, 733–739.

8. SCALE AND PATTERNS
OF COMMUNITY STRUCTURE
IN AMAZONIAN FORESTS

DAVID G. CAMPBELL

Department of Biology, Grinnell College, Grinnell, IA 50112, USA

SUMMARY

Quantitative ecological inventories of *terra firme* (upland) and inundated Amazonian forests provide insights into their community structure, history (including human history) and value. The most rigorous inventories have a minimum threshold of trees ⩾10 cm diameter at breast height and the study sites are of large scale, ranging in size from 1 to 3 ha. The interpretation of these data is dependent on scale. Consistent, and sometimes surprising, patterns are evident:

1   The indices of similarity (Sorensen coefficient) between adjacent 1-ha plots are low, ranging from 0.10 to 0.21 in *terra firme* forest and from 0.16 to 0.47 in *várzea* forest, indicating that there is probably no such thing as a 'representative' small sample for any particular Amazonian forest.

2   All forests have 5–10 very important species (an 'oligarchy').

3   The level of oligarchy of these species is a function of environmental disturbance.

4   The majority of species are rare, represented by only one or two individuals.

5   Some forest types, such as liana forest, indicate long-term perturbation by humans.

6   Quantitative inventory is conducive to the economic mensuration of tropical forest.

These results have enormous significance for the conservation and management of tropical forests.

INTRODUCTION

If the traveller notices a particular species and wishes to find more like it, he may often turn his eyes in vain in every direction. Trees of varied forms, dimensions and colours are around him, but he rarely sees any one of them repeated. Time after time he goes towards a tree which looks like the one he seeks, but a closer examination proves it to be distinct. He may at length, perhaps, meet with a second specimen half a mile off, or may fail altogether, till on another occasion he stumbles on one by accident. (A.R. Wallace 1878)

Wallace's description of tropical nature was typical of the neophyte, and it revealed pattern which is only today being proved quantitatively. One's first impression of a tropical forest, especially when viewed from the edge of a disturbance such as the chaotic jumble of vegetation along a river course, is a wall of inchoate green. The techniques of observation acquired in the temperate north do not apply here. The subtle differences between the species are at first hard to discern. For example, leaf morphology on a single tree may vary as a function of height. On the other hand, many species in disparate families have evolved leaf drip tips to remove excess water, or flaking bark that is adapted to shedding epiphytes, and these convergences conceal the extravagant diversity. Local rarity is also a problem; tropical forest tree species are often solitary – at least on a scale of several square kilometres – and, as Wallace observed, it may take days, or weeks, to find another.

One of the best tools of the ecologist and conservationist to decipher pattern in tropical forests is quantitative ecological inventory, which is defined for the purposes of this chapter as the mapping and enumeration of individuals and species of trees in a sample of forest, the measurement of several important parameters of those individuals, and the analysis of their abundance and distribution as functions of their physical and biotic environments (Campbell 1989).

Quantitative inventory is methodologically simple, employing techniques developed by foresters to measure the amount of saleable wood in a plot of forest. Its parameters of analysis are therefore few: diameter at breast height (dbh), bole height, height of canopy, crown diameter(s), measurements of soil nutrients and hydrology, measurements of light, and a system of Cartesian coordinates to map each tree. Inventories are therefore tedious and expensive to make, and the creativity lies not in the collection, but in the analysis of these data, which is usually accomplished by a computer.

A single quantitative inventory, solitary in space and time, tells us little. But now a critical mass of inventories, dating from the 1950s, is accumulating in Amazonia, and comparisons are beginning to be possible. Striking – and unexpected – patterns are emerging among disparate study sites. Will the patterns someday enable us to model tropical forests, perhaps to reconstruct them? Do they tell us whether the forest is in equilibrium or disequilibrium? Can the patterns be applied to pragmatic concerns, such as extractivism and economic botany? Many of the answers to these questions depend on scale: on whether we can accept extrapolations of time, space and value. This chapter will explore these issues.

MEASUREMENT OF BIOLOGICAL DIVERSITY

It would seem evident that quantitative inventories would be ideally suited to measure the species richness – the alpha diversity – of small plots of tropical vegetation. But even this simple assay is fraught with danger. Many parts of Amazonia are botanically poorly explored. Floras are non-existent or obsolete, monographs spotty in their taxonomic coverage and information on ranges is inadequate. Without such tools, field identification to species of neotropical plants is often impossible. Therefore, the collection of voucher specimens, to be examined by specialists at a later time, is requisite for a rigorous inventory, and it is the collection of these specimens that is the most expensive, time-consuming and dangerous aspect of the research. (To make matters worse, some collectors use destructive methods, such as griffes, to collect vouchers. These injure the trees, increase mortality rates and preclude the long-term study of forest dynamics.) Under circumstances when they are not identifiable to species, the vouchers may be categorized into morphospecies, which can then function as species for purposes of analysis. Regardless, morphocategorization, especially of sterile material, is often a subjective process; whether to split or to lump similar species is often not evident. Workers in quantitative inventory have entered a race to find the most species in a standard sample: the winner is currently about 300 spp. of trees $\geqslant 10$ cm dbh, estimated by a highly competent taxonomist (Gentry 1988). However, I would submit that in most studies the element of subjectivity in morphocategorization induces considerable variance into the counts, and therefore in the estimation of alpha diversity.

Tables 8.1 and 8.2 are summaries of quantitative inventories of Amazonian upland, dry (*terra firme*) and inundated forest conducted in Amazonia. (According to Prance 1979 inundated forests fall into three general categories: those flooded by sediment-rich white water, known in Portuguese vernacular as *várzea* forest; those flooded by waters low in sediment, but rich in tannic and humic acid, known as *igapó*; and those flooded by clear waters that are low in sediment, also known as *igapó*.) The first thing that one notices, of course, is that these forests are, by a factor of 10, more species rich than temperate forests. But within the context of the tropics, two other important trends are evident from these data.

TABLE 8.1. Locations, sampling methods and summary results for studies of *terra firme* forests in Amazonia

Source	Location	Sample size & shape	Criteria for inclusion cm dbh	Trees	Families	Genera	Species
Black *et al.* 1950	Belém, Brazil	1 ha 100 × 100 m	≥10	423	31	65	87
Balslev *et al.* 1987	Añagu, Ecuador	c. 1 ha point-centred-quarter	≥10	728	–	–	228
Boom 1986	Alto Ivon, Bolivia	1.0 ha 10 × 1000 m	≥10	649	28	61	94
Cain *et al.* 1956	Belém, Brazil	2 ha 20 of 10 × 1000 m	≥10	897	39	100	153
Campbell *et al.* 1986	Rio Xingu, Pará	3.0 ha 10 × 3000 m	≥10	1420	39	127	265
Dantas *et al.* 1980	Transamazônica, Brazil	1 ha 40 of 10 × 25 m	≥9.55	504	39	79	120
Gentry 1985	Manu Park, Peru	1 ha 100 × 100 m	≥10	673	43	–	210
Gentry 1988	Yanomomo, Peru	1 ha 10 × 1000 m	≥10	606	58	–	c. 300
Gentry 1988	Cabeza de Mono, Peru	1 ha 10 × 1000 m	≥10	544	40	–	c.185
Gentry 1988	Mishana, Peru	1 ha 10 × 1000 m	≥10	858	50	–	c. 289
Gentry 1988	Tambopata, Peru	1 ha Not stated	≥10	602	42	–	c. 181

Reference	Locality	Area / layout	Size				
Gentry 1988	Cocha Cashu, Peru	1 ha Not stated	≥10	673	48	—	c. 204
Gentry 1988	Neblina Base, Venezuela/Brazil	1 ha Not stated	≥10	513	32	—	c. 102
Hartshorn 1980	Manu Park, Peru	1 ha No layout	≥10	584	—	—	153
Korning et al. 1990	Añagu, Ecuador	1 ha 100 × 100 m	≥10	734	—	46	153
Korning et al. 1990	Añagu, Ecuador	c. 1 ha point-centred-quarter	≥10	728	—	51	239
Mori et al. 1983	Bahia, Brazil*	0.67 ha point-centred-quarter	≥10	600	—	—	178
Mori & Boom (cited in Balslev et al. 1987)	Saul, French Guiana*	c. 1.0 ha point-centred-quarter	≥10	619	—	—	241
Lechthaler 1956	Manaus, Brazil	1 ha No layout	≥8	735	—	—	—
Pires et al. 1953	Castanhal, Brazil	3.5 ha irregular layout	≥10	1482	47	130	179
Pires 1966	Breves, Brazil	1 ha 10 × 1000 m	≥10	516	36	—	157
Prance et al. 1976	Manaus, Brazil	1 ha irregular	≥15	350	43	115	179
Rodrigues 1963	Amapá, Brazil (ridgetop)	1.1 ha 11 or 10 × 100 m	≥15	347	84	63	36
Rodrigues 1963	Amapá, Brazil (ridgetop)	1.5 ha 15 of 10 × 100 m	≥15	307	37	70	96

*Extra-Amazonian, but part of the greater Amazonian biota.
Expanded from Campbell *et al.* (1986).

TABLE 8.2. Locations, sampling methods and summary results for studies of inundated forests in Amazonia

Source	Location forest type*	Sample size & shape	Criteria for inclusion	Trees	Families	Genera	Species
Balslev et al. 1987	Añagu, Ecuador várzea	c. 1.0 ha point-centred-quarter	≥10 cm dbh	417	44	89	c. 149
Black et al. 1950	Belém, Pará várzea†	1.0 ha 100 × 100 m	≥10 cm dbh	564	28	51	60
Campbell et al. 1986	Rio Xingu, Pará várzea‡	0.5 ha 10 × 500 m	≥10 cm dbh	220	17	≥29	40
Campbell et al. 1992	Rio Juruá, Acre 50-year old várzea	1.0 ha 20 × 500 m	≥10 cm dbh	523	37	≥73	106
Campbell et al. 1992	Rio Juruá, Acre 14–50-year old várzea	1.0 ha 20 × 500 m	≥10 cm dbh	420	34	≥52	73
Campbell et al. 1992	Rio Juruá, Acre 14-year old várzea	1.0 ha 20 × 500 m	≥10 cm dbh	777	16	20	20
Gentry 1988	Tambopata, Peru, alluvium	1 ha Not stated	≥10 cm dbh	540	–	41	c. 165
Keel & Prance 1979	Lower Rio Negro igapó	12 plots of 10 × 15 m	≥1 m height	1028	20	34	54
Pires & Koury 1958	Guamá, Pará várzea	3.8 ha 100 × 380 m	≥10 cm dbh	1837	21	79	107
Rodrigues 1961	Lower Rio Negro igapó	350 m² 5 × 70 m	≥1 m height	251	20	–	51

* According to classification of Prance (1979). † Incorrectly described by the authors as igapó. ‡ Although the Xingu is a clear-water river, the authors classified the forest as várzea because of the high sediment load at the study site (O Deserto). Expanded from Campbell et al. (1992).

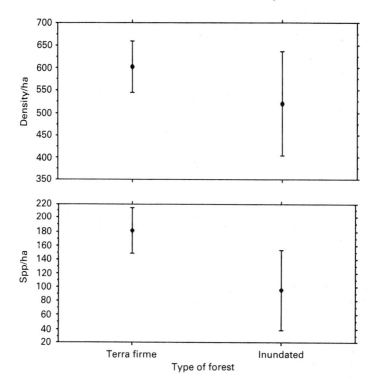

Fig. 8.1. Comparison of stand density per hectare and tree species richness per hectare (for trees ⩾10 cm dbh) in 17 *terra firme* and six *várzea* forests in Amazonia.

The first is that there is no significant difference between the stand densities of *terra firme* and inundated forests (Fig. 8.1). *Terra firme* forests had 602 ± 121 trees/ha (range 423–858; N = 20). Inundated forests had 520 ± 127 trees/ha (range 417–777; N = 8).

Second, the species richness of the *terra firme* plots is significantly greater (P < 0.01) than the inundated plots (Fig. 8.1). *Terra firme* forests had 182 ± 64 (range 87–300; N = 17) species/ha; inundated forests had 96 ± 55 (range 20–165; N = 6). For the analysis of density, inventories greater than 1 ha were divided by their areas to yield the number of trees per hectare, but because species richness does not increase linearly as a function of area, only inventories of exactly 1 ha were used to calculate species richness (Campbell *et al.* 1992).

These data lead to the overwhelming question of Amazonian development. Does the relative paucity of species in inundated forests imply that they could be sacrificed for development and agriculture, instead of

species-rich *terra firme* forest? Consider, as well, the annual deposition
of white-water silt in *várzea* forest, which creates a demonstrable enrichment of cations such as phosphorus, potassium and calcium (Campbell *et
al.* 1992) in marked contrast to the often senile and nutrient-impoverished
soils of *terra firme* forests. Agriculture in the *terra firme*, with few exceptions, rapidly leads to environmental bankruptcy, as the scar tissue of
the Transamazonica demonstrates. It is therefore no coincidence that
subsistence agriculture of corn, beans and peanuts has been conducted
for millennia in Amazonian floodplains, and today experiments in rice
growing in the herbaceous *várzea* have yielded harvests comparable to
other parts of the world (albeit with the heavy application of fertilizer and
pesticides). Indeed, agriculture is not even necessary for *várzea* communities to sustain human populations, if one considers the large populations of tree-eating fish (Goulding 1980), which transmute the otherwise
inedible productivity of the forest into high-quality food. Since pre-
Colombian times, the annual migration of fish to breeding sites after they
have fattened in the *várzea*, known in Portuguese as the *piraçema*, has
provided a protein bonanza for the people who live on the river margin,
and has set the schedule of human activities there.

REPRESENTATIVENESS OF INVENTORY PLOTS

The next question is obvious. How representative are these plots? Are
they sufficiently representative for us to derive conclusions regarding
plant sociology; to design parks and to base policy decisions on them?
Further, do they provide clues as to the assembly of the forest; whether,
for example, replacement of trees follows a random pattern (as postulated
by Hubbell & Foster 1986) or there is succession towards a climax or
another predictable species composition?

The inevitable use of morphocategories in Amazonian forests does not
permit comparative analysis of species similarity between sites, unless, of
course, the morphocollections from different sites are submitted to the
same taxonomic criteria. In other words, quantitative inventory is a
poor source of information on beta diversity. However, a few revealing
inventories have a unified system of morphocategories, enabling analysis
of similarity. Table 8.3 shows such comparisons made within four sets of
plots: three plots in *terra firme* forest on the Rio Xingu, three plots in
várzea on the Rio Juruá, two plots (*várzea* and *terra firme*) in Añagu,
Ecuador, and two *terra firme* plots (liana forest), also on the Rio Xingu.
The results reveal a surprisingly high level of heterogeneity at the species
level among the samples of forest: the maximum similarity (Sorensen's

TABLE 8.3. (a) Indices of similarity of 1-ha samples of Amazonian forest

Forest type	No. of spp. in common	Sorenson's index* for all spp./10 most important
Terra firme, Rio Xingu (Campbell *et al.* 1986) similarity between plots (adjacent)		
1 and 2	24	0.16/0.30
1 and 3	12	0.10/0.30
2 and 3	30	0.21/0.30
Total spp:	265	
Várzea, Rio Juruá (Campbell *et al.* 1992) similarity between plots (adjacent)		
1 and 2	42	0.47/0.00
2 and 3	9	0.16/0.40
1 and 3	10	0.20/0.00
Total spp:	150	
Várzea and *terra firme*, Añagu, Ecuador (Balslev *et al.* 1987) similarity between plots (proximitous)		
	60	0.30/0.38
Total spp:	333	

(b) Indices of similarity of two 1-ha samples of old swidden

Terra firme, liana forest, Rio Xingu (Balée & Campbell 1990) similarity between Araweté and Asurini study sites (90 km apart)		
	44	0.36/0.40
Total spp:	236	

* $H = 2a(2a + b + c)^{-1}$, where a = no. of species common to both plots; b = no. of species unique to area 1; c = no. of species unique to area 2 (Grieg-Smith 1983).
The study sites are: 1, Campbell *et al.* (1986); sequential elongated quadrats (1 km × 10 m, which are in fact more similar to transects than quadrats) in *terra firme* forest at O Deserto (on the Rio Xingu in eastern Brazilian Amazonia). 2, Campbell *et al.* (1992); proximitous parcels (0.5 km by 20 m) of *várzea* forest (near the hamlet of Rodrigues Alves, in western Brazilian Amazonia). 3, Balslev *et al.* (1987); proximitous transects of *várzea* and *terra firme* forest at Añagu, Ecuador. 4, Balée and Campbell (1990): two plots of *terra firme* liana forest, separated from each other by 90 km.

index) is only 0.21 (of a possible maximum of unity) in *terra firme* plots, and 0.47 in the *várzea* forests. The indices of similarity for the 10 most important species (see the definition of importance value, below), at least for the *terra firme* forest plots, are higher, indicating that the forests may be converging on a community that is predictable at least in terms of its most important species, however unpredictable it may be in terms of the majority of species, which are rare (see discussion of oligarchy, below).

Why aren't the *várzea* plots more similar? The disparity in their species compositions is all the more inexplicable because many *várzea* species are wide-ranging, producing fruits that are dispersed over long distances by flotation and by fish. A clue is provided by Table 8.4, which illustrates the niche width (in terms of one dimension: level of annual

TABLE 8.4. The distribution of *várzea* tree species as a function of mean water level during annual inundation. The annual flux on the study sites is as high as 15 m

Species	N	Mean water level (m)	Waller grouping*
Annonaceae sp.	314	4.12	A
Cecropia latiloba	66	4.06	B A
Miconia sp.	167	3.92	B A
Chlorophora tinctoria	53	3.78	B A C
Nectandra immolobilis	50	3.69	B C D
Lueheopsis sp.	15	3.54	C D
Laetia cf. *americana*	105	3.33	D
Triplaris weigeltiana	42	2.24	E
Inga sp.	37	1.72	F
Myrtaceae sp.	26	1.47	F
Sloanea sp.	10	1.24	G H
Cordia nodosa	41	1.03	H I
Astrocaryum juauri	45	0.83	I J
Guarea guidonia	13	0.55	J K
Mimosaceae sp.	22	0.43	J K L
Virola surinamensis	29	0.17	L M
V. elongata	21	0.11	L M
Conceveiba sp.	15	0.08	L M
Pithecellobium sp.	13	0.05	L M
Vismia sp.	17	0.02	M
Alchornea triplinervia	64	0.00	M
Iriartea deltoidea	12	0.00	M
Warscewiczia coccinea	28	0.00	M
Apeiba aspera	12	0.00	M
Casearea sect. *piparea*	15	0.00	M

* Means with the same letter are not significantly different ($P > 0.05$).
Source: Campbell *et al.* (1992).

inundation) for tree species from the three Rio Juruá plots. Obviously, these organisms are quite specific in their requirements and are not exhibiting a random walk of neutral alleles (Hubbell & Foster 1986), at least relative to flooding. The same question pertains here as with the *terra firme* forests. Can the three *várzea* sites be regarded as seral stages marching toward a predictable end? If this is the case, then Campbell *et al.* (1992), knowing the ages of the forests, estimated that they were accumulating species of tree ⩾10 cm dbh at a rate of 1.5 per year.

These data reveal three fundamental lessons for conservation:

1 In species-rich tropical forest there is probably no such thing as a small plot (i.e. less than 10 or 20 ha) that is representative (a concept that implies an equilibrium species composition or an end point in community succession, anyway).

2 Therefore, one cannot reliably extrapolate data from a small plot to the wider forest that surrounds it. Regardless, it is tempting – and often arguably necessary – to do so. The most extreme case of extrapolation may have been Gentry (1982), who used a species–area relationship derived from an abandoned 5.8 ha field in temperate Pickney, Michigan to extrapolate data from 0.10 ha plots to estimate the species richness in imaginary 1.0-ha plots. When the same formula was applied to the known species richness of a 1-ha plot at O Deserto (Campbell *et al.* 1986), it underestimated the species richness in the total 3-ha sample by 60% (106 vs. 265 species), yet the central argument of Gentry's paper was based on these extrapolations.

3 Finally, and this is an ominous conclusion, there may be no such thing as a small park or reserve in species-rich tropical forest that is 'representative' of the greater forest that surrounds it.

These data also show that choice of sampling method for an inventory is probably not critical, in view of the high variance between small plots in a species-rich forest. For example, attempts to evaluate the differences in alpha diversity in adjacent plots of Amazonian *terra firme* forest have been made using quadrats, elongated quadrats and the point-centred-quarter method. Korning *et al.* (1990) compared the results obtained about composition and structure of an Amazonian forest in Añagu, Ecuador using two of these technologies, a quadrat and the point-centred-quarter method. They showed the intuitively logical: that the point-centred-quarter method samples the most species richness and provides the best average values for density and tree size in the area. However, I would submit that the natural variation within the study sites diminishes the sampling differences between sites due to the two methods, and that when researchers choose a sampling method, rather than choosing a method that insures 'representativeness' – a futile goal anyway – they should choose a method for its other attributes, such as facility for nearest-neighbour analysis (quadrat or point-centred-quarter method) or ability to detect light gaps and microhabitats (elongated transect).

IMPORTANCE VALUES:
OLIGARCHIC PLANT COMMUNITIES

Every forest that we have quantitatively inventoried in Amazonia, without exception, has three to five inordinately important species (see definition in Table 8.5) and a predominance of species that are of trivial importance, having densities of one or two in, for example, a 3-ha sample. Forests of this type have been labelled – most aptly – 'oligarchic' (Peters *et al.* 1989b). At first, this may not be much of a surprise, especially in disturbed or early successional communities, where a few hardy, fast-growing species are most conspicuous and are, in fact, often considered to be ecological indicators. This trend is illustrated in Fig. 8.2, which illustrates the importance of the 10 most important species of trees $\geqslant$10 cm dbh for the three 1-ha plots of *várzea* forest in Acre (Campbell *et al.* 1992) along a gradient of inundation. It is evident that the total importance values occupied by the most important species are a function of the amount of disturbance (in this case, inundation), and are an inverse function of species richness. Concomitantly, at least 40% of species on the three plots are locally rare, represented by only one or two individuals (Fig. 8.3).

What is surprising is that this pattern also occurs in species-rich, supposedly mature *terra firme* forest, with no obvious evidence of human perturbation or other recent disturbance. Further, these include forests that were sampled by means of elongated quadrats (typically 10 m × 1 km) and by the point-centred-quarter method, both of which tend to diminish the bias of a square or circular plot of equal area, but a small footprint (100 × 100 m or $r = 56.43$ m, respectively), and which have a sampling resolution about the same as the forest gap mosaic (but see comment on variance, above). A good example is the 3-ha O Deserto *terra firme* study site (Campbell *et al.* 1986). The 10 most important tree species (out of 265) collectively occupy over a third (107 of 300) of the importance values (IVs), whereas the 245 least important species, with frequencies of one or two in the sample occupy a half (149 IVs). Table 8.5 presents the summed IVs for the 10 most important species on quantitative inventories in nine samples of Amazonian *terra firme* as well as five samples of Amazonian *várzea* forests. The mean value for the *terra firme* forests is 115 ± 29 (range 69–158). As would be expected, the

TABLE 8.5. Summed importance values (IVs) for the 10 most important tree species on nine plots of *terra firme* forest and five plots of *várzea* forest

Source	Location	Summed IVs for 10 most important spp.
Terra firme		
Campbell *et al.* (1986)	O Deserto, Rio Xingu, Brazil	124
		93
		124
Balée & Campbell (1990)	Araweté, Rio Xingu	114
	Asurini, Rio Xingu	158
Boom (1986)	Alto Ivon, Bolivia	143
Balslev *et al.* (1987)	Añagu, Ecuador	69
Korning *et al.* (1990)	Añagu, Ecuador	83
		130
Várzea		
Campbell *et al.* (1986)	O Deserto, Rio Xingu, Brazil	222
Campbell *et al.* (1992)	Rodrigues Alves, Rio Juruá, Brazil	116
		150
		285
Balslev *et al.* (1987)	Añagu, Ecuador	95

For trees ≥10 cm dbh; all are exact or approximate 1-ha samples, except the O Deserto *várzea* (0.5 ha).

Importance value is defined by Curtis and Cottam (1962) as: relative density + relative frequency + relative dominance, when

Relative density $= \dfrac{\text{Number of trees of species A}}{\text{Total number of trees in sample}} \times 100,$

Relative frequency $= \dfrac{\text{Number of sampling units containing species A}}{\text{Total number of sampling units for all species}} \times 100,$

Relative dominance $= \dfrac{\text{Basal area for species A}}{\text{Total basal area for all species}} \times 100.$

Therefore the relative density, relative frequency and relative dominance each sum to 100 for all species, and importance value sums to 300 for all species. Goodall (1970) validly criticized relative density and relative frequency as not being commensurable between species with large vs. small individuals, and he further pointed out that comparisons of community types based on composite measures such as importance value are intrinsically weak. In spite of this, importance value, like so many obsolete artefacts embedded in the literature, has persevered. Its single advantage, perhaps, is that it – and its three components – are standardized measures that allow comparisons between study sites. Some authors have used rank order of abundance to characterize tree communities; the results have been similar to the use of IVs.

mean for the *várzea* forests is higher (but not significantly): 174 ± 79 (ranging from 95 to a staggering 285).

Among single species, the patterns of abundance and rarity on small plots are also revealing. For example, *Rinorea juruana* is the twelfth most

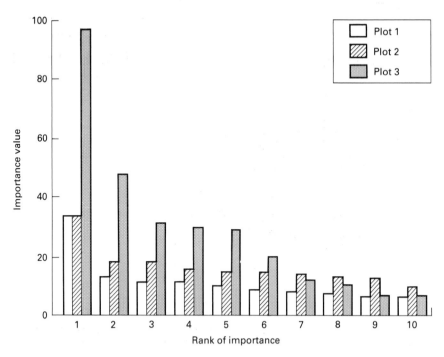

FIG. 8.2. The allocation of importance values as a function of disturbance (level of inundation) on the three plots of *várzea* forest at Rodrigues Alves (see Table 8.4). The plots are along a gradient of inundation: plot 1, which is seldom inundated, is nearly *terra firme* forest; plot 2 is flooded to a mean depth of 1.16 m; plot 3 is heavily inundated to a mean depth of 4.0 m. The plots are also on a gradient of species richness; with 106, 73 and 20 species, respectively.

important species (IV = 4.97) in the O Deserto samples, consisting of 33 individuals. Yet, until this inventory was conducted, the species was almost unknown in herbaria and was considered to be a rare plant. O Deserto may be a source area (Pulliam 1988) for *R. juruana* and previous collections may have been near ecological sinks for the species. It is a matter of scale: just as the inventory samples are too small to detect changes in beta diversity, neither do they have the resolution – that is, the grain size – to detect the mosaic of sources and sinks.

Oligarchies may shed light on the debate as to whether species-rich tropical forests are in a state of equilibrium of disequilibrium. Hubbell and Foster (1986), defending the disequilibrium hypothesis, argued that the species composition in the vicinity of a canopy tree follows an essentially stochastic pattern to which few species can adapt, and thereby

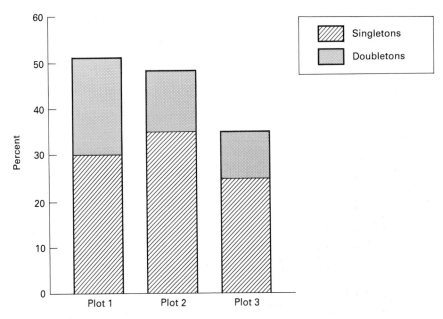

Fig. 8.3. The percentage of tree species represented by a one individual (singletons) and two individuals (doubletons) per hectare on the three Rodrigues Alves *várzea* plots.

decreases the probability that niche differentiation may take place between neighbouring species. I submit that oligarchies – so pervasive in Amazonia – create predictable neighbourhoods, and therefore may foster niche separation. Yet, this is true only if the oligarchies are natural phenomena and are not due to recent human activities, an assumption, as we shall see, that should not be made uncritically.

When, under certain fortunate circumstances the oligarchy happens to be of economically important species, and in the Amazon these are (among others) the babassu palm (*Orbignya phalerata*), camu camu (*Myrciaria dubia*) or the Brazil nut tree (*Bertholletia excelsa*), then the potential for extractivism of forest products may be high (Peters *et al.* 1989a,b). However, it may be no coincidence that the oligarchy is of valuable species. Another way of interpreting these patterns of import-ance is that the oligarchy is an artefact of human interference (although not necessarily enrichment). Take, for example, two plots of liana forest (Table 8.3b) from the Rio Xingu (Balée & Campbell 1990). These plots are not adjacent as in the examples above, but separated by 90 km. Yet their index of similarity is a surprisingly high 0.36. Balée and Campbell argued that liana forest may be due to human perturbation; that liana forests, which occupy about 100 000 km^2 of eastern Amazonia, are very old (perhaps pre-Columbian) late successional swiddens, which would explain their high coefficient of similarity. Stated another way: there may be no such thing as 'pristine' neotropical forest. This is certainly true in much of Central America, which experienced its greatest population density in pre-Columbian, not modern, times, and where, for example, Mayan ruins characteristically support a predictable suite of tree species (i.e. *Achras zapota*, *Brosimum* spp., *Bursera* spp., *Cedrela oderata*, *Orbignya cohune*), all of which were, or are, of economic or ceremonial value.

In summary, quantitative inventories teach us that:
1   A typical forest, be it *terra firme* or *várzea*, has an oligarchy of important tree species.
2   The majority of tree species in a sample are unimportant and rare.
3   The importance of the oligarchy increases as a function of disturbance.
4   Oligarchic forests may already be enhanced (and in the future may be further enhanced) with economically important species.

ECONOMIC MENSURATION OF FOREST

Quantitative inventory is basically little more than the enumeration of species on a small plot, and as such, it is ideally suitable for economic mensuration (which is, or course, what inventory was designed to do in the first place). Yet, tropical forests are so complex, and the human systems of extraction and marketing of forest products so modestly understood, that only recently, with the advent of quantitative inventory, have economic botanists been able to estimate the value of non-timber forest products for local people. The first of these studies were simply estimates of the percentage utilization of tree species in an inventory plot by indigenous tribes. For example, Boom (1989), working with the Chácobo Indians of Alto Ivón, Bolivia, showed that 36% of the tree species on a 1-ha plot were used as food. Typically, as with the measurement of alpha diversity, there has been a competition for high values, and studies of this type have employed co troversial categories of utilization. Boom, for example, considered firewood (15% of the species) as an economic use, and the Chácobo definition of a 'medicinal' plant (25% of the species) may not in any manner conform with the definitions of western medicine (but see below). Other authors have included such suspect categories as plants used for 'spiritual' purposes in order to increase the percentage utilization of the forest.

With the addition of economists to the research teams, valuation has become monetary. Peters *et al.* (1989a), working in Amazonian Peru, calculated that the non-timber forest products (latexes and fruits) in 1 ha of *terra firme* forest were worth $422 per year, while a one-time extraction of timber from the forest was worth $1000. They argued that after a few years of extraction the forest was therefore worth more intact than cut down. However, their analysis assumed unlimited markets for these products and failed to consider the dynamics of supply and demand. In a similar study in Belize, noting that traditional practitioners provide 75% of the primary health care to rural people of that country, Balick and Mendelsohn (1992) calculated that the standing value of traditional medicines in a 1-ha plot of secondary forest submitted to a 30-year rotation was $726/ha, the value of a plot submitted to a 50-year rotation was $3327/ha. The question of the efficacy of these remedies, in terms of the western pharmacopoeia, was not addressed and indeed, for the purposes of these markets (in which value had no correspondence to efficacy), was irrelevant. Further, the markets for these products in urban areas of Belize are far from saturated; indeed, they may be growing as imported western medicines become more expensive. However, it should be noted that these values were extrapolations from 0.25-ha samples, and subject to all the concomitant dangers of scale described above.

Economic mensuration of non-timber forest products has been invoked as a powerful tool for conservation, proving the worth of tropical forest independently of conventional uses such as timber or agriculture. But ultimately it may be perilous to define the value of species-rich tropical forest, which bestows many indirect environmental benefits and which is the principal repository of Earth's terrestrial biodiversity (the full value of which is as yet unfathomed), in such a one-dimensional and short-term measure as money (Ehrenfeld 1988). When forests are measured in this manner, their value becomes a function of the extent of local human knowledge of the biota and the vagaries of the international marketplace, neither of which are intrinsic to the forest itself. As indigenous people disappear, so too does the monetary value of their non-timber forest products. What, for example, if it were demonstrated that, at least in the near time scale, a forest was worth more money converted than intact? Under these circumstances the argument for valuation backfires and economic mensuration may be turned against conservation.

In summary, economic mensuration of species-rich tropical forest is an evocative tool in the battle to conserve both the forest and its indigenous people, but it is subject to parameters of scale and market. It behoves the conservationist to use this tool with caution.

ACKNOWLEDGEMENTS

The author thanks Professors Peter Edwards and Ghillean Prance for helpful comments on the manuscript. This research was partially funded by the John Simon Guggenheim Memorial Foundation and the Henry R. Luce Foundation.

REFERENCES

Balée, W. & Campbell, D.G. (1990). Ecological aspects of liana forest, Xingu River, Amazonian Brazil. *Biotropica*, 22(1), 36–47.

Balick, M.J. & Mendelsohn, R. (1992). Assessing the economic value of traditional medicines from tropical rain forests. *Conservation Biology*, 6(1), 128–130.

Balslev, H.J., Lutyn, J., Øllgaard, B. & Holm-Nielson, L.B. (1987). Composition and structure of adjacent unflooded and flood-plain forest in Amazonian Ecuador. *Opera Botanica*, 92, 37–57.

Black, G.A., Dobzhansky, T. & Pavan, C. (1950). Some attempts to estimate species diversity and population density of trees in Amazonian forests. *Botanical Gazette*, 111, 413–425.

Boom, B. (1986). A forest inventory in Amazonian Bolivia. *Biotropica*, 18, 287–294.

Boom, B. (1989). Use of plant resources by the Chácabo. *Advances in Economic Botany*, 7, 78–96.

Cain, S.A., de Oliveira Castro, G.M., Pires, J.M. & de Silva, N.T. (1956). Application of some phytosociological techniques to Brazilian rain forest. *American Journal of Botany*, 43, 911–941.

Campbell, D.G. (1989). The quantitative inventory of tropical forests. *Floristic Inventory of Tropical Countries* (Ed. by D.G. Campbell & H.D. Hammond), pp. 523–533. New York Botanical Garden, Bronx.

Campbell, D.G., Daly, D.C., Prance, G.T. & Maciel, U.N. (1986). Quantitative ecological inventory of *terra firme* and *várzea* tropical forest on the Rio Xingu, Brazilian Amazon. *Brittonia*, 38(4), 369–393.

Campbell, D.G., Stone, J.L. & Rosas, A. (1992). A comparison of the phytosociology of three floodplain (*várzea*) forests of known ages, Rio Juruá, western Brazilian Amazon. *Botanical Journal of the Linnean Society*, 108, 213–237.

Curtis, M. & Cottam, G. (1962). *Plant Ecology Workbook*. Burgess, Minneapolis.

Dantas, M. & Muller, N.R.M. (1980). Estudos fito-ecológicas do tropico úmido brasileiro: aspectos fitosociológicas de mata sobre latisolo amarelo em Capitão Poço. *Boletím de Pesquisas, Empresa Brasileira de Agropecuaria (EMBRAPA)*, 9, Belém, Brazil.

Ehrenfeld, D. (1988). Why put a value on biodiversity? *Biodiversity* (Ed. by E.O. Wilson), pp. 212–216. National Academy Press, Washington, DC.

Gentry, A.H. (1982). Patterns of neotropical plant species diversity. *Evolutionary Biology*, Vol. 15 (Ed. by M.K. Hecht, B. Wallace & G.T. Prance), pp. 1–84. Plenum, New York.

Gentry, A.H. (1985). Some preliminary results of botanical studies in Manu Park. *Estudios Biológicos en el Parque de Manu* (Ed. by A. Tovar & M. Rios). Ministerio de Agricultura, Lima, Peru.

Gentry, A. (1988). Tree species richness of upper Amazonian forests. *Proceedings of the National Academy of Sciences*, 85, 156–159.

Goodall, D.W. (1970). Statistical plant ecology. *Annual Review of Ecology and Systematics*, 1, 99–124.

Goulding, M. (1980). *The Fishes and the Forest.* University of California Press, Berkeley.

Greig-Smith, P. (1983). *Quantative Plant Ecology*, 3rd edn. Blackwell Scientific Publications, Oxford.

Hartshorn, G.S. (1980). *Forest Vegetation (of Manu Park, Peru).* Tropical Science Center, San José, Costa Rica.

Hubbell, S.P. & Foster, R.B. (1986). Biology, chance, and history and the structure of tropical rain forest tree communities. *Community Ecology* (Ed. by J. Diamond & T.J. Case), pp. 314–329. Harper & Row, New York.

Keel, S.H.K. & Prance, G.T. (1979). Studies of the vegetation of a white-sand and black water igapó (Rio Negro, Brazil). *Acta Amazonica*, **9**, 645–655.

Korning, J., Thomsen, K. & Øllgaard, B. (1990). Composition and structure of a species rich Amazonian rain forest obtained by two different sample methods. *Nordic Journal of Botany*, **11**, 103–110.

Lechthaler, R. (1956). Inventario das árvores de um hectare de terra firme na zona Reserva Florestal Ducke, Município de Manaus. *Instituto Brasileiro de Bibliografia e Documentação, Botânica*, **Publicação No. 3**. Rio de Janeiro, Brazil.

Mori, S.A., Boom, B.M., de Carvalho, A.M. & dos Santos, T.S. (1983). Southern Bahian moist forests. *Botanical Review*, **49**, 155–232.

Peters, C.P., Gentry, A.H. & Mendelsohn, R.O. (1989a). Valuation of an Amazonian rainforest. *Nature*, **339**, 91–93.

Peters, C.P., Balick, M.J., Kahn, F. & Anderson, A.B. (1989b). Oligarchic forests of economic plants in Amazonia: utilization and conservation of an important tropical resource. *Conservation Biology*, **3(4)**, 341–349.

Pires, J.M. (1966). The estuaries of the Amazon and Oyapoque Rivers. *Proceedings of the Decca Symposium* (Ed. by Anonymous), pp. 211–218. United Nations Educational, Scientific and Cultural Organization (UNESCO), New York.

Pires, J.M. & Koury, H.M. (1958). Estudo de um trecho de mata de várzea próximo de Belém. *Boletím Técnico I. A. N.*, No. 36, 3–44.

Pires, J.M., Dobzhansky, Th. & Black, G.A. (1953). An estimate of the number of species of trees in an Amazonian forest community. *Botanical Gazette*, **114**, 467–477.

Prance, G.T. (1979). Notes on the vegetation of Amazonia III, the terminology of Amazonian forest types subject to inundation. *Brittonia*, **31(11)**, 26–38.

Prance, G.T., Rodrigues, W.A. & da Silva, M.F. (1976). Inventário florestal de um hectare de mata da terra firme, km 30 da estrada Manaus-Itacoatiara. *Acta Amazonica*, **6**, 9–35.

Pulliam, H.R. (1988). Sources, sinks and population regulation. *American Naturalist*, **132(5)**, 652–661.

Rodrigues, W.A. (1961). Estudo preliminar de mata de várzea alta de uma ilha de baixo Rio Negro de solo agriloso e úmido. *Instituto Nacional de Pesquisas da Amazônia*, **Publicação No. 10**. Manaus, Brazil.

Rodrigues, W.A. (1963). Estudo de 2,6 hectares de mata de terra firme da Serra do Navio, Territorio de Amapá. *Boletím do Museu Paraense Emilio Goeldi, Historia Natural*, Novo Série, Botânica, **19**.

Wallace, A.R. (1878). *Tropical Nature.* MacMillan and Co., London.

9. HARVESTING SPECIES
OF DIFFERENT LIFESPANS

G. P. KIRKWOOD*, J. R. BEDDINGTON*
AND J. A. ROSSOUW†

Renewable Resources Assessment Group,
Centre for Environmental Technology,
Imperial College of Science, Technology and Medicine,
8 Prince's Gardens, London SW7 1NA, UK and
† MRAG Ltd, 27 Campden Street, London W8 7EP, UK

SUMMARY

The relationship between the potential yield of a species and its demographic characteristics has been recognized in broad outline for many years. In essence, the faster the growth and the higher the level of natural mortality, the higher is the yield. Using a fully age-structured model with density dependence encapsulated in a relationship between the mature stock size and the resulting recruitment, this chapter explores the complications and exceptions to this general rule for fish species. When recruitment is constant and independent of mature stock size, it is shown that the yield as a proportion of unexploited biomass is directly proportional to the natural mortality rate. When recruitment is allowed to vary deterministically with mature stock size, this proportional relationship holds approximately, at least for biologically feasible parameter combinations. Adjustments necessary to deal with the special case of very short-lived species are discussed, as are the potential biases that can result from ignoring age-dependence in natural mortality if it is present. Simulated equilibrium distributions of yield–biomass ratios are presented for cases in which there are different levels of stochastic variability in recruitment. Finally, the ability of species with different lifespans to recover from catastrophic collapses is examined, with longer lived species in this case being more resilient.

INTRODUCTION

It is an intuitively plausible idea that long-lived, slow-growing species have less potential to provide a sustainable yield than short-lived, fast-growing species. This idea was first encapsulated in a simple formula by

Gulland (1971). This formula directly related the potential yield of a species to its natural mortality rate in the equation

$$Y = \tfrac{1}{2}MB_o, \tag{1}$$

where M is the natural mortality rate and B_o is the unexploited population biomass.

Gulland's argument was a simple mix of a theoretical consideration, that the biomass level at which maximum sustainable yield can be obtained occurs at half the unexploited level in a simple logistic model, and an observation from experience of fisheries worldwide that indicated that the maximum yield appeared to occur when the level of fishing mortality was roughly equal to that of natural mortality.

In the context of generalized stock production models, of which the logistic model is one simple example, the Gulland formula was reexamined by Shepherd (1982a). He found that the maximum sustainable yield was usually somewhat less than $\tfrac{1}{2}MB_o$, but variation by a factor of 3 either way about that value was possible. However, Shepherd also noted that his analysis did not specifically take age-structure into account.

Beddington and Cooke (1983) examined the Gulland formula using a more detailed age-structured model. In their analysis, they distinguished between the yield as a proportion of the exploitable biomass (defined as the biomass above the age at which exploitation starts) and the total biomass. Clearly, this distinction is particularly significant in cases where exploitation starts at a relatively advanced age. Results presented in their paper were for a range of lifespans of typical commercially exploited fish. They found the Gulland formula tended to be optimistic, with the potential yield being somewhat lower than it predicted. Their analysis was further restricted by a decision to examine recruitment in a simple way. It was assumed constant above a threshold adult biomass.

The aim of this chapter is to examine in a general way the potential yield of species with different life histories and different stock–recruitment relationships across the whole range of lifespans: from annual species, like squid, to the very long lived, such as whales or the orange roughy. The results obtained here are clearly relevant to studies of mammalian terrestrial organisms, but this issue is not dealt with explicitly here and will be examined elsewhere.

Following the Gulland line, yield is examined as a proportion of the unexploited biomass. There are really two interrelated levels of yield viewed in this way: that as a proportion of the total biomass and that as a proportion of the exploitable biomass. The former is perhaps the more ecologically interesting, in that it sets a limit to the yield that particular species in an ecosystem can provide. The latter lies more in the domain of the fishery manager, who is interested in the limits of commercial exploitation, and has the ability to adjust the age or length at which exploitation commences.

In our examination of the deterministic case, which takes up the bulk of the chapter, we demonstrate that the potential yield measured as a proportion of the unexploited biomass is indeed approximately proportional to the natural mortality rate, at least for biologically feasible combinations of parameters. The constants of proportionality are functions of the key parameters of growth, mortality (lifespan) and the density-dependent response as focused in the variation in recruitment with mature stock size. For a typical case, a simple formula for the potential yield in terms of these key parameters is provided. We also briefly examine the extent to which the presence of age dependence in natural mortality might affect the predictions.

Clearly, a solely deterministic analysis provides only a guide to the potential yield. Two situations are examined where environmental variation is involved. The first follows Beddington and May (1977) by examining the relationship between mean and variance of yield for different parameter combinations and the way this relates to the deterministic results. The second, more unusually, examines the ability of species of different lifespans to respond to and recover from catastrophic events.

POPULATION DYNAMICS MODEL

The dynamics of the fish population are assumed to be described by a relatively simple age-structured model in continuous time. The model is formulated as a set of equations that, by and large, follow the mathematical structure of the dynamics of exploited fish populations described in Beverton and Holt (1957) and used by Beddington and Cooke (1983). The key biological parameters are the rate of natural mortality, the growth rate of the species and the size (or age) at which sexual maturity is reached. The primary variables that can be manipulated to adjust yield levels are the size at which exploitation begins and the mortality rate imposed by harvesting.

Dynamics of a single cohort

We deal first with the dynamics of a single cohort of fish, which initially consists of R fish born at age zero. Let

t_c = the age at first exploitation of fish in the cohort;

l_c = the length at first exploitation;
t_m = the age at sexual maturity;
l_m = the length at sexual maturity;
M = the instantaneous rate of natural mortality (assumed constant for now);
F = the instantaneous rate of fishing mortality;
$N(t)$ = the number of fish of age t years; and
$w(t)$ = the weight of a fish of age t years.
The dynamics of the cohort are described by the following equations:

$$N(0) = R,$$

$$N(t) = R\exp(-Mt) \qquad \text{for } 0 < t \leqslant t_c, \tag{2}$$

$$= R\exp(-Mt - F(t - t_c)) \quad \text{for } t \geqslant t_c.$$

In an unexploited cohort ($F = 0$), the total biomass over the lifespan of the cohort is

$$TB_o = \int_0^\infty R\exp(-Mt)\, w(t)\, dt, \tag{3}$$

and the exploitable biomass is

$$ExB_o = \int_{t_c}^\infty R\exp(-Mt)\, w(t)\, dt. \tag{4}$$

In the exploited case, the equilibrium yield from the cohort under constant fishing mortality F is

$$Y = FR\int_{t_c}^\infty \exp(-Mt - F(t - t_c))\, w(t)\, dt. \tag{5}$$

In common with standard fisheries practice, we assume that growth can be modelled by the von Bertalanffy (1938) growth curve, with the weight $w(t)$ at age t given by

$$w(t) = W_\infty(1 - \exp(-Kt))^3, \tag{6}$$

where W_∞ is the asymptotic weight and K is the growth rate. Lengths and ages are related via

$$l(t) = L_\infty(1 - \exp(-Kt)); \tag{7}$$

however, as it is more convenient to deal with lengths as proportions of the asymptotic length L_∞, from now on the length at first exploitation l_c and the length at sexual maturity l_m should be understood to be measured relative to L_∞.

Within-year components of the population

For the most part, we shall concentrate on deriving equilibrium results. As demonstrated by Beverton and Holt (1957), it then generally suffices to consider only the dynamics of a single cohort. However, even in the equilibrium case it is still important to make clear the relationship between the dynamics of individual cohorts as they age, and the dynamics of the whole population at any single point in time, which of course consists of surviving members of all cohorts born prior to that time.

Assume for convenience that fish are born at the beginning of each year. Then, identifying each individual cohort by the year in which it was born, at the beginning of year T the population consists of fish aged exactly $0, 1, 2, \ldots$, years. Define $N_{T-i}(i) =$ number of fish aged i in the population from the cohort born in year $T - i$, $i = 0, 1, 2, \ldots$ At time t $(0 \leqslant t < 1)$ after the start of year T, the total number of fish in the population is then

$$P_T(t) = \sum_{i=0}^{\infty} N_{T-i}(t + i) \tag{8}$$

and their total biomass is

$$B_T(t) = \sum_{i=0}^{\infty} N_{T-i}(t + i) \, w(t + i). \tag{9}$$

The exploitable biomass at time t within year T is

$$
\begin{aligned}
ExB_T(t) &= \sum_{i=[t_c]+1}^{\infty} N_{T-i}(t + i) \, w(t + i) \quad \text{if } t + i < t_c \\
&= \sum_{i=[t_c]}^{\infty} N_{T-i}(t + i) \, w(t + i) \qquad \text{if } t + i \geqslant t_c
\end{aligned} \tag{10}
$$

where $[x]$ is the integer part of x.

If the population is in unexploited equilibrium, then the dynamics of each cohort follow those described in the preceding subsection with $F = 0$. It follows from equation (9) that TB_o as defined in equation (3) is actually the average total biomass in the unexploited population throughout the year. Similarly, ExB_o is the average exploitable biomass throughout the year in an unfished population. When the natural mortality rate is relatively low, neither the total nor exploitable biomasses will change much during the year, and use of average values causes no problems. However, as will be seen later, if M is very high, as in the case with annual species, some account needs to be taken of within-year changes in biomasses.

Incorporation of density dependence

In the model as described so far, all parameters have been assumed to be constant; they are neither age dependent nor density dependent. We shall examine later what happens if the natural mortality rate varies with age. For most fish populations, it is believed that by far the majority of density-dependent effects occur during the egg and larval stages of their life history. In consequence, most fisheries models incorporate all density dependence in a non-linear relationship between the spawning stock biomass and the subsequent recruitment of juvenile fish to the population. For convenience, we suppress the egg and larval stages, and assume various relationships between the average sexually mature stock biomass during the spawning season of a year and the subsequent recruitment of fish nominally aged zero at the beginning of the following year.

Most fish species have a limited spawning season. In the model, we assume that the spawning season starts at the time of the year corresponding to the onset of sexual maturity, and that the spawning season has duration s. For the numerical calculations, s has been taken to be one tenth of a year. Under these assumptions, the average spawning stock biomass during the spawning season in year T will be

$$SSB_T = \frac{1}{s} \int_{t_m - [t_m]}^{t_m + s - [t_m]} \sum_{i=[t_m]}^{\infty} N_{T-i}(t + i)\, w(t + i)\, dt. \tag{11}$$

As usual, if the population is in equilibrium, the subscripts involving T can be omitted from this expression.

There are various possible stock–recruitment relationships that display different degrees of density dependence. Three frequently used stock–recruitment relationships in fisheries were proposed by Ricker (1954), Beverton and Holt (1957), and Cushing (1973). Shepherd (1982b) has proposed a model with three parameters that can encompass behaviours similar to all of these. Of the three forms of stock–recruitment relationships, the humped form proposed by Ricker (1954) incorporates the greatest degree of density dependence, and thus in a sense represents an extreme. Here, we have chosen to use the Beverton and Holt (1957) form, which has the twin advantages of having a very simple mathematical form and of being relatively conservative in the degree of density dependence it attributes to the population. It can also display behaviour similar to that of the Cushing (1973) form. The effect of incorporating a Ricker (1954) stock–recruitment relationship will be examined in a later publication.

The Beverton–Holt stock–recruitment relationship between spawning stock biomass SSB and recruitment R can be written as

$$R = \frac{a\,SSB}{1 + \dfrac{SSB}{b}}. \tag{12}$$

This relationship is an asymptotic one, with recruitment rising from zero at zero SSB to a maximum of ab as $SSB \rightarrow \infty$. The parameter a represents the slope of the relationship at the origin, and the parameter b is the spawning stock size at which recruitment is reduced to one-half its maximum value.

For the purposes of this chapter, one of the issues we wish to address is the effect of differing degrees of density dependence on yield from otherwise comparable populations. This is achieved by imposing constraints such that these populations always have the same unexploited equilibrium. Two constraints are required. The first is an implicit one resulting from the equilibrium requirement: if the population is in equilibrium, with a constant annual recruitment R_o, then that fixes the value of the spawning stock biomass SSB_o for each value of R_o. If additionally in equation (12) we require that

$$b = \frac{R_o\,SSB_o}{a\,SSB_o - R_o}, \tag{13}$$

then at unexploited equilibrium, the recruitment will be R_o at spawning stock biomass SSB_o.

Under these circumstances, the single parameter a now measures the degree of density dependence. When $a = R_o/SSB_o$ in equation (12), there is no density dependence at all, while as a increases from this value, so does the degree of density dependence. As $a \rightarrow \infty$, recruitment tends to the constant value R_o regardless of the size of the spawning stock biomass. Finally, as it is convenient to express the degree of density dependence on a fixed scale, we put

$$a = \left(\frac{1 + d}{1 - d}\right)^2 \frac{R_o}{SSB_o}, \tag{14}$$

and now the density dependence parameter d varies from 0 to 1. Constrained Beverton–Holt stock–recruitment relations are illustrated in Fig. 9.1, along with a geometric explanation of the transformation in equation (14).

The estimation of stock–recruitment relationships is an exercise fraught with difficulty due to the often highly variable nature of recruitment; see p. 220. However, some insight can be gained from simple observations of where recruitment can be seen to have declined under exploitation.

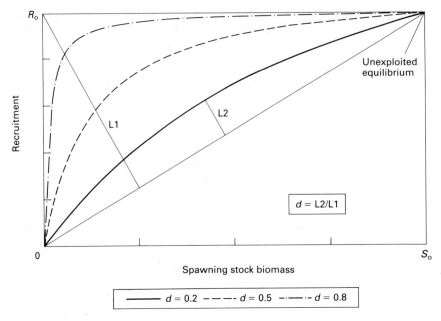

FIG. 9.1. Beverton and Holt stock–recruitment relationships exhibiting differing degrees of density dependence (d), all with the same unexploited equilibrium between spawning stock biomass (S_o) and recruitment (R_o).

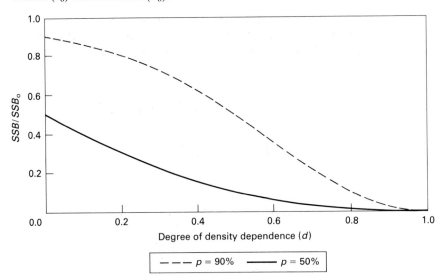

FIG. 9.2. Levels of spawning stock biomass relative to unexploited levels (SSB/SSB_o) at which recruitment has been reduced to a specified percentage p of the unexploited level of recruitment.

Figure 9.2 shows how a decline in recruitment can be related to the parameter d. A guide to the value of this parameter could therefore be obtained by inspection, rather than through detailed statistical fitting. Note that for $d > 0.6$, there is likely to be little hope of discriminating between different values of d from the data. In such circumstances, selection of a conservative value would seem appropriate.

DETERMINISTIC EQUILIBRIUM RESULTS

In this section, we present the relationships found between the deterministic equilibrium maximum sustainable yield (MSY) and the size of the exploitable biomass in a population prior to the commencement of exploitation (ExB_o). The first case considered is that in which recruitment is constant and independent of the size of the spawning stock. This simplest case was examined by Beddington and Cooke (1983), and it is the only one for which any analytic progress can be made. Where recruitment varies according to a stock–recruitment relationship, no analytic results are available, and purely numerical methods must be used.

Constant recruitment

Substituting the expression for the von Bertalanffy growth curve (eqn (7)) into the expression for the yield corresponding to a given fishing mortality rate F and a given length at first capture l_c leads to (following Beverton & Holt 1957)

$$Y = FRW_\infty (1 - l_c)^{M/K} \left(\frac{1}{F + M} - \frac{3(1 - l_c)}{F + M + K} \right.$$
$$\left. + \frac{3(1 - l_c)^2}{F + M + 2K} - \frac{(1 - l_c)^3}{F + M + 3K} \right). \tag{15}$$

Also,

$$ExB_o = RW_\infty (1 - l_c)^{M/K} \left(\frac{1}{M} - \frac{3(1 - l_c)}{M + K} \right.$$
$$\left. + \frac{3(1 - l_c)^2}{M + 2K} - \frac{(1 - l_c)^3}{M + 3K} \right). \tag{16}$$

Taking the ratio of these two quantities and rearranging, it follows that Y/ExB_o has the form

$$\frac{Y}{ExB_o} = Mf\left(\frac{F}{M}, \frac{M}{K}, l_c\right), \tag{17}$$

where $f(.)$ is a complicated-looking function of just three variables: F/M, M/K, and l_c. To obtain an expression for MSY/ExB_o for given M, K, and l_c, it is necessary to maximize equation (17) with respect to F. This cannot be done analytically. However, the form of equation (17) implies that if both l_c and M/K are kept fixed, the function $f(.)$ has now only a single variable, F/M. If the function is then maximized over F, it follows immediately that

$$\frac{MSY}{ExB_o} \propto M, \text{ provided } \frac{M}{K} \text{ and } l_c \text{ are fixed.} \tag{18}$$

It also follows that the fishing mortality rate F_{MSY} that produces the maximum sustainable yield is a fixed multiple of M for given values of l_c and M/K.

In the case of constant recruitment and across species with M/K constant, we have thus shown that at least the proportionality between the yield–biomass ratio and the natural mortality rate proposed by Gulland appears to be correct. At first sight, this finding appears to contradict the results obtained by Beddington and Cooke (1983), whose figures suggest that the relationship between yield–biomass ratios and M is neither proportional nor linear. The difference lies in the requirements in equation (18) that M/K be constant, and that yield–biomass ratios be viewed as a function of l_c, rather than of t_c. The requirement that M/K be constant is not at all restrictive; it simply indicates that the constant of proportionality in the relationship will vary with both the value of M/K and that of l_c.

It is also of interest to note that precisely the same arguments apply to the relationship between M and the ratio of MSY to unexploited total biomass (TB_o), although the constants of proportionality are different, of course. Whether in practice it is more convenient to concentrate on exploitable or total biomasses depends on which of these two quantities is easier to measure.

Having determined that the relationship between yield–biomass ratios and M is one of proportionality when recruitment is constant, it is a simple matter to estimate the different constants of proportionality corresponding to different values of biological parameters. However, before doing so, it is useful to identify likely ranges of these parameters. Turning first to the natural mortality rate, M, we have calculated results for the wide range 0.05–3.0 year^{-1}. This allows consideration of marine species with lifespans as disparate as whales and orange roughy at the low M end

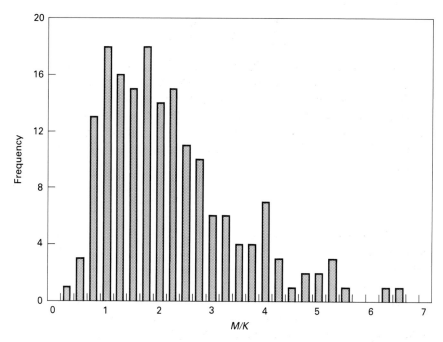

Fig. 9.3. Histogram of estimates of M/K for the 175 fish stocks examined by Pauly (1980).

to annual species such as squid at the other end. Beddington and Cooke (1983) chose to vary M and K separately, but in light of equation (18) above, we shall examine results for differing values of the ratio M/K. Using the large set of estimates of M and K listed in Pauly (1980), we find that typically M/K varies over the range 0.5–4.0. A histogram of M/K values from Pauly (1980) is given in Fig. 9.3.

Table 9.1a lists the constants of proportionality in the relationship using exploitable biomass for a range of values of l_c and M/K. Table 9.1b lists the equivalent constants of proportionality when total biomass is used.

The results in Table 9.1a confirm the results obtained by Beddington and Cooke (1983) that in many cases the constant of proportionality is less than the value of $\frac{1}{2}$ in the Gulland formula. For the parameter values examined, only for high values of M/K and high l_c is the constant $\frac{1}{2}$ or higher. The constant increases monotonically with l_c for constant M/K, but shows a minimum at intermediate values of M/K for each l_c.

When the ratios are calculated in terms of total biomass, Table 9.1b shows that very similar constants of proportionality to those shown in Table 9.1a are obtained for low values of l_c and/or M/K. However, now

TABLE 9.1. Constants of proportionality in the relationship between MSY/ExB_o and M when recruitment is constant, for different values of l_c and M/K

(a)

			M/K		
l_c	0.5	1.0	2.0	3.0	4.0
0.2	0.30	0.25	0.22	0.22	0.23
0.3	0.32	0.28	0.26	0.28	0.30
0.4	0.35	0.32	0.33	0.36	0.42
0.5	0.40	0.37	0.41	0.48	0.55
0.6	0.45	0.44	0.52	0.61	0.68

(b)

			M/K		
l_c	0.5	1.0	2.0	3.0	4.0
0.2	0.30	0.25	0.22	0.21	0.22
0.3	0.32	0.27	0.25	0.26	0.27
0.4	0.35	0.31	0.30	0.30	0.29
0.5	0.39	0.35	0.33	0.31	0.28
0.6	0.43	0.39	0.35	0.28	0.19

the constants generally take their lowest values when both l_c and M/K are high. That the greatest differences between the two sets of results appear for high values of l_c and M/K is of course to be expected, given the differences in the definitions of total and exploitable biomass.

In reality, the differences in the constants of proportionality when both M/K and l_c are high may not be as great as suggested in these tables. The differences arise in part because the ratio between the fishing mortality rate at MSY (F_{MSY}) and M increases as M/K and l_c increase. Pauly and Soriano (1986) have shown that if allowance is made for recruitment to the fishery to occur gradually over a range of lengths, rather than instantaneously on reaching length l_c as assumed here, then the F_{MSY} in the constant recruitment case is lower, as is the maximum yield. This point will be pursued further elsewhere, and for the rest of this chapter we will retain the assumption of knife-edged selection.

Alternative stock–recruitment functions

Given the direct proportionality found between MSY/ExB_o and M when recruitment is constant, it is tempting to hypothesize that the same relationship might hold when recruitment varies as a function of the size of the spawning stock. Unfortunately, this is not true, but for many biologically likely parameter combinations the relationship turns out to be very close to proportional for practical purposes. The reason that proportionality no longer holds is that in both equations (15) and (16), the recruitment R is no longer constant; rather it is also a function of the fishing mortality rate F and equation (17) no longer holds.

As in the preceding section, we will concentrate primarily on investigating the relationship between MSY/ExB_o and M. However, we now also need to specify typical values for the length at maturity relative to the asymptotic length, L_∞. Unfortunately, Pauly (1980) does not cite estimates of the length at maturity, but for the species examined by Beddington and Cooke (1983), it appears that l_m ranges from around 0.2 to 0.8, with a peak around 0.5. We have therefore chosen $l_m = 0.5$ as a typical case.

The result of allowing recruitment to vary with spawning stock size is seen in Fig. 9.4. There, the empirical relationship between MSY/ExB_o

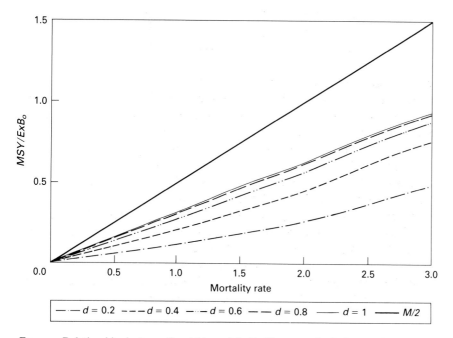

FIG. 9.4. Relationships between the yield–exploitable biomass ratio (MSY/ExB_o) and mortality rate for different degrees of density dependence in the stock–recruitment relationship (d). The straight line with slope $M/2$ is shown for comparison. Other parameters used were $l_c = 0.4$, $l_m = 0.5$, and $M/K = 1$.

and M is shown for differing degrees of density dependence d in the stock–recruitment relationship (see above and Fig. 9.1), in a typical case where the length at first capture is less than the length at maturity, and $M/K = 1$. Two features are obvious. First, while the overall impression is that the relationship is close to linear, in fact for each value of d the actual relationship is best described as being linear with one slope for low values of M, another higher slope for higher values of M, and with a smooth transition between the two linear sections. In this example, the point at which the slope increases is around $M = 2$ for $d = 0.2$, and then at higher values of M for large values of d. Note that the apparent 'waviness' of the lines is an artefact of the time-step used in the numerical calculations.

The second feature to be observed in Fig. 9.4 is that the approximate slope of the lines increases as d increases. This is, of course, as it should be: as the degree of density dependence increases, the population can withstand relatively greater reductions in spawning stock biomass corresponding to larger fishing mortalities, and thus support greater sustainable yields.

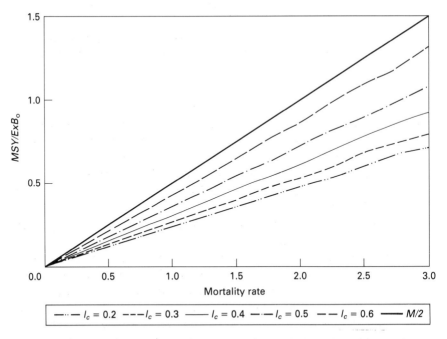

FIG. 9.5. Relationships between MSY/ExB_o and mortality rate for different values of l_c. The straight line with slope $M/2$ is shown for comparison. Other parameters used were $d = 0.8$, $l_m = 0.5$, and $M/K = 1$.

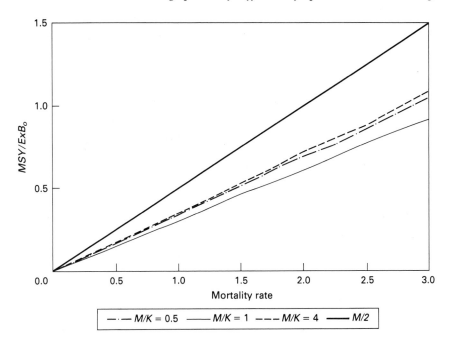

FIG. 9.6. Relationships between MSY/ExB_o and mortality rate for different values of M/K. The straight line with slope $M/2$ is shown for comparison. Other parameters used were $d = 0.8$, $l_m = 0.5$, and $l_c = 0.4$.

Figure 9.5 shows typical relationships between MSY/ExB_o and M when l_c varies, with the relative proportion of the exploitable biomass that can be taken sustainably increasing with length at first capture. Note, however, that while the proportion of exploitable biomass that can be taken increases with l_c, the exploitable biomass itself decreases.

Typical relationships between MSY/ExB_o and M when M/K varies are shown in Fig. 9.6. Somewhat surprisingly, even over a range of M/K of 0.5–4.0 the relative proportions of exploitable biomass that can be taken sustainably are rather similar.

Finally, to round off this illustration of the types of relationships found when the length at maturity is 50% of the asymptotic length, Fig. 9.7 breaks the illusion that the departure from proportionality in the relationship between yield–biomass ratios and M is at most minor. It also demonstrates other non-linearities of the relationships. The line corresponding to $M/K = 0.5$ has been omitted from the figure, as it is virtually indistinguishable from that for $M/K = 1$. The other unusual feature of Fig. 9.7 is the pronounced hump in the uppermost line for M

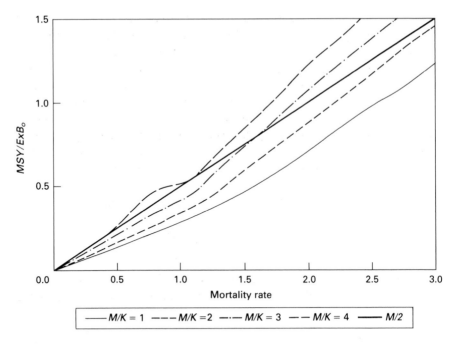

FIG. 9.7. Relationships between MSY/ExB_o and mortality rate for different values of M/K when the length at first capture exceeds the length at maturity. The straight line with slope $M/2$ is shown for comparison. Other parameters used were $d = 0.4$, $l_m = 0.5$, and $l_c = 0.6$.

values between 0.5 and 1.0. This is not an artefact. Rather, it is a real feature arising from the unusual combination of a high M/K, a length at first capture that exceeds the length at maturity, and a relatively small degree of density dependence in the stock–recruitment relation. For these intermediate values of M, it becomes important whether fishing during a year commences before or after the start of the spawning period and whether or not in terms of yield it is worth leaving some spawning stock to survive and spawn an extra year.

Inspection of the full set of empirical relationships obtained indicates that substantial departures from proportionality and non-linearity occur only for quite high values of M (e.g. $M > 1$) and primarily when the degree of density dependence is small. A full investigation of the way in which species occupy different regions of parameter space is beyond the scope of this chapter. However, this particular area of parameter space is biologically the most unlikely: normally one expects that low degrees of density dependence in a stock–recruitment relation will be associated with species that have long lifespans and therefore low mortality rates, and conversely that species with nearly constant recruitment even for small spawning stock biomasses will be short lived and fast growing.

In view of these latter remarks, we attempted to derive an approximate empirical formula expressing the yield biomass ratio as a function of M, l_c, d, and M/K for the case when $l_m = 0.5$. This was done using a log-linear regression in which the dependent variable was the calculated value of MSY/ExB_o when $M = 1$, which was taken to be an approximation for the constant of proportionality, for values of l_c in the range 0.3–0.6, d in the range 0.2–1.0, and M/K in the range 0.5–4.0. After some trial and error, the best fit obtained corresponded to the formula

$$MSY/ExB_o = [1 - (1 - d)^2] \exp\left(-1.33 + 0.64l_c\right.$$

$$\left. - 0.31\frac{M}{K} + 0.80l_c\frac{M}{K}\right). \tag{19}$$

This gives remarkably good predictions for the cited ranges of parameters. The value of R^2 obtained in the regression was 0.93. For only 16 of the

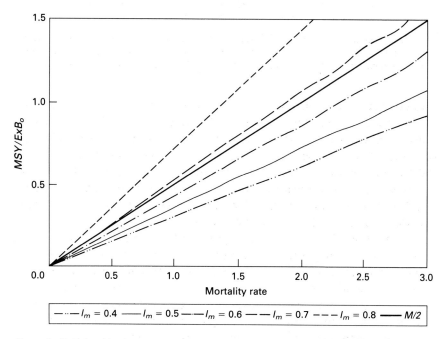

FIG. 9.8. Relationships between MSY/ExB_o and mortality rate for different values of length at maturity (l_m). The straight line with slope $M/2$ is shown for comparison. Other parameters used were $l_c = l_m$, $d = 0.8$, and $M/K = 1$.

124 'observations' did the percentage prediction error exceed 10% and this only occurred for high values of M/K or low d. The maximum percentage prediction error was 23%. While we would not recommend immediate use of this formula at this stage, the good predictions obtained suggest strongly that practically useful predictions of yield in a wide variety of situations could be obtained using a relatively simple empirical relationship.

We have not yet completed a comparably full investigation of cases where l_m takes values different from 0.5, but an indication of the effects of varying the length at maturity is given in Fig. 9.8. In these calculations, it is assumed that the length at first capture and length at maturity coincide, and that $M/K = 1$ and $d = 0.8$. For this more central combination of parameters, the appearance of linearity returns and the approximate constant of proportionality increases with l_m, exceeding $\frac{1}{2}$ for the largest l_m.

Age-dependent mortality

One of the criticisms made of simple fisheries models by those more used to modelling terrestrial mammalian systems is that they make the assumption that the natural mortality rate does not vary with age. For the early stages in the life history of a fish species, the evidence is absolutely unequivocal that there are extreme variations in natural mortality with age. However, most models of exploited fish stocks, and certainly the ones considered here, have been formulated for late juvenile and adult fish, over which an assumption of an approximately constant natural mortality rate may be more tenable. Even so, there is also evidence that natural mortality rates can increase with age for older fish; for example see Beverton and Holt (1957) where approximately linear increases in M with age are shown for a number of fish species. In practice, it is usually difficult enough just to estimate a constant average natural mortality rate for a fish species, so predictions of yield will usually be based on an assumed constant M. In this section, we investigate the extent to which allowing for M to increase with age will affect the results presented for constant M.

To all intents and purposes, the value of M or variations of it with age are irrelevant for all ages younger than the age at first capture, since we are examining ratios between yields and exploitable biomasses. Thus, the cases examined here are for an age-dependent natural mortality schedule of the form

$$M(t) = M_o \qquad\qquad \text{for } t \le [t_c]$$
$$= M_o + \alpha(t - [t_c]) \quad \text{for } t > [t_c]. \tag{20}$$

In order to examine the size of the bias incurred by erroneously assuming a constant M, it is also necessary to calculate an appropriate constant average M that would correspond to a given age-dependent M. An obvious candidate for an equivalent M is

$$EqM = -\ln\left(\frac{\sum\limits_{i=[t_c]+1}^{\infty} N(i)}{\sum\limits_{i=[t_c]}^{\infty} N(i)}\right). \tag{21}$$

This formulation is in fact equivalent to the well-known Heincke (1913) estimate of mortality. For comparability of results with those in earlier sections, it is convenient to examine plots of yield–biomass ratios against

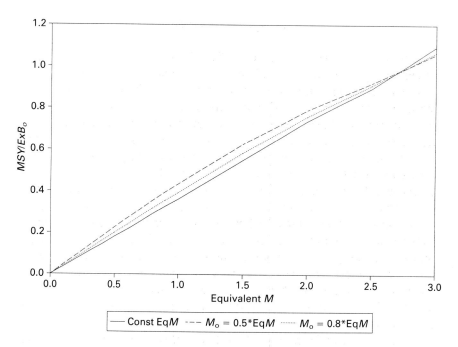

FIG. 9.9. Relationships between MSY/ExB_o and mortality rate when the mortality rate is allowed to vary with age. The dotted lines show the actual relationships for the two different age-dependent mortality schedules with the same equivalent constant Ms (EqM), and the solid line is the result of assuming M is constant at EqM. Other parameters used were $l_c = 0.5$, $d = 0.8$, and $M/K = 1$.

equivalent M for different age-dependent schedules of M. Thus in Fig. 9.9, we present plots for three cases: M constant (so $EqM = M$), M age-dependent with $M_o = 0.8\,EqM$, and M age-dependent with $M_o = 0.5\,EqM$. In the last two cases, the values of α in equation (20) are calculated to give the desired value of EqM.

As can be seen, an erroneous assumption of a constant M leads to a negative bias in the yield–biomass ratios. That bias increases with increasing age-dependency. Interestingly, while it initially appears to increase with increasing M, it then subsequently appears to decrease for higher values of M. No generality is claimed for this latter finding; it could well result from the specific form of age-dependence and assumptions regarding the spawning period for shorter lived fish. What is much more important is the direction of the bias and its relatively small size (the maximum bias is around 25% for $M_o = 0.5\,EqM$). Thus an erroneous assumption of constant M with age is likely to result in a slightly conservative prediction of potential yield when in fact M actually increases with age.

The special case of very short-lived species

On the surface, the results presented above apply equally for all values of M. However, we have already seen that the proportional relationship between yield–biomass ratios and M can break down for high M. In addition, it was intimated earlier that some further difficulties arise in the definition of exploitable biomass when M is large. The argument for treating at least some very short-lived species as special cases is unanswerable when one considers annual species like squid, which die immediately after spawning. Such species can be considered to present extreme cases of age-dependent natural mortality: M is effectively infinite for all ages above the age at maturity.

The difficulty that occurs is not that any of the preceding calculations are incorrect, but rather that the values calculated for the exploitable or total biomass in an unexploited population using equations (2) and (3) no longer correspond to anything really measurable. As noted on p. 203, the biomasses as defined actually represent an average value throughout a year.

When M is relatively small, these biomasses are approximately constant throughout the year, and it is reasonable to treat an estimate of abundance in an unexploited population made at virtually any time throughout a year as an acceptable estimate of the average biomass. However, for annual species not only do these biomasses vary in size considerably throughout the year for ages up to the age at maturity,

TABLE 9.2. Yield–exploitable biomass ratios when $l_m = 0.8$, $M = K = 3$, and $d = 0.8$ for the standard model and for one in which all fish die after spawning once. ExB_o is the average biomass of fish in an unexploited population of length l_c or more, and *Init ExB* is the biomass of fish in an unexploited population of length l_c

	Standard model		Death occurs after spawning	
l_c	*MSY/ExB$_o$*	*MSY/Init ExB*	*MSY/ExB$_o$*	*MSY/Init ExB*
0.2	0.70	1.38	0.92	4.54
0.3	0.77	1.35	1.04	3.05
0.4	0.90	1.15	1.25	1.76
0.5	1.04	0.98	1.49	1.26
0.6	1.24	0.86	1.90	1.00
0.7	1.53	0.82	2.60	0.91

but they are zero for the remainder of the year. This means that interpretation and estimation of average biomasses is rather difficult. In such circumstances, it is more realistic to examine ratios of yield to the biomass in an unexploited population at age t_c (i.e. at the start of what would be the fishing season in an exploited stock). Apart from being easily interpretable, this biomass has the additional virtue of being practically measurable, for example by pre-season surveys of abundance or retrospectively via within-season stock assessments.

Table 9.2 presents numerical results for the case where $d = 0.8$, $M = K = 3$, $l_m = 0.8$, and l_c ranges from 0.2 to 0.7, obtained for two dynamics models: the standard one as described on p. 202, and an alternative where M is infinite for all $t > t_m$. The higher than usual value used for l_m is more typical of very short-lived species. Comparing first the results within the same dynamics model, the different trends with l_c in yield–biomass ratios result directly from the different definitions of exploitable biomass. Of necessity, the average exploitable biomass must decrease with increasing l_c, but for the parameter combinations used in Table 9.2 the maximum unexploited biomass occurs at lengths greater than l_m, so the biomass at the start of the fishing season will increase over the given range of l_c.

Comparing the results across models when the exploitable biomass is that at the start of the fishing season, the yield–biomass ratios are considerably higher when the fish die immediately after spawning, particularly for low values of l_c. This reflects the fact that even for this high M, there are sufficiently many survivors into a second year to influence the size of the exploitable biomass at the start of the season. There are no survivors to a second year if death occurs after spawning.

For the smaller values of l_c, the ratio of *MSY* to the exploitable biomass at the start of the season when death occurs after spawning can be substantially greater than 1.0. This may at first glance appear somewhat bizarre, but it simply reflects the fact that fishing has started well before the population has reached its maximum biomass for the year. This effect has been observed (M. Basson, personal communication) in a real exploited stock of an annual species: the stock of the squid *Illex argentinus* around the Falkland Islands, which have been assessed using the techniques described by Beddington *et al.* (1990).

EFFECT OF A VARYING ENVIRONMENT

While deterministic results are useful, especially to set a scale on yield levels and to indicate how these vary for species with differing biological parameters, the actual yields that will be taken in the presence of a varying environment can differ substantially from those suggested by the deterministic results. For many fish species, the primary focus of variability is in the annual recruitment; for temperate species the recruitment resulting from the same or similar levels of spawning stock can vary by more than an order of magnitude. We therefore examine the effect of a varying environment by allowing for stochastic variability in the stock–recruitment relationship. The same approach was taken by Beddington and Cooke (1983).

Two scenarios are examined. In the first, the annual recruitment arising from given levels of spawning stock biomass is taken to be lognormally distributed. The second examines the ability of species of different lifespans to recover from an extreme case of environmental variability, in which the spawning stock suffers an episode of catastrophic mortality.

Stochastic recruitment

In their examination of the statistical properties of recruitment in some commercial fish species, Hennemuth *et al.* (1980) found that recruitment was highly variable and that its skewed frequency distribution was similar to a log-normal distribution. Similar conclusions have been reached in other studies, by far the most comprehensive and recent being an as-yet-unpublished analysis of stock–recruitment relationships for a very large number of fish stocks by R.A. Myers (personal communication). Accordingly, we investigate the probability distribution of annual yield when the annual recruitment is log-normally distributed with mean equal to that predicted by the deterministic Beverton–Holt stock–recruitment relationship. Based on the analysis of Beddington and Cooke (1983), we have taken the coefficient of variation (CV) of the log-normal errors to be 0.5.

The procedure for generating series of stochastic annual yields and characterizing them for each combination of parameters was as follows. A simulated population was set up at time zero in deterministic equilibrium, harvested at the fishing mortality rate that produced the deterministic *MSY* for that parameter combination. Annual log-normal variation about the stock–recruitment curve was then introduced, and sufficient time was allowed to elapse for the effects of the stochastic recruitment to spread through all age classes in the population. Then the values of annual yield were collected for the next 100 years. This process was repeated 100 times, and the median, 5th and 95th percentiles of the pooled distribution of annual yields were determined. These statistics of the distribution of annual yield were determined for parameter combinations $l_c = 0.4$, $l_m = 0.5$, $d = 0.2$ and 0.8, and M taking selected values in the range 0.05–3.0. Results are presented in terms of the ratios of annual yield to the deterministic exploitable biomass.

Results for the case when $d = 0.8$ are shown in Fig. 9.10. For low values of M, the median annual yield–exploitable biomass ratio is very close to the corresponding deterministic value, but as M increases, the median begins to fall below the deterministic value. This tendency is more marked for the other case examined (not illustrated), in which the degree of density dependence in the stock–recruitment relationship was smaller ($d = 0.2$).

There is a strong tendency for increased skewness in the distribution of annual yield–biomass ratios as M increases. Interestingly, the lower 5th percentile of this distribution stays remarkably constant for M above 0.5. In contrast, the 95th percentile increases rapidly with M. The increasing skewness with increasing M is to be expected; when M is high, the exploitable stock consists of only one or two age classes, each of which has been subject to highly skew log-normal variability, while when M is low, the exploitable stock is the sum of many age classes and its distribution will be more symmetric.

A direct result of the increasing skewness with M is that, while in most cases the annual yield–biomass ratio will be much less than $\frac{1}{2} M$, in an important minority of cases it can exceed $\frac{1}{2} M$. This ability to gain benefits in terms of high yields during good years flows directly from the assumption that fishing mortality (or equivalently fishing effort) is the variable that is used to control the level of harvesting. Figure 9.10 shows clearly that once M gets large, application of a constant fishing mortality

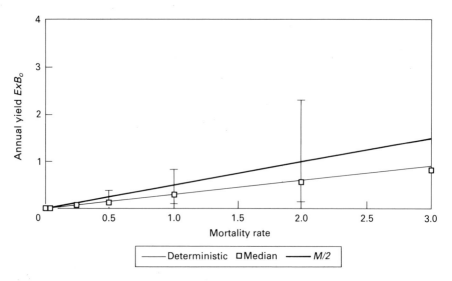

FIG. 9.10. Median and percentiles (5% and 95%) from the distribution of the ratio of annual yield to deterministic ExB_o for different mortality rates when annual recruitment is log-normally distributed about a deterministic Beverton–Holt stock–recruitment relationship. The straight line with slope $M/2$ is shown for comparison. Other parameters used were $d = 0.8$, $l_c = 0.4$, $l_m = 0.5$, and $M/K = 1$.

can lead to yield–biomass ratios considerably in excess of $\frac{1}{2}M$. However, the very high variation in recruitment has tended to be documented for species with relatively low mortality. More detailed examination of these relationships will be explored elsewhere.

Recovery from a stock collapse

The preceding section examined the effect of stochastic variation of recruitment about a stock–recruitment relationship. One feature of that relationship was that recruitment was non-zero for all positive stock biomasses. This would be an optimistic assumption if there existed a critical spawning stock biomass below which recruitment fell to zero; that is if there is critical depensation in the stock–recruitment relationship. Here, we examine the ability of a population to recover from a catastrophic mortality episode.

Specifically, we examine a case in which a population is in deterministic equilibrium at time zero, harvested at the fishing mortality rate that produces the deterministic *MSY*. Just before the start of the spawning

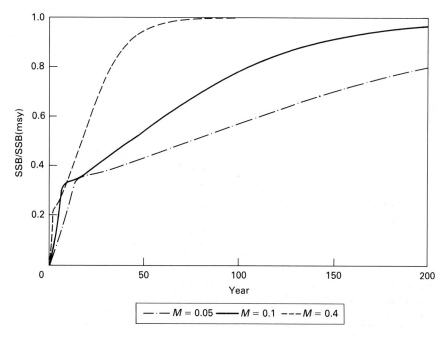

FIG. 9.11. Trajectories of recovery from a catastrophe in which the spawning stock of a stock at exploited equilibrium is killed immediately prior to spawning at year zero, when $l_m = l_c = 0.5$. Trajectories shown are for $M = 0.05$, 0.1 and 0.4. Fishing continues throughout at the deterministic *MSY* level. Other parameters used were $d = 0.2$, and $M/K = 1$.

season in the first year, it is assumed that the entire spawning stock is killed, so that there is no recruitment at all at the beginning of the second year. Subsequently, the population is allowed to recover, if it can, according to the standard deterministic dynamics and deterministic stock–recruitment relationship. During the recovery period, fishing continues at the F_{MSY} rate.

The trajectories taken by the annual spawning stock biomasses for three different natural mortality rates ($M = 0.05$, 0.1 and 0.4) are shown in Figs 9.11 and 9.12. For convenience, we have taken the lengths at first capture and at maturity to be equal. A low degree of density dependence in the stock–recruitment relationship was used ($d = 0.2$), and M/K was set to 1.0. Figure 9.11 examines the case in which $l_c = l_m = 0.5$, and Fig. 9.12 the case in which $l_c = l_m = 0.7$.

Despite exploitation continuing, in each case the spawning stock biomasses recover (eventually), initially at a relatively fast rate and then at a rather slower rate. The key to these recoveries lies in the presence of

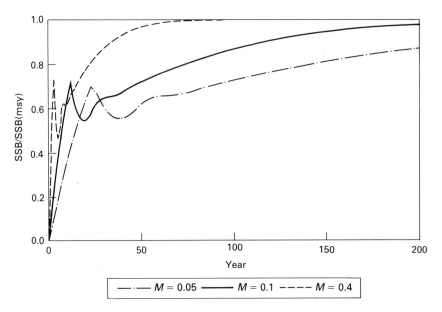

Fig. 9.12. Trajectories of recovery from a catastrophe in which the spawning stock of a stock at exploited equilibrium is killed immediately prior to spawning at year zero, when $l_m = l_c = 0.7$. Trajectories shown are for $M = 0.05$, 0.1 and 0.4. Fishing continues throughout at the deterministic *MSY* level. Other parameters used were $d = 0.2$, and $M/K = 1$.

a buffer in the form of unfished immature age-classes in the population, and in the time lags between birth and the age at maturity. For each of the three mortality rates illustrated, the age at maturity is more than 1 year, so that immediately after the catastrophic mortality suffered by the spawning stock, the population consists of a number of immature unfished age-classes. The number of such age-classes increases with decreasing values of *M*. One year later, the oldest of these immature age-classes becomes mature and the process of rebuilding the spawning stock starts. This process continues until the annual increment to the spawning stock is roughly balanced by the deaths due to natural mortality and fishing. The spawning stock then recovers at a much slower rate that is consistent with the increase in recruitment flowing from an increased spawning stock implied by the stock–recruitment relationship.

Figure 9.12 shows that there can be additional subtleties in the dynamics of recovery. After the initial period of rapid recovery, the spawning stock falls again for a period. During the initial recovery period, the age-classes annually incrementing the spawning stock result from recruitment typical of the spawning stock prior to the catastrophic mortality. The subsequent decline after the initial recovery is a reflection of newly mature age-classes resulting from recruitment after the catastrophe, and thus from a much diminished spawning stock.

The final point to make is that the exclusion of examples with a high M is deliberate. Under the conditions simulated, any species with M sufficiently high that the age at maturity is less than 1 year will by definition be completely extinguished by the catastrophe hypothesized. It may appear at first glance that this last result is an artefact of an extremely unlikely scenario. While in a strict sense that is true, all that is required to bring about conditions nearly matching those simulated here for short-lived species is for them to have a critical spawning biomass below which recruitment is severely diminished. Then a bout of severe overfishing, for example, could well result in a complete stock collapse.

In contrast, the unfished age-classes in longer lived species provide a buffer against such catastrophic events. It follows that in this sense. longer lived species can show a greater resilience than short-lived ones. Further, the greater the number of unfished age-classes, the greater the amount of buffering. Thus extremely long-lived fish species, such as orange roughy, have the ability to withstand occasional major catastrophic events, despite the very low sustainable yields they provide.

CONCLUSIONS

The primary conclusion that may be drawn from the analyses is that although there is a clear approximately proportional relationship between sustainable yield and mortality (lifespan), the constant of proportionality is smaller than had originally been proposed by Gulland (1971). Yield taken as a percentage of unexploited biomass, even in the case of very strong density dependence (recruitment constant regardless of mature stock size), is more likely to be around $0.3M$, rather than $0.5M$ that Gulland suggested. When more realistic levels of density dependence are considered, the percentage yield is further reduced, perhaps to between $0.1M$ and $0.15M$.

When looked at specifically in terms of lifespans, the non-linear relationship between M and lifespan implies that for longer lived species, the sustainable yields are both low and almost independent of lifespan. An exception to this ground rule is the behaviour of very short-lived species. In these species, the details of the life cycle dominate their response to exploitation, such that on sensible and measurable definitions of exploitable biomass, the stock may be capable of producing sustainable yields well in excess of the biomass measured at the start of the fishing season.

Clearly this analysis sets the scene for more detailed investigations of individual species and the way in which the key parameters that determine yields vary. In this context, it is interesting to note that Pauly (1980) found a dependence between natural mortality, growth in size, and mean water temperature for a wide range of fish stocks. Such a relationship affords the possibility of further assessment of the potential yield of species in ecosystems of different types.

The deterministic results are a useful guide to expectations in the real world, particularly for relatively long-lived species. The investigation of stochasticity in recruitment and its implications for the expected variation in yield complements other findings (e.g. Beddington & May 1977) that a policy of constant effort with a target level of fishing mortality can produce substantial benefits in high yields when recruitment is high. The strength of this effect with increasing mortality (reduced lifespan) is particularly marked.

Very short-lived species do appear to be exceptions to the general rules outlined above; they can provide high yields. However, the brief examination of the effect of catastrophic events on the dynamics does indicate that longer lived species retain a resilience to catastrophes that is not available to the short-lived species, which can be particularly vulnerable to a combination of high exploitation and occasional environmental events that devastate the spawning stock.

ACKNOWLEDGEMENTS

The authors are grateful to Mark Bravington for helpful discussions, and to Dr John Caddy and colleagues at the Fisheries Department of FAO, Rome, for commenting on an earlier draft of this manuscript. This research was funded by a grant from the UK Overseas Development Administration, as part of their Fish Management Science Programme.

REFERENCES

Beddington, J.R. & Cooke, J.G. (1983). The potential yield of fish stocks. FAO Fisheries Technical Paper 242.

Beddington, J.R. & May, R.M. (1977). Harvesting natural populations in a randomly fluctuating environment. *Science*, **197**, 463–465.

Beddington, J.R., Rosenberg, A.A., Crombie, J.A. & Kirkwood, G.P. (1990). Stock assessment and the provision of management advice for the short fin squid fishery in Falkland Islands waters. *Fisheries Research*, **8**, 351–365.

Beverton, R.J.H. & Holt, S.J. (1957). On the dynamics of exploited fish populations. *MAFF Fisheries Investigations London*, Series 2, **19**, 1–533.

Cushing, D.H. (1973). The dependence of recruitment on parent stock. *Journal of the Fisheries Research Board of Canada*, **30**, 1965–1976.

Gulland, J.A. (1971). *The Fish Resources of the Ocean.* Fishing News Books, West Byfleet, Surrey.

Heincke, F. (1913). Investigations on the plaice. General report. 1. The plaice fishery and protective measures. Preliminary brief summary of the most important points of the report. *Conseil International pour l'Exploration de la Mer, Rapports et Procés-Verbaux des Réunions*, **16**, 1–67.

Hennemuth, R.C., Palmer, J.E. & Brown, B.E. (1980). A statistical description of recruitment in eighteen selected fish stocks. *Journal of Northwest Atlantic Fisheries Science*, **1**, 101–111.

Pauly, D. (1980). On the interrelationships between natural mortality, growth parameters, and mean environmental temperature in 175 fish stocks. *Journal du Conseil International pour l'Exploration de la Mer*, **11**, 559–623.

Pauly, D. & Soriano, M.L. (1986). Some practical extensions to Beverton and Holt's relative yield-per-recruit model. *The First Asian Fisheries Forum* (Ed. by J.L. Maclean, L.B. Dizon & L.V. Hosillos), pp. 491–495. Asian Fisheries Society, Manila, Philippines.

Ricker, W.E. (1954). Stock and recruitment. *Journal of the Fisheries Research Board of Canada*, **30**, 1965–1976.

Shepherd, J.G. (1982a). A family of general production curves for exploited populations. *Mathematical Biosciences*, **59**, 77–93.

Shepherd, J.G. (1982b). A versatile new stock–recruitment relationship for fisheries, and the construction of sustainable yield curves. *Journal du Conseil International pour l'Exploration de la Mer*, **40**, 67–75.

von Bertalanffy, L. (1938). A quantitative theory of organic growth. *Human Biology*, **10**, 181–213.

10. BLANKET BOGS IN GREAT BRITAIN: AN ASSESSMENT OF LARGE-SCALE PATTERN AND DISTRIBUTION USING REMOTE SENSING AND GIS

E. REID*, G. N. MORTIMER†, R. A. LINDSAY*
AND D. B. A. THOMPSON*

* Scottish Natural Heritage, Research & Advisory Services Directorate,
2 Anderson Place, Edinburgh EH6 5NP, UK and
† English Nature, Geographic Information Unit, Monkstone House,
M2/5, City Road, Peterborough PE1 1UA, UK

SUMMARY

Blanket peatland, amounting to approximately 1.5 million hectares in Great Britain, presents formidable problems for conservation agencies wishing to establish protection or wise-use strategies for the resource. It has poorly defined boundaries, and often covers rugged remote terrain. It also has variety in scale and appearance ranging from the minutely narrow life zones of *Sphagnum* species, through a myriad of hummock/pool complexes, to vast all-encompassing spectacular landscapes.

This chapter employs a combination of a LANDSAT Thematic Mapper (TM) satellite imagery, air photography and field survey to characterize the bog resource in the southern Pennines of England, Great Britain.

Principal Component Analysis (PCA), applied to TM data, highlights blanket peatlands whilst reducing the information in non-peatlands. This identifies natural habitat groupings, with divisions confirmed using TWINSPAN, from ground data survey.

PCA reduced the original five wave bands to three, accounting for 96% of the variation across the 185 km^2 of the satellite scene. Six of the 20 spectral classes were identified as peat vegetation types with greater than 75% ground truth accuracy. A further six were identified as peatland classes which required additional ground data collection and eight were identified as non-peatland classes such as water, forestry and urban areas.

INTRODUCTION

Peatlands represent one of the largest semi-natural habitats in Great Britain, and contain an archive of human and vegetation history stretching back some 5000 years or more. The range of physical and biological situations under which peat is formed gives rise to a variety of peatland habitats that are broadly classed according to their water input and nutrient regime. In terms of conservation and hydro-ecological processes, the peat-forming environment is classed as fen peat or bog peat. However, in appearance bog peats vary considerably in terms of pool–peat configuration, hummock–hollow formation and community and hydro-habitat mosaics.

Fen peats are subject to ground-water flooding, and occur in basins, valleys and flood plains. They are often waterlogged for only part of the year, and are generally dominated by a mixture of tall sedges and herbs. In contrast, bogs occur where peat accumulation remains saturated throughout the year simply through direct rainfall inputs (Moore & Bellamy 1974). Within the bog habitat, two distinct types can be described; raised bogs and blanket bogs. Raised bogs occur as isolated domes of peat in the lowland areas of Britain, rising above a landscape which is not generally dominated by bog peat, but more usually by agricultural land (Lindsay *et al.* in press). Blanket bogs occur in the cooler, wetter, flatter, upland areas of the British Isles and under these conditions peat formation can occur across all parts of the landscape except the steepest slopes (Ratcliffe 1977).

These distinctions in size and scale between raised bog and blanket bog are critical when the pattern and community composition across bog types are to be surveyed and recorded. Surveyed sites within the government conservation agencies form the basis of inventory work and contribute to the local and national overview, and highlight the range of ecological variation, hence allowing areas of interest to be designated. Areas of raised bog can be identified, boundaries can be drawn and the pattern across these sites described by traditional survey methodologies, because individual sites tend to be quite small in total extent, and even large complexes have distinct boundaries encompassing a relatively limited total area. In contrast, however, the very great extent and ill-defined boundaries of the blanket bog environment in Great Britain (Plate 10.1, facing page 244), which amounts to approximately 13% of the global blanket bog resource (Lindsay *et al.* 1988; Ratcliffe & Thompson 1988), means it is much more difficult in such a landscape to draw boundaries around 'sites' and hence to describe their natural heritage importance within local, national or international contexts. Even more fundamentally, however, the resources required to carry out field survey of this entire habitat make traditional survey an unrealistic option for most, if not all, country agency budgets. Current detailed ecological knowledge of the blanket peat habitat is therefore restricted to a relatively small number of sites which lie within a very much larger and undescribed landscape.

The aim of the National Peatland Resource Inventory (NPRI) is to establish a baseline of peatland information from which it is then possible to operate an environmental audit programme. The blanket peatland resource is so extensive that it was considered to lend itself to remote sensing techniques for survey and characterization. Satellite data and aerial photographs were investigated, and a methodology developed for recording, mapping and then characterizing the extent and composition of the resource. This also utilized small-scale recording methods (e.g. quadrat analysis in the field) to qualify the large-scale pattern.

The use of satellite imagery has not yet become widely accepted as a standard tool for ecologists in the mapping of semi-natural vegetation stands. Emphasis instead has been applied to mapping more general land cover types; for example, the Macaulay Land Use Research Institute Landcover Map of Scotland and the Institute of Terrestrial Ecology Landcover Map of Great Britain (Fuller *et al.* 1990).

In order to map and describe vegetation community pattern, our study has developed a heuristic methodology, working with the available algorithms that bring out the maximum amount of information within the blanket peatland areas whilst significantly reducing the information content of non-peatlands.

METHODS
Study site field survey

The NPRI programme is involved with a number of other satellite scenes in Great Britain but for the purpose of this chapter attention is focused on the area covered by TM, row and path number 203/23, 31 May 1985 (Fig. 10.1) – an area which includes the blanket bogs of the southern Pennines.

The British Geological Survey (BGS) 1:50 000 drift series in Great Britain (Fig. 10.2) was used to identify peat soil sites greater than 1 m in depth across the satellite scene. These were digitized in vector format, with kind permission of the BGS. Air photograph cover at an approximate scale of 1:10 000 for selected sites, initially based on the areas highlighted by Ratcliffe (1977) and the Nature Conservancy Council, was identified and the scale across each photograph calculated (Fig. 10.3). Spectral imagery from the visible wavelengths (wave bands 1, 2 & 3) in scene 203/23 was geometrically registered to the National Grid, and a combination of these three wave bands was produced, as hard copy, at

Fig. 10.1. Study area displaying the approximate extent of satellite image 203/23 across the English and Welsh counties.

Fig. 10.2. British Geological Survey 1:50 000 vector boundaries, showing the blanket peatland areas greater than 1 m in depth across Great Britain.

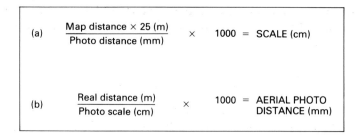

FIG. 10.3. Formula: (a) scale calculation of air photographs and (b) placing of National Grid lines on air photographs.

the same scale as the air photographs. This hard copy was used in the field to collect ground data.

Homogeneous areas of colour and tone – referred to as blocks – were identified on the true colour composite. These were visited on the ground and environmental and vegetation data recorded at that position. Quadrats were recorded using the methodologies adopted for national vegetation surveys, that is, each quadrat contained species information recorded using the 'Domin' scale together with the percentage cover of each vegetation layer (e.g. Ratcliffe 1977; Rodwell 1991). In addition, environmental data such as slope, aspect and modal vegetation height per block were noted. Accurate navigation was achieved with the help of air photographs. The boundaries for each block were digitized in vector format for subsequent use (Plate 10.2, facing p. 244).

Vector and image processing

Areas of non-peat, as defined by the BGS vector data, were physically removed from bands 2–5 and 7 of scene 203/23 using the image processing system. The pixel values from the remaining peat areas were extracted using an algorithm developed by the authors. This was necessary because the image processing system being used was unable to extract pixel values from areas of interest as defined by the BGS peat polygons. A correlation/ covariance matrix was calculated using all of these values, in other words the principal components (PC) were extracted from spectral information for peat areas alone.

PCA resulted in a reduction of the correlation between bands, compression of image variance into fewer dimensions, and hence removal of redundant spectral data (Singh & Harrison 1985; Mortimer 1987). By extracting the spectral information from peat areas alone, PCA highlights the variation in the blanket bog areas and reduces the information in non-peatland areas.

Cluster identification/classification

PC bands 1–3 were used as the base for classification. An unsupervised (ISODATA) algorithm in Intergraph Imagestation Imager was used to identify natural groupings within the PC data. The number of desired clusters within the algorithm was determined from field survey data, and also the National Vegetation Classification (NVC) (Rodwell *et al.* 1991) groups identified within the area covered by scene 203/23. Standard deviation and Euclidean distances were determined after several iterations. Final statistics for each cluster were then used to classify the whole scene using a maximum likelihood algorithm. Examination and merging of classes was carried out by analysing the ground survey descriptions for each class, merging where appropriate, and by looking at the minimum and maximum PCA band pixel values for each class to assess areas of overlap between the classes. Descriptions for each spectral class were drawn and linked to the groups derived from the vegetation analysis.

Vegetation analysis

The 2×2 m randomly located quadrat data from each block were analysed using Two Way Indicator Species Analysis (TWINSPAN) (Hill 1979; Malloch 1988). This data set was divided into a number of homogeneous vegetation groups which could be described in terms of species composition using species given by Rodwell (1991).

Testing

Initial testing involved a desk analysis procedure for the spectral classes in the classification. This established the relationship between the ground data collected and spectral class results.

Subsequent accuracy assessment of peatland classes was based on units of 1 km^2 (referred to as grids), chosen to represent the full range of peat spectral classes but, for simplicity at this stage, to minimize selection of mixed pixels which represent mosaics of those areas which are heterogeneous on the ground. Each 1 km^2 of the classification was plotted at a scale 1:10000. Classes greater than 25 pixels (2.25 ha) were identified on the classification – emphasis was placed on identifying areas that were different from the homogeneous blocks identified in the raw imagery and used to collect the original ground data. These classes were transferred using acetate sheets on to air photographs at the same scale. A 100 × 100 m grid was placed over the 1 km² on the air photograph to allow the area to be thoroughly and easily covered by foot. Squares (100 × 100 m) which contained homogeneous blocks of colour, and which were surrounded by this colour, (i.e. creating a zone of confidence around the test squares) were identified and highlighted for testing. Field testing took place with the aid of Ordnance Survey 1:25 000 maps, vegetation keys and environmental keys; these keys were created from the original ground truth sites. NVC descriptions included additional variables regarded as important from a satellite perspective, such as slope, aspect and vegetation height.

RESULTS

This is a preliminary examination of the results. More detailed publication of the work is planned.

There are several strongly correlated bands within this satellite scene. These correlations were predictable, because the spectral bands recorded by TM were selected largely to highlight the distinctive spectral response of vegetation. Many other variables such as geomorphological features also have distinct spectral response patterns at these wavelengths (Jones 1986). These are accounted for, to a different degree, within each of the bands.

The correlation matrix, resulting from the peat areas alone, reflects the correlation identified across the pixel values in bands 2–5 and 7. This correlation is calculated from the covariance matrix, which is derived from the mean pixel value in a band in relation to the variation of each individual pixel value from this mean. Bands 7, 5 and 3 are highly correlated, as are bands 5, 3 and 2. The eigenvectors (Table 10.1) for the PCs indicate the variance accounted for in the PC band. The PCA

TABLE 10.1. Eigenvectors illustrating the relationship between the satellite bands and three principal components (PC1, PC2, PC3)

	PC1	PC2	PC3
Band 7	0.906	0.358	0.193
Band 5	0.961	0.049	0.239
Band 4	0.801	−0.587	0.104
Band 3	0.954	0.157	−0.224
Band 2	0.95	−0.053	−0.289

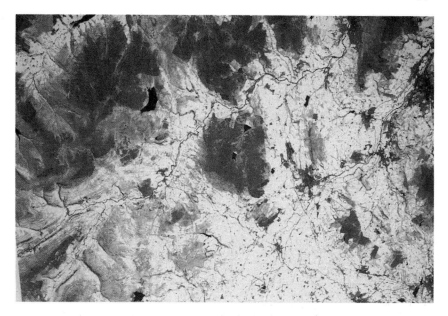

FIG. 10.4. Principal component imagery – highlighting the peatland areas in multiple grey values, while the non-peatland areas show less variation and hence fewer grey values. (Original data © ESA (1985), distributed by Eurimage.)

reduces the original five bands to three bands which between them account for 96% of the image variance. Thus PC1 has variance accounted for in each of the bands, PC2 variance is largely accounted for in band 4, and PC3 in band 2. The variance accounted for by each principal component across the scene shows: PC1 = 83%; PC2 = 10%; and PC3 = 3.1%.

TABLE 10.2. Vegetation types and spectral class link: these peat classes were highlighted as having greater than 75% accuracy when linked to the vegetation type identified through TWINSPAN analysis

Vegetation	NVC type	Spectral class
Eriophorum vaginatum blanket mire (typical)	M20	7
Molinia caerulea dominated community	M20	5
Empetrum nigrum subsp. *nigrum – Vaccinium myrtillus* dominated community	M20	18
Shrub/herb mixed community	M20	16
Calluna vulgaris – Cladonia sub-community	M20b	12
Eriophorum angustifolium bog pools	M3	11

Table 10.3. Constancy table for *Eriophorum vaginatum* blanket mire

NVC:	M20
Variant:	Herb dominated
Spectral class:	7
No. of samples:	133

Species	Constancy
Eriophorum vaginatum	V
Eriophorum angustifolium	V
Empetrum nigrum subsp. *nigrum*	III
Vaccinium myrtillus	III
Calluna vulgaris	II
Rubus chamaemorus	II
Deschampsia flexuosa	II
Molinia caerulea	II
Campylopus pyriformis	II
Campylopus paradoxus	I
Festuca ovina	I
Erica tetralix	I
Vaccinium oxycoccos	I
Polytrichum commune	I
Sphagnum recurvum	I
Drepanocladus fluitans	I
Sphagnum subnitens	I
Lophozia ventricosa	I
Deschampsia setacea	I
Juncus squarrosus	I
Narthecium ossifragum	I
Campylopus atrovirens	I
Campylopus introflexus	I
Pohlia nutans	I
Sphagnum papillosum	I
Marchantia polymorpha	I
Cladonia floerkeana	I
Mean no. of species per quadrat:	4.5

Constancy key: V = species occur in 80–100% of samples; IV = 60–79%; III = 40–59%; II = 20–39%; I = 1–19%.

The PC bands show areas of peatland having grey level values between 0 and 255, whereas the non-peat areas are represented by a very limited number of grey level values, i.e., 250–255 for PC1 (Fig. 10.4).

Analysis of the initial ground-truth data with the classification, before grid testing, identified 20 spectral classes in total. Six have been identified

TABLE 10.4. Constancy table for *Molinia* dominated community

NVC:	M20
Variant:	*Molinia* dominated
Spectral class:	5
No. of samples:	24

Species	Constancy
Molinia caerulea	V
Eriophorum vaginatum	IV
Deschampsia flexuosa	III
Eriophorum angustifolium	III
Calluna vulgaris	I
Vaccinium myrtillus	I
Campylopus pyriformis	I
Empetrum nigrum subsp. *nigrum*	I
Holcus lanatus	I
Sphagnum subnitens	I
Mean no. of species per quadrat:	2.2

Constancy key: V = species occur in 80–100% of samples; IV = 60–79%; III = 40–59%; II = 20–39%; I = 1–19%.

as peat types with greater than 75% ground-truth accuracy, a further six are identified as peat types but require further ground-truth collection, and eight are non-peat classes, for example, forestry, water and agricultural ground (Plate 10.3, facing p. 245).

Examination of the 368 quadrats, environmental data and block descriptions across the scene resulted in six vegetation/environmental classes linked to the spectral classes and three NVC groups (Tables 10.2–10.8). The linkages were created by counting the number of spectral classes found within the TWINSPAN-derived groups, taken to be: *Eriophorum vaginatum* blanket mire and raised mire (M20), *Calluna vulgaris–Cladonia* subcommunity of this (M20b) and *Eriophorum angustifolium* bog pool community (M3).

DISCUSSION

Large-scale, semi-natural vegetation mapping on peatlands and other semi-natural vegetation habitats presents the user of remotely sensed data with a number of problems in interpretation. The reflected solar radiation – which is affected by physiological, ecological and hydrological

TABLE 10.5. Constancy table for *Empetrum* and *Vaccinium* dominated community

NVC:	M20
Variant:	*Empetrum/Vaccinium* dominated
Spectral class:	18
No. of samples:	56

Species	Constancy
Eriophorum vaginatum	V
Eriophorum angustifolium	V
Empetrum nigrum subsp. *nigrum*	V
Vaccinium myrtillus	IV
Calluna vulgaris	II
Rubus chamaemorus	II
Campylopus paradoxus	II
Campylopus pyriformis	I
Deschampsia flexuosa	I
Festuca ovina	I
Lophrozia ventricosa	I
Erica tetralix	I
Juncus squarrosus	I
Narthecium ossifragum	I
Vaccinium oxycoccos	I
Campylopus atrovirens	I
Campylopus introflexus	I
Fissidens adianthoides	I
Hypnum cupressiforme	I
Pohlia nutans	I
Sphagnum fimbriatum	I
Sphagnum papillosum	I
Sphagnum subnitens	I
Cladonia floerkeana	I
Mean no. of species per quadrat:	4.7

Constancy key: V = species occur in 80–100% of samples; IV = 60–79%; III = 40–59%; II = 20–39%; I = 1–19%.

factors, by active management, atmospheric conditions, and by topographical effects – is often heterogeneous and complex across an area, and this makes a simple visual identification of cover-types almost impossible (McMorrow & Hume 1986; Mortimer 1987; Grenon 1989; Fox *et al.* 1992). Attempts have therefore focused on the application of statistical techniques, band rationing (Holben & Justice 1981; Fox *et al.* 1992), supervised and unsupervised classification, and principal component

TABLE 10.6. Constancy table for shrub/herb co-dominant community

NVC:	M20
Variant:	Mixed herb/shrub
Spectral class:	16
No. of samples:	70

Species	Constancy
Eriophorum vaginatum	V
Eriophorum angustifolium	V
Calluna vulgaris	III
Empetrum nigrum subsp. *nigrum*	III
Vaccinium myrtillus	III
Rubus chamaemorus	II
Deschampsia flexuosa	II
Campylopus pyriformis	I
Molinia caerulea	I
Campylopus paradoxus	I
Cladonia chlorophaea	I
Vaccinium oxycoccos	I
Campylopus introflexus	I
Sphagnum papillosum	I
Sphagnum subnitens	I
Juncus effusus	I
Narthecium ossifragum	I
Drepanocladus fluitans	I
Hypnum cupressiforme	I
Lophrozia ventricosa	I
Cladonia bellidiflora	I
Mean no. of species per quadrat:	4.3

Constancy key: V = species occur in 80–100% of samples; IV = 60–79%; III = 40–59%; II = 20–39%; I = 1–19%.

analysis to identify and reduce this spectral heterogeneity (Belward *et al.* 1990). Equally, accepting the limitations of per-pixel classifiers' research has concentrated on extracting the proportions of cover types contributing to the spectral response from a single pixel. This technique is known as spectral mixture modelling; however, the software to apply this is not widely available (Foody & Cox 1991). Blanket peatlands that occur in the flatter areas of the uplands are therefore, by their nature, more suited to the application of satellite image processing than are most other semi-natural habitats, although habitats which are less rugged and steep,

TABLE 10.7. Constancy table for *Calluna* dominated community

NVC:	M20b
Variant:	*Calluna* dominated
Spectral class:	12
No. of samples:	75

Species	Constancy
Calluna vulgaris	V
Eriophorum angustifolium	V
Eriophorum vaginatum	V
Vaccinium myrtillus	III
Empetrum nigrum subsp. *nigrum*	II
Campylopus pyriformis	II
Campylopus paradoxus	I
Cladonia chlorophaea	I
Deschampsia flexuosa	I
Pohlia nutans	I
Molinia caerulea	I
Campylopus introflexus	I
Lophozia ventricosa	I
Cladonia floerkeana	I
Betula seedling	I
Carex nigra	I
Juncus effusus	I
Juncus squarrosus	I
Nartheciun ossifragum	I
Campylopus atrovirens	I
Dicranum scoparium	I
Hypnum cupressiforme	I
Sphagnum fimbriatum	I
Sphagnum papillosum	I
Sphagnum subnitens	I
Cladonia bellidiflora	I
Mean no. of species per quadrat:	4.5

Constancy key: V = species occurs in 80–100% of samples; IV = 60–79%; III = 40–59%; II = 20–39%; I = 1–19%.

such as estuaries, salt marshes, lakes, flood plains, mountain plateaux, and woodlands may also be characterized by such techniques.

The heuristic methodology adopted here, by first utilizing the satellite image to guide the ecologist and subsequently building on this knowledge for a more refined classification, has proved, so far, to be highly successful. This first principle is most valuable when used in conjunction with the

TABLE 10.8. Constancy table for *Eriophorum angustifolium* dominated communities

NVC:	M3
Variant:	Bog pools
Spectral class:	11
No. of samples:	10

Species	Constancy
Eriophorum angustifolium	V
Rubus chamaemorus	I
Vaccinium myrtillus	I
Sphagnum cuspidatum	I
Calluna vulgaris	I
Empetrum nigrum subsp. *nigrum*	I
Molinia caerulea	I
Vaccinium oxycoccos	I
Campylopus pyriformis	I
Polytrichum commune	I
Mean no. of species per quadrat:	2.2

Constancy key: V = species occur in 80–100% of samples; IV = 60–79%; III = 40–59%; II = 20–39%; I = 1–19%.

BGS digital data to highlight the areas of importance and to reduce any confusion which traditionally has been recorded and characterized by previous studies (Palylyk & Crown 1984; Mortimer 1987; Fox *et al.* 1991; Federal Geographic Data Committee 1992). It has also highlighted the importance of spectral homogeneity in the design of a rationalized and innovative approach to field techniques.

By utilizing small-scale recording methodologies, such as quadrat analysis, micro-topographical identification, NVC keying and simple environmental parameters together with satellite imagery, it is possible with the aid of GIS and image processing to quantify, for the first time, the entire extent of community types within the blanket peat resource. However, it should be noted that detailed site-based survey will always prove more valuable when assessing specific characteristics of a 'site', because the remotely sensed data restrict the resolution of what it is possible to identify to a pixel of 30 × 30 m.

A further paper will describe the technical analysis of the variance accounted for in each of the principal components, as well as describing the resulting accuracy assessment from the grid field testing technique. This will form the basis for the production of a baseline map of the blanket peatland communities of the southern Pennines and, as such, will represent the launching pad for a longer term programme of environmental audit. The original challenge was that of finding a means of mapping an immense semi-natural resource. The result has been the development of a technique which not only maps the resource, but offers the potential for repeated survey on a time scale which is so significantly short as to make regular environmental audit a realistic option.

In terms of the scale of detail possible with this technique, it should first be noted that the vegetation represented by the 368 ground-truth quadrats is almost entirely NVC *Eriophorum vaginatum* blanket mire (M20). The use here of satellite imagery reveals that a finer distinction, than those described by the NVC subcommunities, (Rodwell 1991) is possible. In this study, M20 can be subdivided into *Molinia caerulea* dominated herb community, *Empetrum nigrum* subsp. *nigrum* and *Vaccinium myrtillus* dominated community, and a shrub/herb mixed community (shrubs are a mixture of *Vaccinium myrtillus*, *Calluna vulgaris*, *Empetrum nigrum* subsp. *nigrum*, 'herbs' are a mixture of *Eriophorum vaginatum* and *Eriophorum angustifolium*).

This refined classification, describing the large-scale pattern across 185 km^2 of the satellite image, is of considerable use in the conservation and management of the blanket peatland resource. It provides a means of quantifying resource quality, and of analysing and monitoring widespread but often diffuse and poorly documented impacts such as heavy grazing, fire, drainage, recreation, and afforestation.

Once a broad picture of the quality and quantity of the resource has been generated, this can assist in the adoption of strategic approaches to planning and management by local, national and international colleagues, for Britain's internationally important blanket bog systems. In addition, this picture may also contribute to the detection of even larger scale dynamic aspects of environmental change such as global warming, and perhaps in doing so will help to promote the ideal of sustainability.

ACKNOWLEDGEMENTS

This chapter describes the work initiated under the Chief Scientist Directorate of the Nature Conservancy Council. It is now managed by the Uplands and Peatlands Branch of the Research and Advisory Services Directorate of Scottish Natural Heritage. The GIS/image analysis was mainly undertaken in English Nature. We express our gratitude to Dr J. Budd and Dr J. Hellawell for their patience and understanding, and Marcus Poley, for his time in commenting on the drafts. We would

PLATE 10.1. Blanket peatland stretching across a vast landscape, surrounding the smaller units of agricultural land.

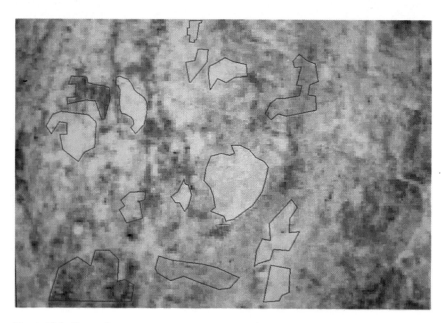

PLATE 10.2. True colour contrast stretched satellite image, showing the vectorized ground collection sites. (Original data © ESA (1985), distributed by Eurimage.)

[facing page 244]

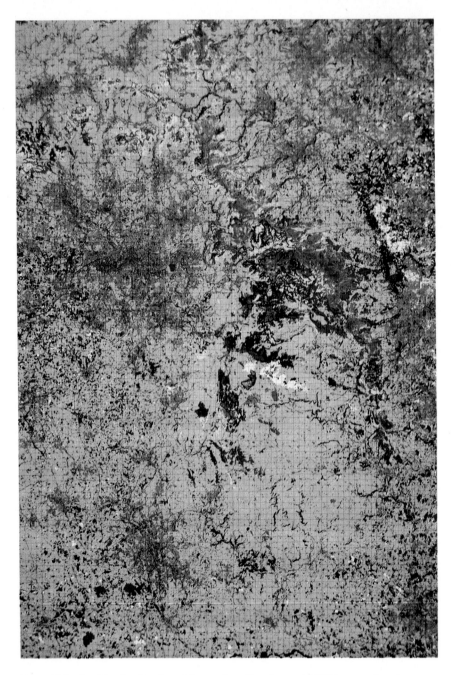

PLATE 10.3. Satellite classification of the southern Pennines, with BGS peat soil vector data. The peatland areas are clearly highlighted in purple and pink, agricultural land and the cities are in orange and blue (scale 1:200000). (Original data © ESA (1985), distributed by Eurimage.)

especially like to thank present and past members of the NPRI team: Sarah Ross, Stuart Gillies, Penny Mayer, Fiona Everingham, Ylva Fanden-Lilja, Sarah Haynes and Sue Rogers. Many thanks also to the English Nature Bakewell Office for their help in field preparation. We acknowledge ESA and Eurimage for the original satellite data and British Geological Survey for peatland data.

REFERENCES

Belward, A.S., Taylor, J.C., Stuttard, M.J., Bignal, E., Mathews, J. & Curtis, D. (1990). An unsupervised approach in the classification of semi-natural vegetation from Landsat Thematic Mapper data – A pilot study on Islay. *International Journal of Remote Sensing*, **11**(3), 429–455.

Federal Geographic Data Committee (1992). Application of satellite data for mapping and monitoring wetlands – fact finding report No1. US Geological Survey, Wetlands subcommittee, 590 National Center, Reston, Virginia.

Foody, G.M. & Cox, D.P. (1991). Estimation of sub-pixel land cover composition from spectral mixture models. Proceedings of a symposium. Spatial Data 2000 conference, Christ Church, Oxford University, September 1991.

Fox, D., Hindley, D. & Power, C. (1992). The determination of areas of peat in the lowland wetlands of North West England using satellite imagery – WP106. National Remote Sensing Centre Limited, Farnborough, Hampshire.

Fuller, R.M. & Parsell, R.J. (1990). Classification of TM imagery in the study of land-use in lowland Britain: Practical considerations for operational use. *International Journal of Remote Sensing*, **11**, 1901–1917.

Grenon, A. (1989). *Peatlands and Remote Sensing – Inventory of peatlands in Quebec*. Service de la cartographie centre, Quebec de co-ordination de la teledetection, Ministrie de L'énergie et des Resources, Sector Tries, Sainte-Foy, Quebec.

Hill, M.O. (1979). *TWINSPAN – a FORTAN program for arranging multivariate data in an ordered two-way table by classification of the individuals and attributes*. Cornell University, Ecology and Systematics.

Holben, B. & Justice, C. (1981). An examination of spectral band rationing to reduce the topographical effect on remotely sensed data. *International Journal of Remote Sensing*, **12**, 115–133.

Jones, A.R. (1986). The use of Thematic Mapper imagery for geomorphological mapping in arid and semi-arid environments. Symposium on Remote Sensing for Resource Development and Environmental Management. ITC, Enschede, Netherlands.

Lindsay, R.A. & Andrews, J. (1993). *Lowland Raised Bogs in Great Britain*. Scottish Natural Heritage and the Joint Nature Conservation Committee (in press).

Lindsay, R.A., Charman, D.J., Everingham, F., O'Reilly, R.M., Palmer, M.A., Rowell, T.A. & Stroud, D.A. (1988). *The Flow Country – The Peatlands of Caithness and Sutherland*. Nature Conservancy Council, Peterborough.

Malloch, A.J.C. (1988). Vespan II – A computer package to handle and analyse multivariate species data and handle and display species distribution data. Institute of Environmental and Biological Sciences, University of Lancaster, Lancaster.

Manley, B.F.J. (1986). *Multivariate Statistical Methods – A Primer*. Chapman & Hall, London.

McMorrow, J. & Hume, E. (1986). Problems of applying multispectral classification to upland vegetation. *Proceedings of a Symposium held by Commission IV of the International Society for Photogrammetry and Remote Sensing (ISPRS) and the Remote Sensing Society (RSS)*, Edinburgh, 1986, pp. 610–620.

Moore, P.D. & Bellamy, D.J. (1974). *Peatlands.* Paul Elek Scientific Books, London.

Mortimer, G.N. (1987). *The use of Landsat Thematic Mapper imagery for monitoring blanket peat erosion in the Southern Pennines.* MSc Thesis, Cranfield Institute of Technology–Silsoe College, Bedford.

Palylyk, C.L. & Crown, P.H. (1984). Application of clustering to Landsat MSS Digital data for Peatland Inventory. *Canadian Journal of Remote Sensing*, **10(2)**, 201–209.

Ratcliffe, D.A. (1977). *A Nature Conservation Review.* Cambridge University Press, Cambridge.

Ratcliffe, D.A. & Thompson, D.B.A. (1988). *The British uplands: their ecological character and international significance.* In: *Ecological Change in the Uplands* (Ed. by M.B. Usher & D.B.A. Thompson), pp. 9–36. Special Publication Series of the British Ecological Society, No. 7, Blackwell Scientific Publications, Oxford.

Rodwell, J.S., Pigott, C.D., Ratcliffe, D.A., Malloch, A.J.C., Birks, H.J.B., Proctor, M.C.F., Shimwell, D.W., Huntley, J.P., Radford, E., Wigginton, M.J. & Wilkins, P. (1991). *British Plant Communities. Vol. 2. Mires and Heaths.* Cambridge University Press, Cambridge.

Singh, A. & Harrison, A. (1985). Standardized principal components. *International Journal of Remote Sensing*, **6**, 883–896.

11. TSETSE DISTRIBUTION IN AFRICA: SEEING THE WOOD *AND* THE TREES

DAVID J. ROGERS* AND BRIAN G. WILLIAMS†

* Department of Zoology, South Parks Road, Oxford OX1 3PS, UK and
† THEU, London School of Hygiene and Tropical Medicine,
Keppel Street, London WC1E 7HT, UK

SUMMARY

After an introduction outlining the differences between the biological and statistical approaches to understanding the distribution and abundance of organisms, this chapter gives two examples of dimension-reducing statistical techniques whereby large amounts of environmental data can be processed and sifted to extract useful correlates of the distributional ranges of animal species. These techniques are illustrated using the tsetse fly, *Glossina morsitans*, as an example.

The first technique is that of linear discriminant analysis which predicts the past and present distribution of *G. morsitans* in Zimbabwe, Kenya and Tanzania with an accuracy of >80%. Conclusions from the statistical analysis coincide with previous biological interpretations of the distribution of this species in Africa. The message to emerge from the analysis is that Global Circulation Models (GCMs) will need to achieve a greater degree of accuracy than at present if they are to be useful in making predictions about changing vector distributions with global climate change.

The second technique is temporal Fourier analysis of a series of Normalized Difference Vegetation Indices (NDVIs) of Africa derived from the Advanced Very High Resolution Radiometers (AVHRR) of Earth-orbiting meteorological satellites of the National Oceanic and Atmospheric Administration (NOAA) series. The analysis captures the important characteristics (i.e. the average, amplitude and phase) of the major annual and biannual cycles of vegetation growth. Examples are given of features (the Gezira irrigation project in Sudan) and processes (the timing of the peak vegetation growth along the Nile from Uganda to the Mediterranean) which are revealed clearly by Fourier analysis. A strong association is demonstrated between the amplitude of the first term of the Fourier expansion (= the amplitude of the annual cycle of vegetation growth) and savannah woodland areas of Africa, and a similar close association is shown between the same features of the analysis and the areas infested with the tsetse *G. morsitans*.

Both discriminant analysis and Fourier analysis achieve dimension-reduction without the obfuscation of the underlying biological processes that is often associated with statistical processing of biological data (e.g. as with principal components analysis). Finer resolution of biologically important features may be possible in both space and time using satellite sensor information at a finer spatial resolution.

INTRODUCTION

The distribution and abundance of organisms can be studied in two ways. The first involves a biological approach, in which demographic rates are measured and related to obvious biotic and abiotic factors that might determine or influence them. The aim is to construct models based on the biology of the organism that describe changes in the population over time, and these models can be used to investigate the impact of proposed interventions on population size. While this approach is sound, it requires data for several generations of the study species before any sensible analysis of species' dynamics can be carried out. Resources are often limited and, as a result, such studies cover restricted areas. If the resulting biological models are applied to much larger areas, it must then be assumed that the relationships of the organisms to their abiotic and biotic environments are the same throughout their range, an assumption that is rarely tested and seldom justified.

The second approach to studying the distribution and abundance of organisms is based on a statistical analysis of the relationship between population data (presence or absence of a species, or records of its abundance) and environmental factors that are often measured for other purposes. These factors are generally abiotic (meteorological variables) or environmental (vegetation), and rarely include any measure of the distribution and abundance of natural enemies, parasites or predators. The statistical approach requires extensive data sets that encompass a wide range of environmental conditions, some suitable and some unsuitable for the study species. In the analysis one tries to define a set of statistically sound rules for predicting the presence or absence of species and their spatial abundance. Problems with the statistical approach often hinge upon the non-linear response of the predicted variable to the set of predictors (Ter Braak & Prentice 1988; Hill 1991). More sophisticated mathematical techniques can be used to overcome these problems, but are generally very difficult to interpret biologically.

Whilst the biological approach can investigate experimentally the links between demographic variables and the species' environment, the statistical approach relies on correlations only. Nevertheless, in an ideal world, the two approaches should lead to the same conclusions and should identify the same limiting factors. Though past attempts to unite these two approaches have had limited success, there are several reasons to be optimistic about the future. First, the increased interest in global change has highlighted the biologists' lack of knowledge of the critical determinants of the distribution and abundance of many key species (either those threatened with extinction, or those, such as pests or vectors, whose increasing ranges might increase the spread of crop failures or diseases). Second, an increasing number of ground-based environmental data sets that can be used for statistical analyses are becoming available. Third, there is now a substantial archive of satellite sensor data, a major advantage of which is its extensive and uniform coverage of large areas of the tropics where other, ground-based information is often patchily or erratically recorded. Finally, the development of Geographical Information Systems provides a means of storing and processing spatial data in ways hitherto unavailable to the general biological community.

Unfortunately many of these opportunities come with costs, in particular those of storing the large amounts of data collected, and of detecting important patterns within the data sets, the latter becoming more difficult as the volume of data increases. The risk is that we will soon be overwhelmed with data which, in their raw, unprocessed form, are worthless. The immediate need, therefore, is to develop analytical techniques that can be used as a filter to find the useful information in the available data sets and to present them in a way that makes biological sense. What we require is mathematical simplification without biological corruption. In this chapter we describe two techniques for extracting significant information from large data sets by finding a small number of linear combinations of the original variables that contain most of the relevant information. Such 'dimension reduction' is used to extract biologically useful information from environmental and satellite data sets relevant to tsetse flies, *Glossina* spp., in Africa.

BIOLOGICAL BACKGROUND

There are 22 species of tsetse, a genus currently restricted to Africa. Both sexes of all species live only on vertebrate blood. Three ecological groups of flies are recognized; the forest-dwelling *fusca* group, the forest and riverine *palpalis* group, and the savannah *morsitans* group. The life cycle of the fly is relatively straightforward. Approximately every 9 or 10 days mature female flies produce fully grown larvae viviparously and these burrow into the soil and pupate within a few minutes of larviposition. Three or more weeks later the teneral adults emerge from the soil and the life cycle continues (Buxton 1955).

The diseases transmitted by tsetse are caused by trypanosomes, flagellate Protozoa in the genera *Trypanosoma*, *Nannomonas* and *Duttonella* (Hoare 1972). The host-range of most trypanosome species is quite wide (Molyneux & Ashford 1983). Species in all three trypanosome groups infect domestic animals, with variable consequences for the hosts, but only two subspecies of *T. brucei*, *T.b. gambiense* and *T.b. rhodesiense*, affect humans, causing the generally endemic and milder West African ('Gambian') and the generally epidemic and more acute East African ('Rhodesian') sleeping sickness respectively. Animal trypanosomiasis is more widespread than human trypanosomiasis in Africa and, because most large vertebrates (including man) are more abundant in the savannah areas of Africa than elsewhere, the *morsitans* and *palpalis* groups of flies are of greater economic importance than is the *fusca* group.

ANALYSING THE DISTRIBUTION OF TSETSE

Tsetse survival in the laboratory is related to the humidity and temperature at which the flies are kept (Buxton & Lewis 1934) and fly mortality rates in the field are correlated with saturation deficit and temperature (Rogers & Randolph 1985). The biological studies leading to these conclusions unfortunately have not been carried out over large areas so that the predictions of the pan-African distributional range of tsetse arising from them (Rogers 1979; Rogers & Randolph 1986) must be treated with caution. Arising from such studies are data sets for fly mortality rates (or correlates of these rates, such as the physical size of flies) and relative population estimates in both space and time and these data sets have recently been correlated with satellite data derived from the National Oceanic and Atmospheric Administration (NOAA) series of meteorological satellites (Rogers & Randolph 1991). The NOAA data were processed to give Normalized Difference Vegetation Indices (NDVIs), calculated from two of the five channels of the satellite's Advanced Very High Resolution Radiometer (AVHRR),

$$NDVI = (Ch_2 - Ch_1)/(Ch_2 + Ch_1),$$

where Ch_1 = AVHRR Channel 1 reading (= radiance at 580–680 nm wavelength, visible red) and Ch_2 = AVHRR Channel 2 reading (=

radiance at 725–1100 nm wavelength, near infra-red). AVHRR data, which have many environmental applications (Prince *et al.* 1990), are now routinely processed to produce NDVIs, which are the most widely available of a number of vegetation indices (Jackson & Huete 1991). The NDVI is directly related to the photosynthetic activity of plants (Tucker & Sellers 1986) and has also been correlated with vegetation type on a continental scale (Tucker *et al.* 1985), vegetation biomass (Huete & Jackson 1987) and seasonal crop production (Bartholomé 1988). The spatial resolution of the AVHRR sensors is at best 1.1 km, but when the data are stored on-board the satellites the limited tape storage facilities reduce this to *c.* 4 km. Later image registration to a map co-ordinate system results in NDVI picture elements ('pixels') with a spatial resolution of *c.* 8 km, considerably poorer than the 10 m spatial resolution of the SPOT satellite's panchromatic sensor or the 30 m spatial resolution of the LANDSAT's Thematic Mapper (TM). However, AVHRR data are available on a daily basis from each of two satellites (Huh 1991), whilst the higher spatial resolution data are available only every 16 days from LANDSAT or nominally every 26 days from SPOT (the receptors of which can be directed from the ground to review particular sites up to seven times during this interval) (Hugh-Jones 1989; Cracknell & Hayes 1991). The high temporal frequency of AVHRR data allows several images taken in a short period of time to be combined in order to eliminate clouds from the final image product. Clouds have very low NDVI values and automated cloud removal is usually achieved by selecting the maximum NDVI value for each site in each 10-day period ('maximum value compositing').

Our current view is that NDVIs effectively integrate a variety of environmental factors that are important for tsetse fly survival. For example, vegetation growth is related to both rainfall and temperature, and active vegetation provides conditions of relatively high humidity in which flies thrive. Satellite imagery therefore provides a link between the intensive biological approach and the extensive statistical approach, providing both an interpretation of the biological result and a data layer for use in the statistical analyses.

In the absence of extensive biological information on tsetse, alternative statistical techniques have been used to 'fit' tsetse distributions to environmental data sets derived from ground-based meteorological recordings and from satellite sensors (Rogers & Randolph 1993; Rogers & Williams 1993; Gareth Staton, personal communication). The meteorological records are interpolated to calculate values of meteorological variables within cells of a geographical data base and the satellite sensors

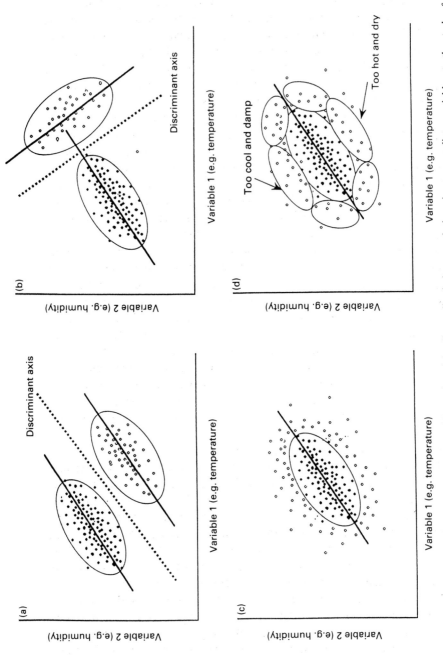

FIG. 11.1. Diagrammatic representation of the problems of applying linear discriminant analysis, using two predictor variables, to the study of animal distributions. (a) Areas of presence and absence have the same variances and co-variances and the discriminant axis easily separates them. (b) and (c) show two complications that arise when variances and covariances are not equal. It may still be possible to use the same procedures of discriminant analysis (as in (b)) or modifications of it (as in (d)). From Rogers & Randolph (1993).

provide further data layers for the analysis. Of the techniques tested we have found that linear discriminant analysis, whilst not the most accurate (Gareth Staton, personal communication), provides considerably more biological insight than many of the alternatives. The theory of this technique is shown diagramatically in Fig. 11.1. Using discriminant analysis a new observation is assigned to one of several categories that have previously been characterized in terms of means, variances and co-variances of a number of predictor variables which previous experience suggests may be important in discriminating between the categories. In the examples in the literature (e.g. Marriott 1974) the categories are classes such as species or species' groups and the discriminating variables are measurements of different parts of the organisms concerned. In the present case there are only two categories, presence and absence of tsetse (there are very few extensive data sets containing information on fly abundance), and the predictors are environmental variables such as temperature, rainfall, etc. Each category is characterized by a group mean in the n-dimensional space of the predictor variables and the way in which the sample points cluster around this group mean is described by the covariance matrix of the predictor variables. Given a new point in the n-dimensional space, it is possible to determine the probability with which it belongs to each group. This is analagous to the univariate case where both the difference from the group mean and the standard deviation around the group mean are used to determine in which percentile an observation lies. In the multivariate case co-variances between the predictor variables are also important in determining category characteristics and therefore assignment rules, since an observation may be well within the cluster of points around one group mean, but physically closer (in n-dimensional space) to another group mean.

One of the assumptions of discriminant analysis is that the variances and co-variances of the predictor variables are the same around each of the group means (as in Fig. 11.1a). Observations from all groups are combined to estimate the within-group co-variance matrix of the predictor variables, and this is used to determine the probability that an unclassified/new point belongs to each group in turn (Green 1978). There are reasons to believe that this assumption may not be strictly valid for distributional data (since species presumably select, or are selected by, a rather well-defined and non-random subset of environmental conditions), but discriminant analysis is relatively robust to violations of the initial assumptions (Marriott 1974). Alternatively assignment can be made on the basis of the co-variance matrix of each category separately (Tatsuoka 1971).

The output of discriminant analysis for a single observation is a set of simple probabilities of belonging to each group in turn, and all probabilities within the set sum to 1.0 (i.e. it is assumed that the observation is drawn from the populations that gave the co-variance matrix or matrices used to make the prediction). Probabilities of presence are worked out for each site in turn (i.e. each cell within the geographical data base), using that site's set of predictor variables, and a map of the result is produced. (Two-group discrimination can be treated as a linear regression between the distributional data, coded 0 for absence and 1 for presence, using the same set of predictor variables. The resulting predictions of group membership are identical.)

In the present examples, the within-group co-variance matrix was calculated for a subsample of points (a 'training set') chosen randomly from the observed data set. Whilst a record of fly presence is indisputable evidence of the (at least temporary) suitability of that site for tsetse, a record of fly absence may arise because of the genuine unsuitability of that site, or because the site has not been surveyed. These unsurveyed sites, if they contain tsetse, will incorrectly contribute to the group of 'tsetse absence' and there will be a tendency to underestimate areas of fly presence. If possible, therefore, only sites known to have been surveyed should be included in the training set. In the absence of good 'tsetse absence' data we might resort to determining the tsetse's environmental envelope using packages such as BIOCLIM (Nix 1986).

Table 11.1 and Fig. 11.2 show the results of applying linear discriminant analysis to the distribution of the tsetse *Glossina morsitans* in Zimbabwe, Kenya and Tanzania. Table 11.1 includes the rank order of the variables as judged by the Mahalanobis distances (Marriott 1974), and determined in a step-wise fashion by first choosing the single variable that gave the greatest separation in multivariate space between sites of tsetse presence and absence (i.e. the highest squared Mahalanobis distance) then selecting a second variable, using the same criterion, to add to the first, and so on. The map of fly distribution in Kenya and Tanzania was compiled by Ford and Katondo (1977), but that for Zimbabwe is based on records prior to game elimination by European colonizers, the rinderpest panzootic at the end of the last century, and the activities of the Tsetse Control Division of the Department of Veterinary Services, Zimbabwe, in the present century, each of which has contributed to the eradication of flies from areas they previously inhabited (Ford 1971).

Several conclusions can be drawn from these analyses (Rogers & Randolph 1993). First the averages of the key predictor variables may differ by rather small amounts between areas of fly presence and absence.

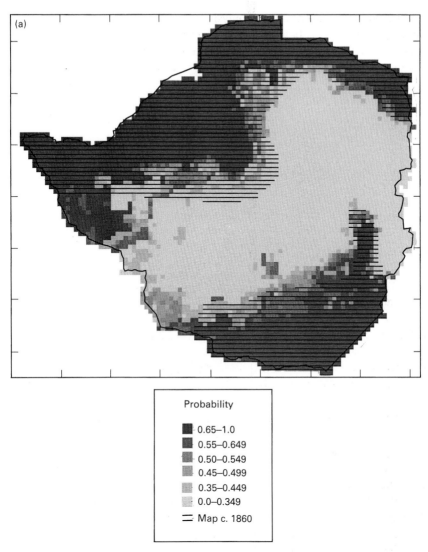

FIG. 11.2. Predicted distributions of *G. morsitans* in (a) Zimbabwe and (b) Kenya and Tanzania based on linear discriminant analysis using a subset of the observed distributions (also shown). The fly map for Zimbabwe (from Ford 1971) pre-dates a rinderpest panzootic that killed many vertebrate hosts of tsetse at the end of the last century, after which flies disappeared permanently from much of their previous range. The fly map for Kenya and Tanzania is the current distribution for this species, from Ford & Katondo (1977). Predicted distributions are on the probability scale shown in the figure. From Rogers & Williams (1993). (a) is drawn in the Plate Carrée projection and (b) in the Hammer–Aitoff projection (Snyder 1987).

Continued on p. 256

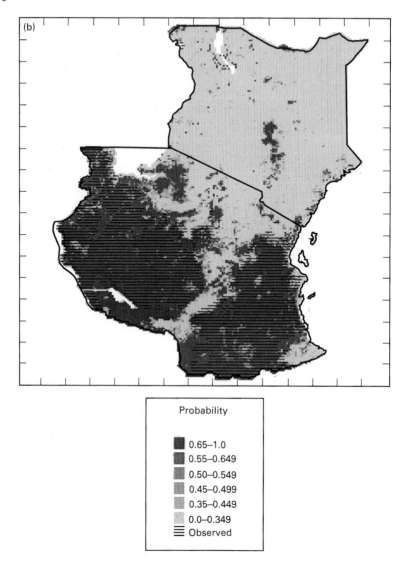

FIG. 11.2. *Continued.*

For example the average difference between *Txmm*, the maximum of the mean monthly temperature (the most important predictor), in areas of fly presence and absence in Zimbabwe is about 3°C whilst for the four key temperature variables for the same species in Kenya and Tanzania it is less than 0.4°C (details in Table 11.1). Making any predictions of changing

TABLE 11.1. Average values of the predictor variables used in the discriminant analysis of the distribution of *Glossina morsitans* in Zimbabwe, Kenya and Tanzania

	Elevation*	NDmean	NDmin	NDmax	NDran	Txmx	Txmm	Tmmm	Tnmm	Tnmm
Zimbabwe										
Abs†	1097.1	0.29	0.19	0.38	0.19	30.51	23.38	19.98	6.0	14.66
Pres	790.0	0.29	0.16	0.40	0.24	33.09	26.32	22.23	8.36	16.94
Rank	5	10	3	4	8	9	1	6	7	2
% correct	84	85	83	85	86	85	82	86	86	84
Kenya and Tanzania										
Abs	953.7	0.24	0.02	0.48	0.46	30.78	24.63	22.72	15.49	20.95
Pres	963.4	0.44	0.1	0.7	0.6	30.47	24.56	22.58	14.59	20.53
Rank	4	7	10	1	8	5	6	2	3	9
% correct	78	83	84	69	84	81	82	75	77	84

* Variables (in order) are elevation (metres), Normalized Difference Vegetation Index (NDVI), monthly mean (NDmean), minimum (NDmin), maximum (NDmax) and range (NDran), maximum of the monthly maximum temperature °C (Txmx), maximum of the monthly mean temperature (Txmm), mean of the monthly mean temperature (Tmmm), minimum of the monthly minimum temperature (Tnmm) and minimum of the monthly mean temperature (Tnmm).

† Abbreviations: Abs, sites where the vectors are absent; Pres, sites where the vectors are present; Rank refers to the order of importance of each variable (as determined by the analysis) for the production of the predicted distribution maps; % correct, the percentage of sites where the analysis makes a correct prediction of presence or absence when all variables of this and all higher ranks are included.

vector distributions under different scenarios of global change will therefore require a degree of accuracy of Global Circulation Models (GCMs)
that few presently possess. Second, the analysis suggests that the major
limiting variable for a species may change from place to place (for tsetse
it is temperature in Zimbabwe but Vegetation Index in Kenya and
Tanzania). It may be important, therefore, to carry out separate analyses
for different parts of a species' range to check for changes in the order of
importance of the variables included in the analysis. Third, at the very
edge of the range of tsetse, as in the case of Zimbabwe which is near the
southern continental limits of tsetse in Africa, a single predictor variable
may be sufficient to describe fly distribution, whilst throughout the continental range of flies several variables may play an important role. In
Zimbabwe, adding all the other variables to *Txmm* improves the fit from
82% to only 85% correct. In Kenya and Tanzania adding temperature
and elevation to *NDmax* (the maximum NDVI, the most important
predictor for these two countries) improves the overall fit from 69%
correct to 84% correct. It is likely that different variables are relatively
more important in different areas of these two countries. Fourth, the
rather small proportion of false negatives, an incorrect prediction of
absence, to false positive results, an incorrect prediction of presence,
suggests that whilst the analysis has correctly identified the major environmental constraints, the present tsetse distribution maps may underestimate the actual distribution of vectors. (However, the smaller
proportion of false negatives to positives may arise from the fact that a
smaller proportion of Kenya and Tanzania is inhabited by the flies rather
than uninhabited by them.) False positive areas should be targeted by
survey services since they may reveal the presence of vectors at low
density. At the very least they represent 'ecological corridors' along
which tsetse could move into new areas. Fifth, the statistical procedure
of linear discriminant analysis and the potential biological significance of
the results are transparent to the user in ways that 'black-box' techniques
for analysing vector distributions are not. Some of these other techniques
involve fitting a large number of arbitrary parameters. The final values of
the parameters are chosen to give the best fit, but they may not, in the
end, give a significantly better fit to vector distributions than does linear
discriminant analysis (Rogers & Randolph 1993).

In this initial analysis those variables that are most useful in making
statistical predictions about the distribution of tsetse also make biological
sense. It has long been known that Zimbabwe represents the cold-
temperature limits of tsetse, and fly distribution limits very often follow
elevation contours. However, many of the temperature variables are

highly correlated with each other, and the one that appears most important may be acting as a surrogate for another temperature variable. In Zimbabwe, for example, both *Txmm* and *Tnmm*, the maximum and minimum of the mean monthly temperature respectively, are higher in tsetse than in non-tsetse areas; *Tnmm* is less variable than *Txmm* and so statistically may be a slightly less accurate predictor.

Future developments of this technique will be used to compare linear discriminant analysis with non-linear approaches to predicting tsetse distributions, treading a delicate line between statistical accuracy and biological realism. What we wish to avoid is a statistically perfect prediction of a fly distribution map that we know is inaccurate. Applied carefully, these new techniques should continue to throw light on the environmental conditions on which tsetse depend.

SATELLITE SENSOR IMAGERY AND VEGETATION CLASSIFICATION

NDVIs provided several of the data layers used in the discriminant analysis of tsetse (maximum, minimum, average and range of NDVIs for each site), and are clearly of at least regional importance. In the past the information content of temporal sequences of NDVIs has been extracted using techniques such as principal component analysis (Tucker *et al.* 1985; Townshend & Justice 1986). As in other uses of this approach (e.g. Jeffers 1978) it is quite difficult to interpret biologically the principal axes obtained. The first principal axis contains more than 80% of the information in the 12 monthly NDVIs for an average year (D.J. Rogers, unpublished) and is strongly correlated with the annual average NDVI, whilst the second axis appears to be related to seasonal changes in the vegetation index (Tucker *et al.* 1985). In an attempt to extract biologically useful information from sequences of NDVIs, temporal Fourier analysis (Chatfield 1980) was performed on the 36 monthly images for the years 1987–89 (spatial Fourier analysis is often used in image processing packages to filter out periodic noise in the images arising from faults in the satellite sensors (Cracknell & Hayes 1991), but has previously been used, as here, for temporal analysis of NDVIs by Menenti *et al.* 1991). Fourier analysis describes the seasonal NDVI as the sum of sinusoidal components with frequencies of one to six cycles per year. The first term in the Fourier expansion (hereafter the 'first component') gives the best fit of a sinusoidal wave to the annual cycle of vegetation growth, the second term (the 'second component') gives the biannual cycle, and so on. The analysis gives both the phase and the amplitude of each term,

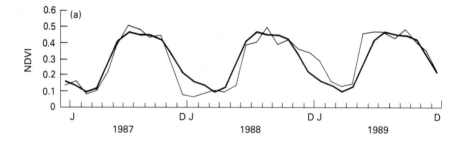

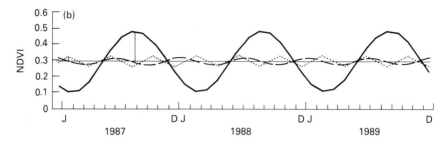

Fig. 11.3. An example of temporal Fourier analysis of the Normalized Difference Vegetation Index from a site in southern Sudan, east of Jonglei on the White Nile (31.6°E, 7°N), for the years 1987–89. The upper graph shows the observed NDVI (thin line) and the fitted Fourier curve, the sum of the first three terms ('components') of the Fourier expansion (thick line). Details of these components are shown in the lower graph. The first component (the solid line), with a frequency of one cycle per year, has a much larger amplitude than the second (dashed line) or third (dotted line), with frequencies of two and three cycles per year respectively, and so makes a major contribution to the overall fit (all three components are drawn around the mean NDVI). Thin vertical lines indicate the phase (= timing) of the first peak of each of the Fourier components.

and the final fit to the observed data can be judged by visual comparison of the observed NDVI and the NDVI predicted using the parameters derived from the Fourier analysis. An example is shown in Fig. 11.3, from a site in southern Sudan, where a very strong annual cycle explains most of the variation in the observed NDVI.

Fourier analysis provides a convincing description of NDVIs from a wide range of ecological zones within Africa; a sequence from the forest zone of Liberia to the edge of the Sahara desert in Mali is shown in Fig. 11.4. In the forest zone the annual average is high, and there is little seasonal variation; in the desert zone there is also little seasonal variation, but around a much lower average. In between, both annual and biannual components may be important. What is especially noticeable is the high

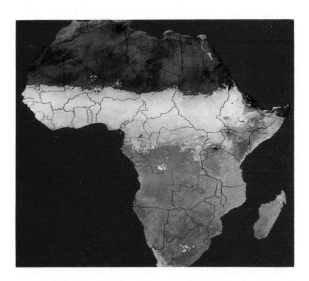

PLATE 11.1. Fourier analysis of monthly NDVIs for the whole of Africa for the period 1987–89 inclusive. Analysis is based on the 36 monthly images, each the average of three, 10-day images produced by selecting for each pixel the maximum NDVI during the period (= maximum value compositing), to eliminate clouds. The average NDVI is put in the red gun of the computer screen, the phase of the first Fourier component in the green gun and the amplitude of this component in the blue gun. Values in each gun of the colour image were then stretched across the full range of intensities within the image processing system. International boundaries are also shown (Hammer–Aitoff projection).

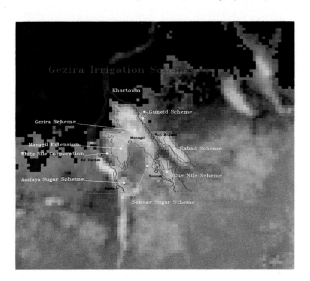

PLATE 11.2. Close-up of part of Plate 11.1 showing the Gezira irrigation project area between the White and Blue Niles, just south of Khartoum. Different parts of the irrigation project are labelled (the green parallelogram marks the limits of the map. Scale: Managil to Kosti is c. 130 km) (from RIM 1987).

[facing page 260]

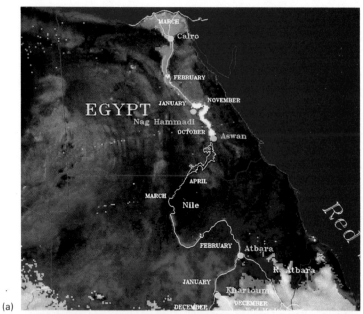

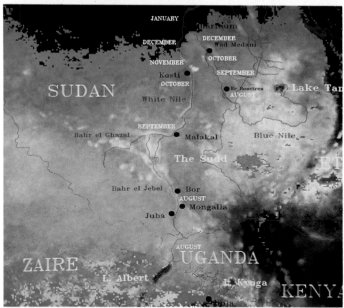

PLATE 11.3. Enlargement of part of Plate 11.1 showing (a) the Upper and (b) the Lower Nile basin. Courses of the White and Blue Niles and the months of peak NDVI along the river, probably associated with the seasonal growth of vegetation associated with seasonal river flow, are shown. The long time interval between peak growth at the point of entry into Lake Nasser and peak growth just down river of the Lake is presumably determined by the irrigation schedule of the Aswan dam (scale: Cairo to Khartoum is *c.* 1600 km).

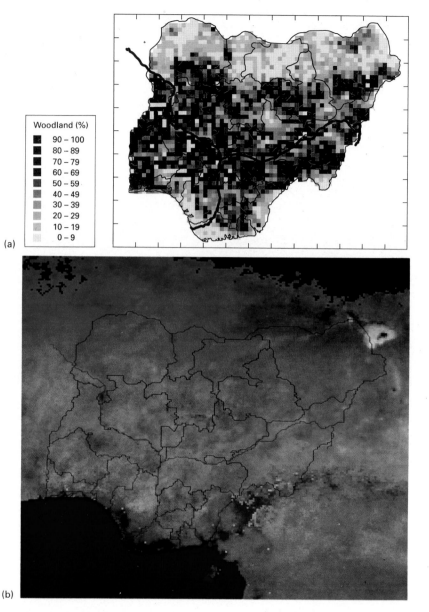

(a)

(b)

PLATE 11.4. (a) Results of an aerial survey of woodland in Nigeria carried out by the Environmental Resource Group Oxford in 1991 (RIM 1992). Each grid square is *c*. 20 km on a side. (b) enlargement of part of Plate 11.1 for Nigeria, with the same state boundaries as in Plate 11.4a. Notice the correspondence between the heavily wooded areas of Plate 11.4a and the blue areas of Plate 11.4b, i.e. areas where the amplitude of the first Fourier component is particularly pronounced (Plate Carrée projection). (The different colour shading of Plates 11.2, 11.3 and 11.4 in comparison with Plate 11.1 is due to colour stretching of only parts of the image, to reveal the features of interest.)

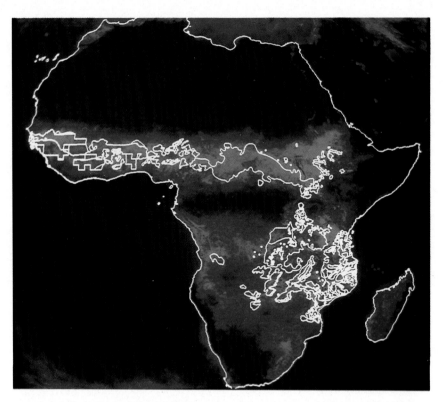

PLATE 11.5. The amplitude of the first Fourier component is shown on a map of Africa (produced by turning off the red and green guns of the computer screen that gave Plate 11.1) and the pan-African distribution of the tsetse *G. morsitans* is shown (within the white continental boundary lines). Plate 11.4 suggests that the brighter blue areas are associated with seasonal woodlands, the habitat of the savannah species of tsetse. This conclusion is here supported on a pan-African scale (Hammer–Aitoff projection).

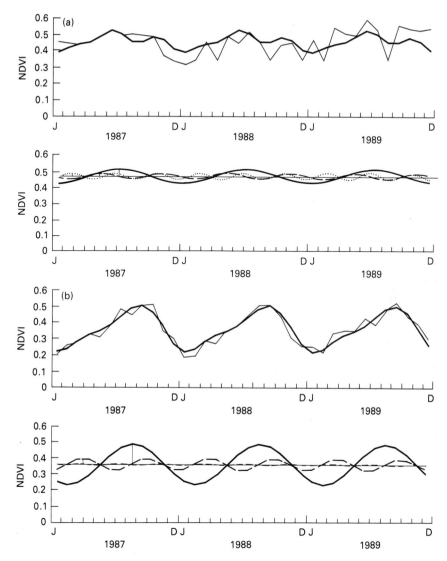

Fɪɢ. 11.4. Examples of Fourier analyses of a series of NDVIs for sites from the coast of West Africa, inland to Mali for 1987–89. (a) 9°W, 6°N, in the Liberian rainforest, (b) 6°W, 10°N, near Korhogo, Côte d'Ivoire, (c) 1°W, 13°N, near Ougadougou, Burkina Faso and (d) 0°E, 16°N, near Gao, Mali. Notice changes in the mean NDVI and the changing relative importance of the first and second Fourier components (details as in Fig. 11.3).

Continued on p. 262

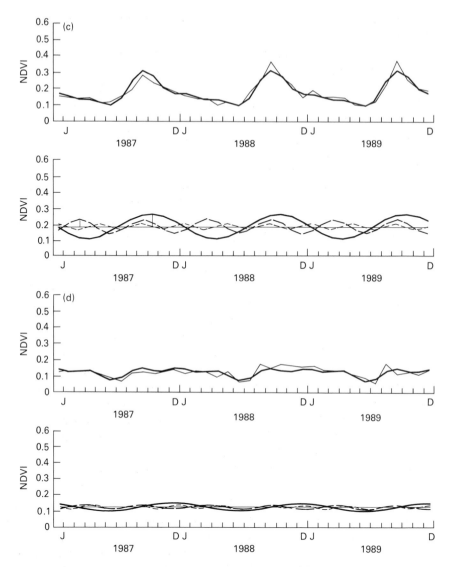

Fig. 11.4. *Continued.*

amplitude fluctuations in the savannah zone of Africa (Fig. 11.4b and c). Thus this analysis confirms, but presents in a different way, the conclusions drawn from principal components analysis of similar images (Tucker *et al.* 1985). The annual cycle of vegetation growth appears to be well captured by Fourier analysis, as can be seen from the examples in Fig. 11.4.

The Fourier analysis of the NDVIs for the whole of Africa for 1987–89 is shown in Plate 11.1 (facing p. 260), where the average NDVI (one of the outputs of the analysis) has been put in the red gun of the screen, the phase of the first component in the green gun and its amplitude in the blue gun. Areas of the image in Plate 11.1 which are predominantly red indicate sites with a large, but unvarying NDVI (= 'forests'); areas which have a blue tinge have seasonally highly variable NDVIs, whilst those which are bright green have a late peak of maximum vegetation growth. The fact that they are green means that neither the average NDVI nor the amplitude of the annual cycle of vegetation growth is very pronounced in such areas. Areas with a high average NDVI plus a late peak of vegetation growth appear either yellow in Plate 11.1 (since red plus green = yellow in colour monitors), or white if the seasonality is also very pronounced (red + green + blue = white).

The Fourier image in Plate 11.1 reveals information about both patterns and processes connected with vegetation types in Africa. One example of a pattern is shown in Plate 11.2 (facing p. 260), a close-up of the Gezira irrigation scheme in Sudan, between the White and Blue Niles just before they join at Khartoum. This scheme is one of the oldest large-scale irrigation projects in the continent and is supplied mostly by the pure waters of the Blue Nile which, through a series of irrigation canals, move down a gradient towards the White Nile. (The history of the Gezira project is described briefly in RIM 1987.) The original Gezira scheme was initially expanded into the Managil Extension, and other schemes were later added around these. Much of the cultivated area is presently under a 4-year crop rotation (of cotton, wheat, legumes such as ground nuts, and fallow), but some of the schemes are more permanently cropped with sugar cane. Some areas, such as the White Nile Corporation Schemes, have much lower yields than others, and this can be partly attributed to poorer irrigation. The non-cropped areas outside the irrigation scheme provide grazing for the considerable numbers of livestock owned by the workers within the scheme, and the whole is a rather intricate, inter-dependent patchwork (RIM 1987). A map of the Gezira scheme is super-imposed on the image in Plate 11.2, and a strong correspondence can be seen between the map and the underlying image. The brightest areas in Plate 11.2 are those that are best irrigated, while the sugar schemes have a speckled appearance.

An example of an ecological process revealed by Fourier analysis is shown in Plate 11.3a and b (facing p. 260) for the upper and lower Nile basin respectively. The White Nile and the main Nile can be detected in these images (the map overlay shows the course of the river) through a combination of differences between the riverine vegetation and the surrounding vegetation. These differences are of several types. In southern Sudan the Sudd can be picked out because it has a high and constant average NDVI throughout the year whereas the surrounding vegetation has a lower average and shows strong seasonality. Further north it is the riverine vegetation which shows strong seasonal variation whilst the surrounding semi-desert or desert shows little seasonal change. The seasonal variation in the NDVI along the course of the river is presumably related to the seasonal flow of water along the river (Shahin 1985; Howell et al. 1988), determined by regional rainfall patterns. Of particular interest is the phase (= month) of peak vegetation growth of the annual cycle of vegetation change, which is indicated along the course of the river in Plate 11.3. The outflow of water from Lake Victoria shows little seasonality and there appears to be little seasonality in the NDVI along the Victoria Nile. Beyond Lake Albert, however, the river develops a seasonal flow because of water received from various rain-fed tributaries. Peak vegetation growth occurs during July/August. Continuing along the Nile, the peak of vegetation growth occurs later in the year the further along the river it is measured, so that at Khartoum it occurs around December of the same year. Here the highly variable influx of the Blue Nile contributes significantly to the total volume of water in the main Nile (Shahin 1985). Peak seasonality nevertheless appears to follow a reasonable timing up to Lake Nasser where all signs of a seasonality different from that of the surrounding vegetation disappear. However, beyond the Aswan dam there is a remarkably strong seasonal signal of vegetation growth with a peak in September/October, approximately half a year later than the peak at the entry point into Lake Nasser. This is associated with the timing of the release of waters by the Aswan dam authorities (subject to international agreements with other countries along the Nile). Another dam at Nag Hammadi appears to delay the seasonal peak still further, after which the peak vegetation growth along the remainder of the Nile follows a sensible sequence, ending in the Nile delta in March, approximately 20 months after it was first detected near Lake Albert, almost 3000 miles away but only 750 m higher. It appears therefore that a ('statistical') drop of water from Lake Victoria takes almost 2 years to reach the Mediterranean!

It is possible to look at the Fourier analysis of the NDVI for points along the course of the Nile, and some examples are shown in Fig. 11.5. The slight difference in seasonality of the first Fourier component detected

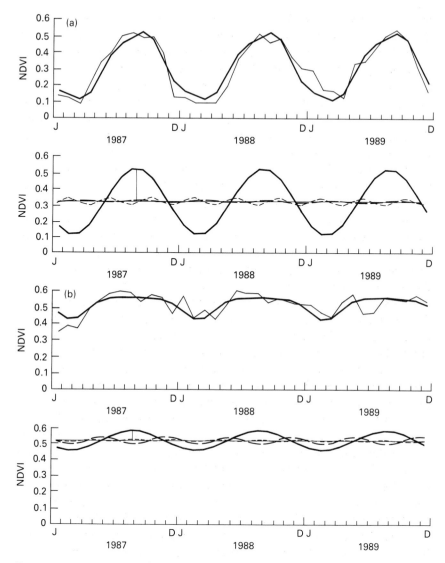

FIG. 11.5. Fourier analysis from selected points near to, or on, the Nile River. (a) West of Jonglei (i.e. just downriver from Bor) (30.6°E, 7°N), (b) on the Nile near Jonglei (30.8°E, 7°N), i.e. in the Sudd, (c) in the desert east of Qena (32.95°E, 25.85°N), on the sharp bend in the Nile between the Aswan Dam and Nag Hammadi, (d) at Qena (32.75°E, 25.85°N) on the Nile and (e) at Tahta (31.45°E, 26.75°N) just downriver from Nag Hammadi (for most place names see Plate 11.3).

Continued on p. 266

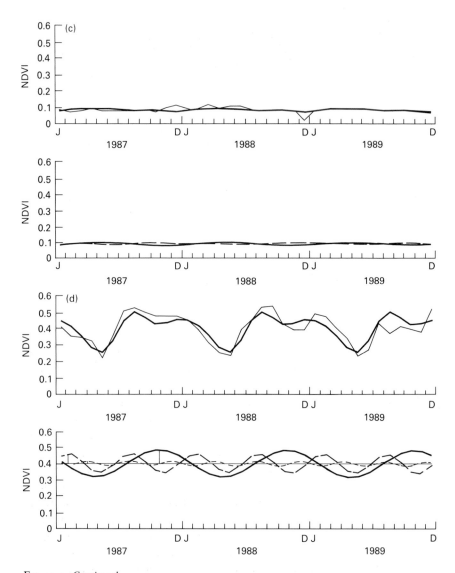

FIG. 11.5. *Continued.*

between the Aswan dam and Nag Hammadi and beyond Nag Hammadi itself can be seen in Fig. 11.5d and e to hide a much more dramatic difference between these two sites revealed by looking at both the first and second Fourier components. Whereas the annual signal dominates at the first site, the biannual signal is stronger than the annual signal at the

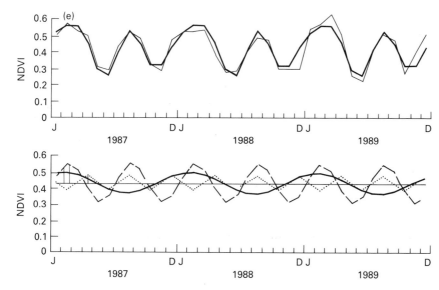

FIG. 11.5. *Continued.*

second site. This suggests either a quite different irrigation schedule at the two dams, or a quite different cropping pattern, or some combination of the two.

Finally we return to the problem of predicting tsetse distributions. The seasonal variation of vegetation growth picked out by the amplitude of the first Fourier component appears to be strongly correlated with savannah woodland areas of Africa. Plate 11.4 (facing p. 260) shows the comparison for Nigeria where the Environmental Resource Group Oxford (ERGO) has recently carried out an aerial survey of livestock and land-use categories for the Federal Government of Nigeria (RIM 1992). Plate 11.4a shows a map of the percentage woodland seen in the 20-km grid squares of the country-wide survey whilst Plate 11.4b shows an enlarged part of Plate 11.1, with the Nigerian state boundaries from Plate 11.4a superimposed. The correspondence between the heavily wooded areas of Plate 11.4a and the blue areas of Plate 11.4b is remarkable. *G. morsitans* is a species of tsetse associated with savannah woodland and Plate 11.5 (facing p. 261) shows the agreement between its pan-African distribution and regions of the continent showing strong annual seasonality of vegetation growth. Not only are the northern limits of this species in West Africa well described by the Fourier analysis, but so too are the southern limits in this region. It has long been acknowledged that the northern limits are strongly correlated with rainfall (or vegetation determined by rainfall), but the southern limits have been much more difficult to predict (Nash 1948).

CONCLUSIONS

It is relatively easy to extract biological conclusions from the two methods of analysis presented here although, as emphasized in Chapter 7, it must be remembered that correlations do not necessarily imply causation. When correlations are consistent, however, and when they are supported by intensive biological studies, we may be more certain of their biological significance.

The statistical analyses presented here should help to refine research questions to be addressed by future field work, the need for which becomes more urgent as indications of long-term climatic change are more widely demonstrated and accepted. The stimulus to discover the determinants of animal (and plant) distributions is ironically greatest at the moment when long-term distributions are on the point of disruption by human activity, in much the same way that diseases stimulate the study of the healthy human body, or mechanical breakdowns show the need to understand how machines work properly. Statistical analysis works best in a constant world and, in a changing environment, the same methods of analysis may mislead more than they direct. Nevertheless the option to investigate the biological determinants of distribution and abundance is no longer open, because of a lack of time to carry out such studies. Answers are required quickly and they will need to be based on statistical analysis.

This study has shown how satellite sensor imagery can have an important role in detecting patterns and processes in biological systems. Temporal Fourier analysis appears to be a novel and exciting way to extract hitherto unused aspects of NDVI information. The output of Fourier analysis is more easily interpreted in terms of the seasonality and amplitude of vegetation growth and is therefore more understandable biologically than the output from other methods of analysing NDVIs. The images used in this study had a spatial resolution of *c.* 8 km, much coarser than the maximum of 1.1 km of which the NOAA series of satellites is capable. Ground-based receivers near to, or beneath the track of, the orbiting satellites can receive the full resolution 1.1 km data in all five wave bands of these satellites. This would allow the processing of the NDVI Channels 1 and 2 in different ways to give a variety of vegetation indices, some of which are probably more useful for heavily vegetated areas than is the NDVI, which was originally formulated to detect changes in the sparsely vegetated regions of the Sahelian zone of Africa (Jackson & Huete 1991). Other potentially useful information is available in the non-NDVI infra-red bands that have been used in the past to study wetlands (Van de Griend *et al.* 1985; Xue & Cracknell 1992). Novel combinations of the spectral signals and the much finer spatial resolution of the 1.1 km data at the high temporal frequency of the NOAA satellites must be explored if full advantage is to be taken of these satellite sensor data for biology and conservation. Problems of sensor calibration, atmospheric aerosol effects and view angle must also all be overcome (Goward & Huemmrich 1992) before such imagery can usefully complement the higher spatial resolution LANDSAT and SPOT imagery.

New space platforms are currently being planned which increase both the spatial and spectral resolution of remotely sensed information. The wealth of data that these promise to provide could overwhelm biologists unless a start is made soon on developing techniques of dimension reduction that provide useful, large-area information for each of the many problems of research and conservation that will need to be addressed in the very near future. Literally and metaphorically we would like to see both the wood *and* the trees.

ACKNOWLEDGEMENTS

We thank Dr William Wint of the Environmental Resource Group Oxford (ERGO) for help and advice with the interpretation of the Fourier images, for the loan of slides used during the presentation of the talk at the British Ecological Society Symposium, for reading and commenting on the manuscript and for permission to use information in two reports from Resource Inventory and Management Ltd (RIM 1987 and 1992), for the map of Plate 11.2 and for Plate 11.4a.

The climatic data for Zimbabwe were kindly provided by Trevor Booth, CSIRO, Australia, and for East Africa by Dr Tim Robinson. Satellite sensor data were provided by the FAO Remote Sensing Centre, Rome, courtesy of Jelle Hielkema and by USAID/FEWS NASA GSFC, courtesy of Barry Henricksen.

Financial support was provided by the UNDP/World Bank/WHO Special Programme for Research and Training in Tropical Diseases, project 860010, and all image processing was carried out using equipment provided by the NERC, project GR3/7524 (to DJR).

REFERENCES

Bartholomé, E. (1988). Radiometric measurement and crop yield forecasting: some observations over millet and sorghum experimental plots in Mali. *International Journal of Remote Sensing*, **9**, 1539–1552.

Buxton, P.A. (1955). *The Natural History of Tsetse Flies.* Memoir of the London School of Hygiene and Tropical Medicine No. 10. H.K. Lewis, London.

Buxton, P.A. & Lewis, D.J. (1934). Climate and tsetse flies: laboratory studies upon *Glossina submorsitans* and *G. tachinoides*. *Philosophical Transactions of the Royal Society, London, Series B*, **224**, 175–240.

Chatfield, C. (1980). *The Analysis of Time Series: an Introduction.* Chapman & Hall, London.

Cracknell, A.P. & Hayes, L.W.B. (1991). *Introduction to Remote Sensing.* Taylor & Francis, London.

Ford, J. (1971). *The Role of the Trypanosomiases in African Ecology: a Study of the Tsetse Fly Problem.* Clarendon Press, Oxford.

Ford, J. & Katondo, K.M. (1977). *The Distribution of Tsetse Flies* (Glossina) *in Africa 1973.* Organization of African Unity – Scientific and Technical Research Commission. Cook, Hammond & Kell, London.

Goward, S.N. & Huemmrich, K.F. (1992). Vegetation canopy PAR absorptance and the normalised difference vegetation index: an assessment using the SAIL model. *Remote Sensing of Environment*, **39**, 119–140.

Green, P.E. (1978). *Analyzing Multivariate Data.* The Dryden Press, Hinsdale, Illinois.

Hill, M.O. (1991). Patterns of species distribution in Britain elucidated by canonical correspondence analysis. *Journal of Biogeography*, **18**, 247–255.

Hoare, C. (1972). *The Trypanosomes of Mammals.* Blackwell Scientific Publications, Oxford.

Howell, P., Lock, M. & Cobb, S. (Eds) (1988). *The Jonglei Canal: Impact and Opportunity.* Cambridge University Press, Cambridge.

Huete, A.R. & Jackson, R.D. (1987). Suitability of spectral indices for evaluating vegetation characteristics on arid rangelands. *Remote Sensing of Environment*, **23**, 213–232.

Hugh-Jones, M. (1989). Applications of remote sensing to the identification of the habitats of parasites and disease vectors. *Parasitology Today*, **5**, 244–251.

Huh, O.K. (1991). Limitations and capabilities of the NOAA satellite advanced very high resolution radiometer (AVHRR) for remote sensing of the Earth's surface. *Preventive Veterinary Medicine*, **11**, 167–183.

Jackson, R.D. & Huete, A.R. (1991). Interpreting vegetation indices. *Preventive Veterinary Medicine*, **11**, 185–200.

Jeffers J.N.R. (1978). *An Introduction to Systems Analysis: with Ecological Applications.* Edward Arnold, London.

Marriott, F.H.C. (1974). *The Interpretation of Multiple Observations.* Academic Press, London.

Menenti, M., Azzali, S., Verhoef, W. & van Swol, R. (1991). Mapping agro-ecological zones and time lag in vegetation growth by means of Fourier analysis of time series of NDVI images. Report No. 32, SC-DLO. Winand Staring Centre, Wageningen, The Netherlands.

Molyneux, D.H. & Ashford, R.W. (1983). *The Biology of* Trypanosoma *and* Leishmania, *Parasites of Man and Domestic Animals.* Taylor & Francis, London.

Nash, T.A.M. (1948). *Tsetse Flies in British West Africa.* HMSO, London.

Nix, H. (1986). A biogeographic analysis of Australian elapid snakes. In: *Atlas of Elapid Snakes of Australia.* (Ed. by R. Longmore), pp. 4–15. Australia Flora and Fauna Series No. 7. Australian Government Publishing Service, Canberra, Australia.

Prince, S.D., Justice, C.O. & Los, S.O. (1990). *Remote Sensing of the Sahelian Environment: a Review of the Current Status and Future Prospects.* Technical Centre for Agricultural and Rural Cooperation, Commission of the European Communities.

RIM (1987). *Gezira Livestock Integration Study, Final Report Vol. 4. Livestock, Human and Environmetal Resources.* Devco, Dublin.

RIM (1992). *Nigerian Livestock Resources.* Resource Inventory and Management, St Helier, Jersey.

Rogers, D.J. (1979). Tsetse population dynamics and distribution: a new analytical approach. *Journal of Animal Ecology,* **48**, 825–849.

Rogers, D.J. & Randolph, S.E. (1985). Population ecology of tsetse. *Annual Review of Entomology,* **30**, 197–216.

Rogers, D.J. & Randolph, S.E. (1986). The distribution and abundance of tsetse flies (*Glossina* spp.). *Journal of Animal Ecology,* **55**, 1007–1025.

Rogers, D.J. & Randolph, S.E. (1991). Mortality rates and population density of tsetse flies correlated with satellite imagery. *Nature,* **351**, 739–741.

Rogers, D.J. & Randolph, S.E. (1993). Distribution of tsetse and ticks in Africa: past, present and future. *Parasitology Today,* **9**, 266–271.

Rogers D.J. & Williams, B.G. (1993). Monitoring trypanosomiasis in space and time. *Parasitology,* **106**, S77–S92.

Shahin, M. (1985). *Hydrology of the Nile Basin.* Elsevier, Amsterdam.

Snyder, J.P. (1987). *Map Projections – a Working Manual.* US Geological Survey Professional Paper 1395. US Government Printing Office, Washington.

Tatsuoka, M.M. (1971). *Multivariate Analysis: Techniques for Educational and Psychological Research.* John Wiley, New York.

Ter Braak, C.J.F. & Prentice, I.C. (1988). A theory of gradient analysis. *Advances in Ecological Research,* **18**, 271–317.

Townshend, J.R.G. & Justice, C.O. (1986). Analysis of the dynamics of African vegetation using the normalized difference vegetation index. *International Journal of Remote Sensing,* **7**, 1435–1445.

Tucker, C.J. & Sellers, P.J. (1986). Satellite remote sensing of primary production. *International Journal of Remote Sensing,* **7**, 1395–1416.

Tucker, C.J., Townshend, J.R.G. & Goff, T.E. (1985). African land-cover classification using satellite data. *Science,* **227**, 369–375.

Van de Griend, A.A., Camillo, P.J. & Gurney, R.J. (1985). Discrimination of soil physical parameters, thermal inertia and soil moisture from diurnal surface temperature fluctuations. *Water Resources Research,* **21**, 997–1009.

Xue, Y. & Cracknell, A.P. (1992). Thermal inertia mapping: from research to operation. *Proceedings of the 18th Annual Conference of the Remote Sensing Society* (Ed. by A.P. Cracknell & R.A. Vaughan), pp. 471–480. Remote Sensing Society, Nottingham.

12. MONITORING SPECIES PERFORMANCE OF COMMON DOMINANT PLANT SPECIES

PALMER NEWBOULD

Dromore House, 88 Coolyvenny Road, Coleraine,
Northern Ireland BT51 3SF, UK

SUMMARY

Dominant plant species set a structural pattern for the ecosystem within which other plant and animal species live. *Phragmites australis* and *Cladium mariscus* are dominant over considerable areas of marsh in the Parc Natural S'Albufera in northern Majorca. Above-ground biomass, leaf area index and standing dead material were estimated for a number of stands and dates in 1991.

The growth and structure of both species are influenced by the time since last burn. Burning is accidental or at least unpredictable. Above-ground net primary production was estimated as 376 and 694 g dry matter/m^2/year for two stands of *Cladium* and 855 and 1515 g dry matter/m^2/year for two stands of *Phragmites*. However, *Phragmites* is characterized by rapid translocation of stored photosynthate in the spring, and the reverse in the autumn, in contrast with the steadier progress of the *Cladium* with its perennial leaves.

Leaf area index peaks in July at a little over 3 for two *Cladium* and one *Phragmites* stand, but is considerably lower (1.6) and earlier for the other *Phragmites* stand. Some of the *Phragmites* stands suffer high levels of caterpillar damage to the larger shoots, damaged shoots being replaced by two or three smaller lateral shoots. Standing dead material increases from zero after a fire to an equilibrium value between two to three times the net primary production after about 5 years or more.

Closer understanding of the patterns of growth and biomass accumulation in *Cladium* and *Phragmites* would contribute significantly to the proper management of the Albufera marsh for conservation but will require considerable research commitment.

INTRODUCTION

The UK government signed the Biodiversity Convention at the Rio Conference in June 1992. Biodiversity is commonly assessed in terms of genetic diversity, often as the number of species occurring together in an area. But another facet of biodiversity is the structural diversity represented in a population of a single plant species. The pattern or structure imposed on the plant community by the growth of one or more dominant species was the central theme of A.S. Watt's masterly Presidential Address to the British Ecological Society in 1947, 'Pattern and Process in the Plant Community'. Watt covered a range of community types, notably woodland, dwarf shrub heath, bogland, bracken and grassland. His work on dwarf shrub heath influenced the subsequent work on growth and nutrient cycling in heather moorland carried out by Gimingham (1972) and many others which established the sequence of development of heather stands from pioneer through building and mature phases to the degenerate phase. This research was especially appropriate to Britain since heathland and heather moorland are (or were) especially well represented in Britain and Ireland in relation to their world distribution, are important for nature conservation and landscape protection and are also managed for grazing by sheep and grouse. Even where management is not deliberate, most areas of heathland are burnt fairly regularly. Watt himself was subsequently able to study a few sites where *Calluna* had remained unburnt for a long period of time and had assumed an uneven-age structure so that pioneer, building, mature and degenerate phases formed a mosaic in a single area (Watt 1955).

Watt's hypothesis, linking pattern in space with process in time, was based on a considerable volume of detailed and precise information, including the rerecording of permanent quadrats. Recent research on heather has increasingly availed of modern techniques, reinforcing but not challenging Watt's conclusions. The ability of the heather, for example, to withstand heavy grazing depends upon the vigour and age of the heather before the grazing occurs. Felton and Marsden (1990) suggest that in well-managed heather moor if less than 40% of the current season's growth is used, grazing is not damaging. Tolerance of grazing increases through the building stage and declines rapidly thereafter. Felton and Marsden recommend appropriate sheep stocking rates for heather moorland. MacDonald (1990) and Armstrong and MacDonald (1992) suggest how existing utilization rates can be measured. Performance indicators of heather can alert the conservation manager to situations where the heather is being overgrazed before the situation is irreversible. Essentially these indicators involve measuring the current year length growth of long shoots and relating this to the extent to which long shoots have been grazed. Careful study of the 'condition' of heather allows precise diagnosis of overgrazing where this is occurring, and the active management of the site for conservation can be adjusted appropriately. Better understanding of the growth pattern of the dominant species is valuable for fine-tuning management practice. Heather moorland is a good example of the connection between basic ecological research and applied research leading to land use and land management policy. The boundaries and management prescriptions of some Environmentally Sensitive Areas declared by departments of agriculture are based on this research.

Another example of using plant performance to confirm past management practice is the work of Kassas (1951) at Chippenham Fen. He studied the width of past growth rings of ash trees (*Fraxinus excelsior*) relating to the past 150 years and was able to relate accelerated growth to clearance of drainage channels and reduced growth to periods of neglected drainage.

This chapter reports some preliminary findings relating to the growth of two dominant species in a marshland area in Majorca.

SITE, SPECIES AND METHODS

Project Albufera, based in the Parc Natural S'Albufera, in northern Majorca, was initiated by Max Nicholson in 1989 and has been organized by Earthwatch Europe. The project is on-going. Annual reports of progress are available from Earthwatch at their Oxford office. A concise account of the history and limnology of the Albufera is given by Martinez-Taberner *et al.* (1990). The primary objective of the project is to carry out baseline studies which could be repeated from time to time and which would allow the detection and measurement of environmental change. A secondary objective is to provide information to the park management authority (Conselleria d'Agricultura i Pesca, Govern Balear) which would be useful in the management and interpretation of the park. The studies are mainly biological, meteorological and hydrological. The biological studies involve the recording of plant and animal species, including vegetation studies.

The park has an area of 1700 ha and the predominant vegetation type is marsh, occupying a salinity gradient with *Cladium mariscus* (L.) Pohl at the freshwater end and *Arthrocnemum* spp. at the saline end. *Phragmites australis* (Cav.) Trin. ex Steudel appears to have a wide ecological tolerance and occurs almost throughout. It seems likely that *Phragmites* and *Cladium* between them contribute a high proportion of the herbaceous plant biomass on the Albufera and it is hoped to assess this using satellite imagery. Normal measures of frequency or abundance applied to these two common species are not very revealing. Their phenology and growth forms are quite different and it seemed desirable to measure biomass, productivity and other performance indicators in different sites, and use these data as the baseline for assessing both environmental heterogeneity and environmental change. The data will also contribute to an understanding of the structure and functioning of the marsh ecosystem. It may, for example, be possible to define the habitat requirements of the moustached warbler (*Acrocephalus melanopogon*) in terms of the performance parameters of *Phragmites* and *Cladium*, allowing management to provide for the protection of this important species.

Much general information about the ecology of *Phragmites* and *Cladium* has been summarized by Haslam (1972) and Conway (1942) respectively in their Biological Flora accounts of these species. In both cases these accounts refer to earlier research papers.

At Albufera both species are subject to occasional and irregular burning which may be accidental or deliberate, often spreading into the park from the adjacent agricultural land. There is a cycle of regrowth following burning. Hence one factor influencing performance is probably time since last burn. The above-ground growth of *Phragmites* is essentially annual; the shoots die back each autumn and new shoots arise from the rhizomes each spring. At Albufera, as also reported by Haslam (1972) from Malta, some of the shoots remain viable throughout the winter and produce lateral shoots in the spring, but this is not especially common. A proportion of stems produce a terminal inflorescence. In some areas the stems are widely attacked by caterpillars, probably of the moth *Archanara geminipuncta*, which bore into the stem and feed on the vascular tissue within the internode resulting in the death of the shoot above that internode. This removes apical dominance and buds are formed below the affected internode, producing two or three side shoots. Although these are much thinner than the original shoot they may still produce inflorescences. In some cases the reed bunting (*Emberiza schoeniclus*) uses its powerful bill to open up the internode and eat the caterpillar. The incidence of caterpillar damage varies in different parts of the marsh. There is some indication that old stands (in the sense of time since last burn) are more vulnerable than young ones; caterpillar populations build up from zero after burning. After burning *Phragmites* produces a high density of small shoots, sometimes with just a few larger shoots. In successive years the density becomes lower, and shoot diameter comes to have a bimodal distribution, a few thick shoots and a large number of thin ones, a phenomenon discussed by Haslam (1972).

In contrast to *Phragmites*, *Cladium* has perennial leaves, each leaf reaching an age of 3–4 years before becoming brown and senescent.

Months Days Site no.	February 48	April 111	July 201	October 292
2		*		
5		*		
11		*		
1	* ——➤	* ——➤	* ——➤	*
7	* ——➤	* ——➤	* ——➤	*
9	* ——➤	* ——➤	* ——➤	*
10	* ——➤	* ——➤	* ——➤	*
12		*		
13		*		
14		*		

FIG. 12.1. Albufera. Sampling strategy 1991.

Each year new leaves are produced in the centre of the shoot and old leaves die off at the outside. It does not produce inflorescences as freely as *Phragmites*, and a new shoot may not flower until its fourth year. This does not always apply after burning, since some shoots flower in the year directly after a burn. Once a *Cladium* shoot has flowered, it normally dies.

Where horse or cattle grazing has been introduced, *Phragmites* is heavily grazed and *Cladium* would seem to be a fodder of last resort, seldom grazed. However, fresh green *Cladium* regrowth after a burn may be grazed. The present project has not addressed the impact of grazing on *Phragmites*.

Sampling methodology was tested in 1990 and a full but practical sampling programme was devised for 1991. Four sites, two dominated by *Phragmites* and two by *Cladium*, were selected for four sampling dates. A further six sites were selected for sampling during April 1991. Accessibility was one important criterion in the choice of sites. Approximately half the marsh was burnt over at the end of September 1990, and so sampling sites were distributed between sites burnt and unburnt on that occasion. The sampling strategy is shown in Fig. 12.1, the locality of the sites in Fig. 12.2 and information about the sites is listed in Table 12.1. No chemical

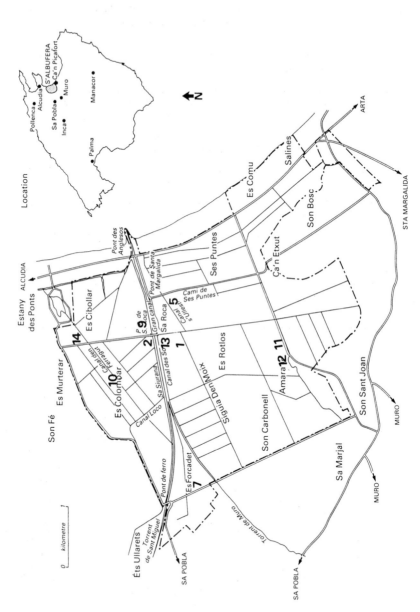

Fig. 12.2. Albufera. Location map.

TABLE 12.1. Sample site descriptions

1 *Cladium* dominant. South of stepping stones leading to Watkinson hide. Usually 10–20 cm water. Burnt 1989 and 1990

2 *Phragmites* dominant. Es Colombar. Across two bridges and then in on the left. 'Mature' *Phragmites*, unburnt for a long time. Usually *c.*50 cm water. A few cattle grazing October 1991

5 *Phragmites* dominant. Enter from bridge over the aqueduct. Burnt 1988. Few large stems but dense stand of small stems. Water level fluctuating

7 *Phragmites* dominant. Es Forcadet, western margin of park. Tall, vigorous *Phragmites*. Regrowth started (out of season) immediately after fire (27/9/90) but many of the thin shoots died before February. 10 cm water February, below ground level April

9 *Cladium* dominant. NE of Grand Canal. Unburnt for many years. A rather mixed stand. Probably near tolerance limits of *Cladium*. Water near or just above soil surface

10 *Phragmites* dominant. Es Colombar. Scattered *Tamarix*, tall 'mature' *Phragmites*, much standing dead. Unburnt for several years. Water up to 50 cm. Brackish

11 *Phragmites* dominant. Amarador. September 1990 fire was patchy here, so Site 11 was burnt, Site 12 not. Water below soil surface. Light horse grazing

12 *Phragmites* dominant. As 11, but not burnt September 1990. Probably burnt 1987 or 1988

13 *Cladium* dominant. 'Mature' *Cladium* prior to September 1990 fire. Vigorous, difficult to penetrate prior to the fire. Usually 10–20 cm water. Access only by boat

14 *Phragmites* dominant. Fairly near and similar to Site 10, but drier. Water +10 cm to −10 cm. Brackish

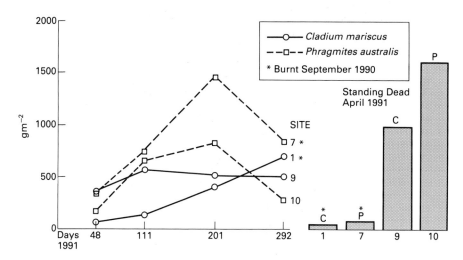

FIG. 12.3. Albufera. Above-ground biomass in grams dry weight per square metre at four sample sites on four sample dates in 1991 (days 48, 111, 201, 292). Standing dead in same units for the 111-day sample date.

TABLE 12.2. Above-ground dry weight and net primary production

	Site 1				Site 7				Site 9				Site 10			
	F	A	J	O	F	A	J	O	F	A	J	O	F	A	J	O
Above-ground dry weight (g m⁻²)																
Phragmites	1		5	7	330	746	1464	843	4	45	139	49	172	655	824	282
Cladium	74	140	400	687	1				360	571	514	507				
Calystegia						17	26	17						63	31	10
Other, *Scirpus, Juncus*						4	25	7	52	84	139	85				
Total	75	140	405	694	331	767	1515	867	416	700	792	641	172	718	855	292
Net primary production above-ground (g m⁻² year⁻¹)	36				66				982				1591			
Standing dead	694				1515				>376				855			

TABLE 12.3. Biomass and other data, April 1991

Year burnt: Site No.:	1989, 1990 1	 2	1988 5	1990 7	 9	 10	1990 11	 12	1990 13	 14
Biomass (g dry matter m^{-2})										
Phragmites	140	550	594	746	45	655	278	387	28	592
Cladium		18	133	21	571		2	70	176	
Other					84	63		3	1	12
Total	140	568	727	767	700	718	280	460	205	604
Standing dead (g dry matter m^{-2}) (% total biomass)	36 (26)	734 (129)	564 (78)	66 (9)	982 (140)	1591 (222)	175 (63)	846 (184)	48 (23)	1040 (172)
Leaf area index (m^2 leaf m^{-2} ground)	0.92	1.32	1.93	1.55	3.31	2.52	0.96	1.16	0.96	1.59
% large *Phragmites* stems attacked by caterpillar	NA	69	84	0	0	7	0	96	NA	48

NA = Not applicable.

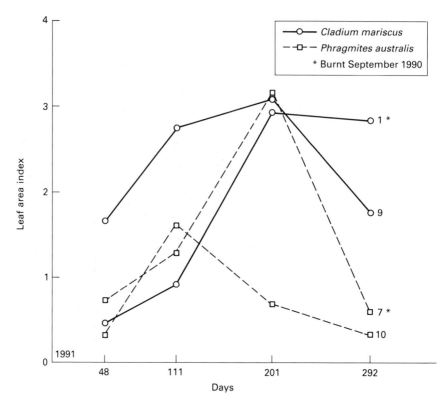

FIG. 12.4. Albufera. Leaf area index (leaf area per unit ground area) at four sample sites on four sample dates in 1991 (days 48, 111, 201, 292).

analyses of soil or water were carried out, though this omission can be rectified later. At each sample site and date, five 1 m × 1 m quadrats were harvested at ground level, sorted, the plant shoots were measured and weighed and subsamples were dried to allow the results to be converted to dry weight. Leaf area index was measured as the product of the dry weight of leaf material per square metre and the surface area per gram dry weight of those leaves. On the occasion of the April sample only, standing dead material was harvested and weighed.

Results are presented in Tables 12.2 and 12.3 and Figs 12.3–12.5 and are discussed below.

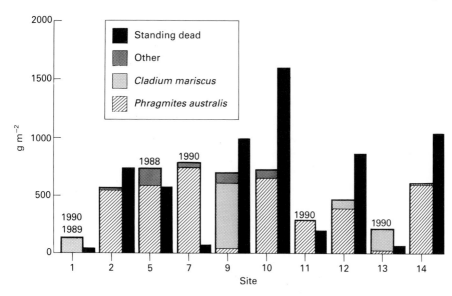

FIG. 12.5. Albufera. Above-ground biomass and standing dead in grams dry weight per square metre at 10 sample sites in April 1991.

RESULTS

Net primary production (Fig. 12.3, Table 12.2)

Four samples during the year is not really enough for the estimation of net primary production (NPP). However, in the case of Site 1 and Site 7 all the above-ground biomass had been formed since the beginning of October 1990, so that a minimum estimate of above-ground NPP for these two sites is $694\,\mathrm{g\,m}^{-2}$ for Site 1 and $1515\,\mathrm{g\,m}^{-2}$ for Site 7. Indeed on Site 7 an additional $60\,\mathrm{g\,m}^{-2}$ of small shoots had grown after the fire but died before the February sample date.

The rapid growth of *Phragmites* early in the year, and while the leaf area index is still low, is probably attributable to translocation of stored photosynthate from the rhizomes to the stems. This phenomenon, which is in contrast with the pattern in *Cladium*, is reviewed by Hara *et al.* (1993). It means that true NPP may actually be less than the maximum above-ground biomass.

The sharp decline in above-ground biomass of *Phragmites* between July and October is partly explained by translocation of photosynthate back to the rhizomes and the death of shoots, in the sense that they no longer bear green leaves or sheaths.

In the case of *Cladium*, on Site 1 the above-ground biomass continued to increase up to the October sample indicating NPP of 694 g m^{-2}. On Site 9, the minimum estimate of above-ground NPP is the July figure minus the February figure, 376 g m^{-2}. While a few first year shoots of *Cladium* on Site 1 flowered, their density was low and none was recorded in the sample quadrats. The initial population of *Cladium* had a shoot density of about 42 shoots per square metre but an additional 11 new shoots per square metre were added between July and October. On Site 9, five *Cladium* shoots per square metre (out of 90) flowered and subsequently died and 13 new shoots per square metre were recruited between July and October.

Standing dead

Dead material accumulates until either it is destroyed by fire or equilibrium is attained between production and decomposition. Both Site 9 (*Cladium*) and Site 10 (*Phragmites*) give the impression of maturity, as if they had reached a steady state. This suggests that in a steady state above-ground standing dead is two to three times the dry weight of above-ground NPP, (Table 12.3). Presumably some of the standing dead material subsides into the water and contributes to the formation of organic sediments or export. Where the *Phragmites* stem has been penetrated by a caterpillar, with or without additional reed bunting damage, this forms a point of weakness and a number of stems break or bend at this point. The physical structure of the habitat is probably important for its bird and invertebrate populations but this has not yet been investigated.

Leaf area index

The development of the leaf area index from February to October is shown in Fig. 12.4. In estimating the leaf area index of *Phragmites* only the leaf blades are included and the leaf sheaths are omitted. Their inclusion would add about 20% to the figure. They usually appear pale green as opposed to the deeper glaucous green of the leaf blades.

In estimating the leaf area index of *Cladium*, leaves or parts of leaves which have gone brown are omitted as are the fine leaves from the centre of the shoot which are white in colour. In a 'mature' stand of *Cladium* such as Site 9, the leaf area index includes leaves probably varying in age from o to 4 years or perhaps older. In this site the *Cladium* leaf area index peaked in July at 3.09 but the total leaf area index at this sample time was 4.80 due to significant contributions from *Juncus maritimus* and *Carex extensa*.

In Site 7, the *Phragmites* grew vigorously after the September 1990 fire and peaked at a leaf area index, for *Phragmites*, of 3.15, to which can be added 0.4 contributed by other species, mainly *Calystegia sepium* and *Scirpus maritimus*.

April results from 10 sites

Table 12.3 sets out data from the April 1991 samples. Sites 2, 5, 7, 9, 10 and 14 had above-ground biomass within the range $568-767 \, \text{g m}^{-2}$. Of the remaining sites the regrowth of the two *Cladium* sites burnt in September 1990 (1, 13) was quite slow getting under way, probably because there was less photosynthate stored below-ground than in corresponding *Phragmites* sites. The two *Phragmites* sites, 11 and 12, were also less vigorous than the other *Phragmites* sites, probably for reasons of soil or water chemistry. The two sites are directly comparable with each other, except that 11 was burnt in September 1990, whereas 12 has probably not been burnt for 3 or 4 years. The relatively high standing dead in 11 was due to the survival of a number of charred shoot bases. With this exception the standing dead in the recently burnt sites ranged from 9 to 26% of the living biomass. In Site 5, burnt in 1988, it was 78% of living biomass and in sites unburnt for longer periods, standing dead ranged from 129 to 222% of living biomass. Burning a mature site releases significant amounts of plant nutrients via the ash. However, the September 1990 fire was followed by extreme floods so that most released nutrient probably ended up in the Bay of Alcudia where the extensive *Posidonia* beds may have benefited.

Caterpillar damage

It seems likely that most or all the caterpillar damage to the *Phragmites* stems on the Albufera is caused by *Archanara geminipuncta*. Tscharntke (1990) worked on *Phragmites* stands along the river Elbe near Hamburg where *Archanara geminipuncta* caused damage to the *Phragmites* shoots similar to that occurring on the Albufera and this species was also responsible for damage to *Phragmites* in the Netherlands (van der Toorn & Mook 1982). Van der Toorn and Mook also record damage by the larvae of another moth species *Rhizedra lutosa*, which also damages the rhizomes. This has not been observed from the Albufera.

Caterpillar damage on large *Phragmites* stems was zero on recently burnt sites (7, 11) and high on Sites 2, 5 (burnt 1988), 12 and 14. Strangely it was quite low (7%) on Site 10 which gave the impression of maturity. A considerable proportion of the standing dead stems had been subject to caterpillar damage so it seems possible that the caterpillar population on this site had peaked and crashed, as described by Tscharntke (1990) and also mentioned by Mook and van der Toorn (1982). The eggs of *Archanara geminipuncta* overwinter under the leaf sheaths of reed shoots formed in the previous year, usually 20–40 cm above the ground or water level and are therefore prone to destruction by fire.

DISCUSSION

Data presented

Westlake (1963) reviewed estimates for the biomass and productivity of a variety of plant communities, including a number of aquatic or wetland communities. His general figure for organic productivity of temperate reedswamp is $45 \, t \, ha^{-1} year^{-1}$. The maximum figure recorded here ($1515 \, g \, m^{-2}$) is equivalent to $15 \, t \, ha^{-1} year^{-1}$. Even allowing for below-ground production, it seems likely that the productivity of the Albufera marsh is low by comparison with similar ecosystems elsewhere. Wheeler and Shaw (1991) report on above-ground biomass for a wide range of fen vegetation in lowland England and Wales. Taking all their stands together, the mean change in standing crop between April and September was $721 \, g \, m^{-2}$, to be compared, crudely, with Albufera NPP. April litter averaged $854 \, g \, m^{-2}$ and September litter $729 \, g \, m^{-2}$, indicating rather over 100% of above-ground NPP.

More precise comparison can be made between their *Cladio-Molinietum* (April–September mean increment $680 \, g \, m^{-2}$) and the Albufera *Cladium* (NPP 376, $694 \, g \, m^{-2}$); and between their *Phragmites* consociation (April–September mean increment $1300 \, g \, m^{-2}$) and the Albufera *Phragmites* (NPP 855, $1595 \, g \, m^{-2}$). These figures are remarkably similar, despite considerable differences in climate including length of potential growing season. The Albufera *Phragmites* gives an impression of vigour, based on height growth and thick stems. But leaf area indices are relatively low. Satellite imagery may help to indicate the amount of radiant energy absorbed. Vigorous height growth may carry penalties of low light penetration, early death of lower leaves, etc. But this would not affect the production of *Cladium* which also seems no higher than comparable stands in Britain.

Leaf area index is generally correlated with biomass. For all the *Cladium* samples $r = 0.90$, which is not surprising since most of the tissue is leaf material. The figure for all *Phragmites* samples is 0.81. It is not clear that leaf area measurements supply sufficient extra information to justify the effort.

The overall importance of caterpillar damage is not certain. Mook and van der Toorn (1982) cite values for August/September shoot biomass of wet undisturbed *Phragmites* (probably the closest to Albufera Sites 7 and 10) of $900-1200\,\mathrm{g\,m}^{-2}$ in different years. Severe infestation by *Archanara geminipuncta* may lower the maximal shoot biomass by 25–35%. This is mainly due to reduced growth and leaf area in June, after which the growth of the new side shoots may allow some element of 'catching-up'. Mook and van der Toorn attempt to characterize damage using the parameter leaf area duration (LAD) up to 7 July. Their wet undisturbed site shows a much lower LAD (around 50 leaf area days) in the 2 years of maximum *Archanara* damage compared with values around 100–170 in years with less damage. Growth on the Albufera starts earlier in the year and the comparable LAD values are 270 for Site 7 and 160 for Site 10. If LAD, for sake of comparison is measured simply between April and July the values would be 200 for Site 7 and 104 for Site 10. Neither of these sites suffered significant caterpillar damage in 1991 and LAD figures are not available for those sites which did.

Mook and van der Toorn (1982) explain variations in yield partly by damage factors (burning, late frost, caterpillar damage) which operate through effects on shoot density and self-thinning. Shoot density and dimensions of *Phragmites* are also influenced by the events of the previous summer. The growth dynamics of shoot populations of *Phragmites* have been further elucidated by Hara *et al.* (1993).

It is clear that the performance of *Phragmites* and *Cladium* varies from place to place, and presumably also from year to year, within the Albufera marsh. A more precise analysis of this variation would be valuable in determining the importance of variation in the structure of the ecosystem for other species, the impact of environmental conditions, burning, grazing and insect damage on the performance of these dominant species, and how best to manage the marshes. However, the sampling procedures are relatively laborious, and a good deal more work is needed to elucidate, for example, the sequence of regrowth after burning. The sampling procedures inevitably cause some damage to the habitat, not so much due to the material removed, but due to trampling and access.

Additional data required

The model of the growth and production of *Cladium* and *Phragmites* needs to be refined by more data sets of the same type; to be relevant to management it needs to be broadened to include grazing, and probably also mowing, though this management technique is not currently employed. It is also necessary to combine the type of ground truth work described here with the interpretation of satellite imagery, which has been on the agenda throughout and on which a start has now been made (Riddiford 1993). This should indicate the approximate extent and biomass of *Cladium* and *Phragmites* present on the Albufera.

Relevance to ecological theory and conservation management

If satellite imagery, from different seasons of the year, were combined with additional data sets of the general type presented here what use could be made of this? Here speculation runs ahead of data, but in ecology it often does.

The data could contribute both to ecological and conservation theory, and also to practical countryside management in northern Majorca. It could contribute at three distinct scales which can be termed the landscape scale (say $500\,km^2$), the ecosystem scale ($20\,km^2$) and the management compartment scale ($<1\,km^2$).

Landscape scale

The landscape scale involves a number of units such as an extensive upland catchment yielding relatively pure water, a lowland catchment area including the town of Sa Pobla and a zone of intensive horticulture yielding enriched and polluted water, the marsh itself, the coastal strip with its beach, dunes and tourist hotels and the Bay of Alcudia into which the outflow from the marsh discharges. Within the Bay there are extensive beds of *Posidonia oceanica*. Each spring a mound of dead *Posidonia* leaves is removed from the beach, to facilitate tourism, and is used by the vegetable growers as compost. This represents nutrient recycling on a macro scale. The landscape-scale model could be described in a number of units such as energy flow, carbon, mineral nutrients, water, money or employment. These socioeconomic and ecological models could provide a basis for sustainable development in northern Majorca in the medium and long term. The park has a crucial role, and marsh is its predominant ecosystem. A description of the functioning of the marsh, based on satellite imagery and the sort of ground truth described here is crucial for the formulation of these models which involve the interaction of the marsh, which is essentially a conservation area, with the horticultural industry and the tourist industry. Co-ordinated management on the landscape scale will involve a number of different agencies and will be difficult to achieve, so that initially the models may be of more theoretical than management value. Ultimately they concern the importance and proper management of Mediterranean wetlands in tourist areas.

Ecosystem and compartment scale

At the ecosystem scale, treating the Albufera marsh as a unit, the opportunities for co-ordinated management are much greater. There is a management plan for the whole park, which defines overall management strategy and is implemented via management prescriptions for the 28 individual compartments (Mayol 1991). As in all conservation management, clear definition of objectives is important and difficult. The overall objective is the maintenance and enhancement of biological diversity in what is a highly man-modified ecosystem. Examples of the history of modification, the industrial archaeology of the site, should also be conserved. The natural history and human history must be attractively presented to visitors including tourists, local inhabitants and school children. The park must seek local, Balearic and international support so as to acquire the resources needed for imaginative management. It must be a 'good neighbour' both to the farmers inland and the tourist industry along the coastal strip. All this is happening to a greater or lesser degree. The question to be addressed here is how far data on the biomass and productivity of *Phragmites* and *Cladium* can contribute to the management of the individual compartments and of the park as a whole.

The three main management tools at present are fire, grazing and manipulation of the hydrology. It is very difficult to prevent the occurrence of fire which can spread in from outside the park, though the frequency of burning could be *increased* by controlled fires. Also some parts of the park are more vulnerable to fire spreading from outside than others.

Some areas of marsh are subject to relatively intensive grazing by cattle or horses; the marsh is usually burnt first to make it more accessible to the stock but thereafter the grazing itself should keep it open. The impact of grazing on the biomass and productivity of *Phragmites* has not been studied in this project though work on *Phragmites* grazing marsh has been carried out in the Netherlands (e.g. van Deursen & Drost 1990) and elsewhere. It would be a natural extension of the Albufera work described here.

The hydrology can be manipulated in various ways, such as reedcutting and dredging in existing canals, the creation of new canals and the manipulation of water levels by the raising and lowering of sluices. It would be feasible for the marsh to act as a treatment plant for the runoff from Sa Pobla and the horticultural area, though this would be more effective if the resultant biomass were harvested. Management of the marsh by mowing or cutting is not current practice at Albufera, but at the time when there was a paper factory at Sa Roca, within the marsh, the reeds and sedges were cut as a source of fibre for paper manufacture. If a cutting regime were reinstated in some compartments, it might be more controllable than fire, and could be used to shift the balance between *Phragmites* and *Cladium* if this were thought desirable. The crop could be used as fuel, and the nutrient removal function of the vegetation would be enhanced.

In a situation where the diversity of higher plant species is relatively low, structural diversity may become even more important. The habitat requirements of important bird and insect species need to be better known and the simplified biomass and production data presented here need to be extended to provide a better description of the structure of the ecosystem. This involves the demography of the shoot populations of the two populations, numbers and dimensions. The data collected do include much of this information but they require further analysis.

Phragmites sometimes gives the impression of supporting greater biological diversity than *Cladium*. Conversely *Phragmites* is ubiquitous whereas *Cladium* has a more restricted distribution. In terms of the conservation of Mediterranean wetlands, the *Cladium* marsh on the Albufera is of outstanding importance.

CONCLUSION

It has long been a hope, embodied for example in the International Biological Programme, that ecosystem management, especially for amenity and conservation, could be based on scientifically valid ecosystem models. But these models, typically based on energy flow and mineral cycles, have proved elusive and expensive. Populations of single species, whether plant or animal, are simpler. In a situation where a single species, or in this case two species, dominates a considerable area, it should be easier to develop a functional population model as a basis for management. One bonus in the case of *Phragmites* is the increasing amount of ecological work on the species which, although it covers a considerable geographical range, appears to present a coherent description of its autecology.

In making a comparison, in the introduction to this chapter, with work on heather, it is worth considering the volume (scientist-years, journal pages, etc.) of research on heather. If there were a comparable volume of work on *Phragmites* (especially) and also *Cladium*, understanding and management of the Albufera would be much easier. But the habitat is less user-friendly and more prone to damage by the researcher.

ACKNOWLEDGEMENTS

I was introduced to both the Albufera and Earthwatch by Max Nicholson. Contributions to the cost of this research project were made by Earthwatch, the Oleg Polunin Memorial Fund from Charterhouse and the BES Small Projects Fund (Coalburn Trust) to all of whom I am most grateful. Logistic support was provided by Joan Mayol, Biel Perello and other park staff, and also by Toni Martinez and Enrique Descals from Universitat de les Illes Balears. The support of Pat and Dennis Bishop meant more than they realized. I was helped with the actual fieldwork by my wife, Dinah McLennan, Nick Riddiford, Frank Perring, Jon King and Earthwatch volunteers too numerous to mention (but a special mention is due to Sharon Bey who came back for a second time).

This chapter is contribution No. 8 in the Parc National S'Albufera series.

REFERENCES

Armstrong, H.M. & MacDonald, A. (1992). Tests of different methods for measuring and estimating utilisation rate of heather (*Calluna vulgaris*) by vertebrate herbivores. *Journal of Applied Ecology,* **29,** 285–294.

Conway, V.M. (1942). *Cladium mariscus* (L.) R.Br. Biological Flora of the British Isles. *Journal of Ecology,* **30,** 211–216.

Felton, M. & Marsden, J. (1990). Heather regeneration in England and Wales. A feasibility study for the Department of the Environment by the Nature Conservancy Council, Peterborough.

Gimingham, C.H. (1972). *Ecology of Heathlands.* Chapman & Hall, London.

Hara, T., van der Toorn, J. & Mook, J.H. (1993). Growth dynamics and size structure of *Phragmites australis,* a clonal plant. *Journal of Ecology,* **81,** 47–60.

Haslam, S.M. (1972). *Phragmites communis* Trin. Biological Flora of the British Isles. *Journal of Ecology,* **60,** 585–610.

Kassas, M. (1951). Studies in the ecology of Chippenham Fen. II. Recent history of the fen from evidence of historical records, vegetational analysis and tree-ring analysis. *Journal of Ecology,* **39,** 1–32.

MacDonald, A. (1990). *Heather Damage: a Guide to Types of Damage and their Causes.* Research and Survey in Nature Conservation. Conservation Management of the Uplands No. 28. Nature Conservancy Council, Peterborough.

Martinez-Taberner, A.M., Moya, G., Ramon, G. & Forteza, V. (1990). Limnological criteria for the rehabilitation of a coastal marsh. The Albufera of Majorca, Balearic Islands. *Ambio*, **19**, 21–27.

Mayol, J. (1991). Plan d'us i gestio del Parc Natural de S'Albufera de Mallorca. Documents tecnics de Conservacio 3. SECONA. Palma de Mallorca.

Mook, J.H. & van der Toorn, J. (1982). The influence of environmental factors and management on stands of *Phragmites australis*. II. Effects on yield and its relation with shoot density. *Journal of Applied Ecology*, **19**, 501–518.

Riddiford, N. (1993). Monitoring for environmental change, The Earthwatch Europe S'Albufera project. A summary report of the fourth season's work 1992. Earthwatch, Oxford.

Tscharntke, T. (1990). Fluctuations in abundance of a stem-boring moth damaging shoots of *Phragmites australis*: causes and effects of overexploitation of food in a late-successional grass monoculture. *Journal of Applied Ecology*, **27**, 679–692.

van der Toorn, J. & Mook, J.H. (1982). The influence of environmental factors and management on stands of *Phragmites australis*. I. Effects of burning, frost and insect damage on shoot density and shoot size. *Journal of Applied Ecology*, **19**, 477–500.

van Deursen, E.J.M. & Drost, H.J. (1990). Defoliation and treading by cattle of reed *Phragmites australis*. *Journal of Applied Ecology*, **27**, 284–297.

Watt, A.S. (1947). Pattern and process in the plant community. *Journal of Ecology*, **35**, 1–22.

Watt, A.S. (1955). Bracken versus heather, a study in plant sociology. *Journal of Ecology*, **43**, 490–506.

Westlake, D.F. (1963). Comparisons of plant productivity. *Biological Reviews*, **38**, 385–425.

Wheeler, B.D. & Shaw, S.C. (1991). Above-ground crop mass and species richness of the principal types of herbaceous rich-fen vegetation of lowland England and Wales. *Journal of Ecology*, **79**, 285–301.

13. AN INVESTIGATION INTO METHODS FOR CATEGORIZING THE CONSERVATION STATUS OF SPECIES

GEORGINA M. MACE

Institute of Zoology, Zoological Society of London, Regent's Park, London NW1 4RY, UK

SUMMARY

New proposals for methods of categorizing the conservation status of species in The World Conservation Union (IUCN) Red Lists have been applied to a range of vertebrate taxa over the last 2 years. The data arising from this process provide a means of assessing the applicability and consequences of the new quantitative criteria. A detailed analysis is presented in this chapter of 779 species and 643 subspecies in 10 higher order taxa (orders and families) of vertebrates. Overall, 43% of species and 51% of subspecies were recorded as 'threatened', a substantial increase over recent estimates for similar groups. This increase appeared to result from the application of quantitative criteria in a comprehensive manner. In an assessment of the threat category classification for waterfowl, population size was the most significant operational variable. Declining populations and restricted geographical range area were also likely to contribute significantly to threatened status, but alone these did not correlate closely with threat level. Alternative approaches to identifying indicators of extinction that can be readily measured on natural populations are discussed. These approaches hold promise for developing more specific predictions with better prospects for directing effective conservation action.

INTRODUCTION

The purpose of threatened species categories is to highlight species at risk of extinction. The regular publication by IUCN, The World Conservation Union, of Red Data Books, engenders public interest, and also serves to focus attention on the species listed. Conservationists and parks planners use the lists to direct their work and they also serve as a basis for many educational programmes. They have an important role in publicity for, and planning of, species-based conservation programmes.

The Red Data Books first appeared in 1966 (Munton 1987) and were initially general compendia about endangered mammals and birds. Since that time there have been numerous other publications from IUCN, more recently including complete listings for animals, the IUCN Red Lists (IUCN 1990) which are now published every 2 years. The present system for classifying species in Red Data Books has a long history (Munton 1987; Scott *et al.* 1987) and is based upon the categories Extinct, Endangered, Vulnerable, Rare, Indeterminate and Insufficiently Known, with other categories appearing in some volumes (Fitter & Fitter 1987; IUCN 1990).

Several problems with these categories and the lists that are based upon them have recently emerged, some of which are caused by difficulties over interpretation of the subjective definitions. The definitions are merely stated in terms of '. . . in danger of extinction . . .'. Without any reference to a time scale or the estimated probability of extinction within this time frame, the definition can be interpreted in different ways by different authorities. With the proliferation of threatened species lists of national, regional and taxonomic bases, which may disagree over the status of a particular species, the problem has become more evident, since under the existing definitions there is no good way to resolve these differences. Many of the recent publications, especially the Action Plans produced by IUCN/Species Survival Commission (SSC) Specialist Groups, also use the threat category for selecting high priority projects. If conservation action is to be based upon them, threatened species categories will require some better validated approach. A more objective system that can be challenged and judged against an accepted set of rules is needed.

The publication of Red Data Books and Lists has other implications. Diamond (1987), among others, has discussed how this immediately puts a limit on the number of taxa that are recognized as being under threat of extinction because only described species can be listed, and once a list exists conservationists and survey expeditions focus especially on those species. Little attention is paid to taxa that are not listed, either because they have not been described or are not listed as threatened, and their plight worsens. In addition, once a taxonomic or regional list is produced, there is an implicit message that all described taxa in that group or region have been evaluated. This is often far from the case. Even among the vertebrates only a proportion of all the described species have been evaluated by IUCN. The birds have been fully reviewed; but only about 50% of mammals, less than 20% of reptiles, 10% of amphibians and 5% of fish (WCMC 1992). Despite these difficulties, threatened species categories are widely used and recognized, especially at an international

TABLE 13.1. Proposed definitions for IUCN threatened species categories from Mace and Lande (1991)

Critical	50% probability of extinction within 5 years or two generations, whichever is the longer
Endangered	20% probability of extinction within 20 years or 10 generations, whichever is the longer
Vulnerable	10% probability of extinction in 100 years

level. If the existing level of credibility given to them is to continue, they need to be more soundly based and more consistently applied.

In 1990, IUCN began a process to redefine the categories of threat used in Red Data Books. Mace & Lande (1991) put forward for discussion a set of definitions for the categories (see Table 13.1), as well as some population criteria for assigning taxa to the categories. The categories are reduced in number compared to existing IUCN categories, and describe the probability of extinction within a specified time frame. With increasing threat levels the probabilities increase and the time frame shortens. This makes sense both in biological terms as well as in practical terms by focusing attention especially on the most threatened forms. The term 'threatened' is used to describe species that fall into any of the categories Critical, Endangered or Vulnerable.

The development of these new definitions was based around a set of objectives to ameliorate the problems described above. Fundamental population characteristics that should be applicable across broad taxonomic groups were at the core of the proposal, which aimed to make maximum use of however many or few population data were available. We (Mace & Lande 1991) recommended that the criteria were especially appropriate for large vertebrates and that further development was needed for other groups. However, the definitions, because of their very funda-mental nature, should be appropriate across all life forms.

We also stressed that there should be a distinction drawn between the assessment of the severity of threat, and the setting of conser-vation priorities. Threatened species categories should simply present an assessment of how likely a species is to go extinct within a certain period of time. In contrast, conservation priorities will need to be based upon a set of additional criteria as well as this. For example, the likelihood that restorative action can be successful, species densities and the density of other threatened species occupying the same habitat, taxonomic uniqueness, and financial and logistical considerations will all be important.

Over the last couple of years these new definitions and the population criteria have been applied quite widely in various activities of the IUCN/ SSC, although the precise proposal is still under review. A new set of draft criteria for IUCN has recently been published, based around the same general principles (Mace *et al.* 1992) and is under consultation within the IUCN/SSC. In this chapter I use the data that have been generated from application of Mace and Lande (1991) definitions and criteria to examine the utility of the system, the extent to which it will alter the number of species listed as threatened and the kinds of population data that are employed in the analysis. Some more recent developments in identifying significant and easily measured population variables for predicting extinction are discussed at the end.

APPLICATION OF THE NEW CATEGORIES

The process

Since 1990, various specialist groups of the IUCN/SSC have been applying the Mace and Lande (1991) criteria in a set of workshops aimed at providing strategic guidance for the application of intensive management techniques to threatened taxa. The workshops are called Conservation Assessment and Management Plans (CAMPs) and are generally a collaborative undertaking by the Captive Breeding Specialist Group (CBSG) and the taxon specialist group for the taxon under consideration (see Seal *et al.* 1993a,b). Each workshop covers a higher taxonomic level (usually an order of family) and a number of experts including field workers, taxonomists, wildlife managers, population biologists and *ex-situ* managers are assembled. Using a previously agreed taxonomy, the group then considers the status of each species or subspecies, and makes recommendations for conservation action under a number of different categories (e.g. taxonomic research, field surveys, husbandry research, etc.). The assessment of status is made for every species or subspecies using the criteria published in Mace and Lande (1991). The output from each CAMP includes a table listing all taxa considered and the category of threat assigned, as well as information on estimates of population size, the direction of recent trends in population size, the number of known subpopulations and the size of the known geographical distribution, that has been used to assign a threat category. These reports form the basis of the analysis presented here (CBSG 1992).

By September 1992, eight CAMPs had been completed and published – waterfowl, cranes, parrots, Asian hornbills, primates, felids, canids and hyenas and antelope, though the number is increasing at a rapid rate and by May 1993 the total number of CAMP workshops held had risen to 21, covering over 3500 taxa (species and subspecies) (Seal *et al.* 1993a). Of these, 3150 taxa had been classified using Mace and Lande (1991) criteria and 1345 of these were threatened (43%). This is a much higher proportion of threatened species than is seen in the IUCN Red List, which gives the highest frequencies of threatened status among well-studied groups such as the mammals and birds, which have respectively 11.7 and 10.6% threatened (WCMC 1992). This increase in the frequency of listing by the CAMPs could be a result of the new criteria and definitions, or of the complete analysis of all members of a higher order taxon, or could be due to the increased attention given to subspecies by CAMPs, which routinely aim to consider all subspecies. The CAMP data allow detailed examination of the basis for species listing which is analysed in this chapter.

There is a very strong emphasis by the CAMP workshop organizers on numerical estimates being given for as many population criteria as possible. This is felt to be important both to make assessments transparent and to open the process and data to review and improvements by other workers. In some cases this policy has caused problems as it leads implicitly to the publication and dissemination of some poor quality data which have been through only a limited review process. The analysis presented here is therefore limited to a very general investigation of the results, especially focusing on how well the criteria have met the objectives laid down, and to guide the development of a system which can be formally recommended.

Assessing the qualities of the system

In general CAMP workshops have been successful in classifying the vast majority of species and subspecies they consider. In many cases, all taxa were classified, usually by making inferences from, for example, published species densities for related forms in similar habitats, range maps, habitat maps and anecdotal observations. On average over 90% of species and subspecies considered have been classified (Seal *et al.* 1993a), though the proportion varies widely with the group under consideration. Many CAMPs have successfully classified all species and subspecies. Several workshops did, however, append an additional threat category at a lower level of threat than the Mace and Lande 'Vulnerable'. This was called variously 'High Anxiety' or 'Vulnerable II' and suggests that there were taxa about which field biologists were concerned but which did not trigger the criteria for the 'Vulnerable' category. Thus the system seems to meet the objective of being generally applicable (at least to most large vertebrates) but the criteria for the lowest level of threat (and possibly also the definition) may need further consideration. The criteria could also clearly be applied at a variety of levels since species, subspecies and geographical populations were routinely classified in many of the CAMP reports.

Perhaps one evident failing so far has been to include some statement about uncertainty in either the population parameters or the classification resulting from these. This means that categories for species with excellent population, range and historical data are presented in lists alongside classifications based on the most tenuous kind of inference. The inferences are doubtless important for developing a better process and better data to feed into it, but perhaps need to be indicated more clearly.

The number of threatened species

Since the CAMP process is exhaustive in its analysis of a particular higher taxonomic group it should be possible to review the effect that this has upon the total numbers of threatened species.

This analysis was based on CAMP reports for 10 families or orders of reptiles, birds and mammals (see Table 13.2). These groups were selected because the data tables were relatively complete and it had not proven necessary to add categories for any of them. In addition most of these have now been through a review subsequent to the initial workshop. The data were extracted from summary tables published by CBSG (1992).

An examination of the numbers of threatened species here is complicated by the fact that the CAMPs make many assessments below the species level, generally for subspecies, but sometimes also including geographically distinct populations. The extent of this varies but could artificially inflate the frequency of threatened status if subspecies are more likely to be threatened than species. This analysis was therefore done separately for species and subspecies. Where all subspecies of a particular species were categorized the species was excluded from the analysis (generally it was not classified at all at the species level). Where only a subset of subspecies was categorized, both species and subspecies data were included in the analysis. Other subunits (e.g. geographical populations, island races, etc.) were excluded. In the case of the Canidae, all subspecies were listed in the CAMP report but only a small number were classified, often just those believed to be threatened. For this group therefore, the subspecies classification was not used, but species categories taken or inferred from the report to provide data for analysis. In total 779 species and 643 subspecies were included in the analysis of the 10 higher order taxa.

Table 13.2 shows the number and percentage of species and subspecies classified into the three categories of threat, 'Safe' and 'Unknown' for each of the 10 taxa. The percentage in each threat category is calculated only from the number classified (excluding unknowns). In two of the groups there was a significant difference between species and subspecies in the numbers threatened, though in opposite directions. In boid snakes, relatively more species than subspecies were threatened ($\chi^2 = 4.50$, $P < 0.05$) and in psittacines there were relatively more threatened subspecies than species ($\chi^2 = 24.9$, $P < 0.001$). Although it might be anticipated that subspecies would be more susceptible as a result of their more limited distribution, this does not appear to be borne out by the results. In fact, for many taxa, subspecies are recognized commonly among widespread abundant species and many monotypic species have very limited distributions. Thus there does not seem to be any difference between species and subspecies in their threatened status. Overall 93% of species and 86% of subspecies could be classified by the new criteria.

In all groups studied Vulnerable was the most common category among the threatened taxa, and Critical the rarest. The most seriously threatened group studied here was the Asian hornbills (Bucerotidae) with over 80% of species and subspecies threatened. The waterfowl (Anseriformes) were the least threatened with just over 30% of both species and subspecies. Overall, the average level of threat was 43.2% of species and 51.1% of subspecies (Table 13.2).

These levels can now be compared with the IUCN Red List classification for the same groups. In Table 13.3, the numbers listed as threatened by IUCN (1990) (i.e. Endangered, Vulnerable, Rare, Indeterminate, Insufficiently Known) are compared with numbers listed by the CAMP documents as Critical, Endangered or Vulnerable for each of the groups studied. Except for species of deer, cranes and canids (Cervidae, Gruidae and Canidae) more species were consistently listed as threatened by the CAMP, and the difference is further seen in comparisons of subspecies listing.

This change in number of threatened species from 211 to 313 represents an overall increase of 48%. This may partly result from the completeness of surveys by CAMPs, but many of the groups studied here are well known and have been comprehensively reviewed by appropriate authorities (Collar & Andrew 1988; IUCN 1990) so this seems unlikely to be a complete explanation. It seems likely that much of the increase also results from the application of the new criteria, which should tend to pinpoint species that had not previously been considered to be at risk, perhaps because no active field survey or research was focused upon them. It is also likely to be the case that where species data are being

TABLE 13.2. Summary of CAMP results for selected groups. The table lists the number and percentage of species and subspecies in each of the categories Critical, Endangered, Vulnerable and Safe. The number and percentage that could not be classified (Unknown) is given, and percentages for threatened taxa are calculated as the percentage of those species and subspecies that were classified (total known). The probabilities at the end of each line are for a χ^2 value to reject the null hypothesis that there is no significant difference in the proportion of threatened species and subspecies within each group

Taxon	Name	Species No.	Critical		Endangered		Vulnerable		Safe		Unknown		Total	Total known	% Threatened
			No.	%	No.	%	No.	%	No.	%	No.	%			
Reptilia	Boidae	37													
	Species		1	5.88	2	11.76	6	35.29	8	47.06	0	0.00	17	17	52.94
	Subspp.		1	1.45	4	5.80	11	15.94	53	76.81	0	0.00	69	69	23.19[*]
	Varanidae	45													
	Species		0	0.00	1	3.45	10	34.48	18	62.07	5	0.17	34	29	37.93
	Subspp.		0	0.00	1	5.00	6	30.00	13	65.00	5	0.25	25	20	35.00[ns]
	Iguanidae	39													
	Species		1	4.00	2	8.00	14	56.00	8	32.00	2	0.08	27	25	68.00
	Subspp.		2	6.90	8	27.59	13	44.83	6	20.69	8	0.28	37	29	79.31[ns]
Aves	Anseriformes	154													
	Species		5	4.59	9	8.26	22	20.18	73	66.97	0	0.00	109	109	33.03
	Subspp.		5	4.07	14	11.38	22	17.89	82	66.67	0	0.00	123	123	33.33[ns]
	Gruidae	15													
	Species		1	16.67	0	0.00	3	50.00	2	33.33	0	0.00	6	6	66.67
	Subspp.		8	32.00	7	28.00	4	16.00	6	24.00	0	0.00	25	25	76.00[ns]

Psittaciformes	356									
	Species	22 7.28	25 8.28	71 23.51	184 60.93	12 0.04	314	302	39.07
	Subspp.	9 8.33	17 15.74	47 43.52	35 32.41	7 0.06	115	108	67.59†
Bucerotidae	24									
	Species	1 10.00	3 30.00	4 40.00	2 20.00	0 0.00	10	10	80.00
	Subspp.	4 9.30	10 23.26	22 51.16	7 16.28	0 0.00	43	43	83.72^{ns}
Mammals Marsupialia	210									
	Species	6 3.35	20 11.17	60 33.52	93 51.96	31 0.17	210	179	48.04
	Subspp.	0 –	0 –	0 –	0 –	0 –	0	0	–
Canidae	34									
	Species	2 5.88	4 11.76	7 20.59	21 61.76	0 0.00	34	34	38.24
	Subspp.	0 –	0 –	0 –	0 –	0 –	0	0	–
Cervidae	58									
	Species	4 28.57	4 28.57	3 21.43	3 21.43	4 0.29	18	14	78.57
	Subspp.	17 12.41	28 20.44	23 16.79	69 50.36	69 0.50	206	137	49.64^{ns}
Total										
	Species	43 5.93	70 9.66	200 27.69	412 56.83	54 6.93	779	725	43.17
	Subspp.	46 8.30	89 16.06	148 27.61	271 48.92	89 13.84	643	554	51.00
	Total	89 6.96	159 12.43	348 27.21	683 53.40	143 10.06	1422	1279	47.00

$* P < 0.05. † P < 0.001.$
ns = Not significant.

TABLE 13.3. Comparison of numbers of threatened species and subspecies from CAMP analyses using Mace and Lande (1991) definitions and criteria and existing IUCN (1990) classifications for the same higher order taxa

	CAMP			IUCN		
	Species	Subspp.	Total	Species	Subspp.	Total
Boidae	9	16	25	9	3	12
Varanidae	11	7	18	4	1	5
Iguanidae	17	23	40	15	1	16
Anseriformes	36	41	77	21	–	21
Gruidae	4	19	23	7	–	7
Psittaciformes	118	73	191	79	–	79
Bucerotidae	8	36	44	7	–	7
Marsupialia	86	–	86	37	–	37
Canidae	13	–	13	19	–	19
Cervidae	11	68	79	13	17	30
Total	313	283	596	211	22	233

assembled for this kind of exercise, and there is a paucity of high-quality information, the tendency will be for participants in the workshop to be cautious and give the benefit of the doubt to listing. Overall, however, it seems likely that the increase in threatened species numbers results from the detailed process and the application of quantitative criteria.

The increase in the number of subspecies listed (22 to 283) is certainly an overstatement of changes in threat levels arising from application of the new criteria. IUCN lists rather few subspecies, unlike the CAMPs which routinely aim to list all. However, it is worth noting that the procedure used in this analysis, to exclude species from CAMP listing where all subspecies were listed, should increase the IUCN count of threatened species relative to CAMPs.

In conclusion, the much higher level of threatened species documented in CAMPs is not simply a product of comprehensive reviews of the taxa, nor of an increased emphasis on analysis at the level of subspecies.

POPULATION AND DISTRIBUTION VARIABLES IN CLASSIFICATION

Introduction

The criteria in Mace and Lande (1991) were essentially based around four different kinds of population variable: population size, population subdivision, rates of population decline and the impact and frequency of catastrophes. Any one of these factors operating at a certain level can potentially trigger listing, though more usually two are required. In fact, the different variables have very different population dynamic properties so that, for example, the effects of a deterministic population decline can be more reliably predicted than the consequences of small and stable population size. In spite of this, population size dominates much of the thinking about extinction risk, especially in empirical studies (Soule 1987; Pimm *et al.* 1988; Berger 1990). Population size may be deceptively hard to estimate and for many populations its effects are confounded by environmentally driven fluctuations which may render single census estimates meaningless. Metapopulation structure too has complex interactions with extinction risk, and is hard to deduce for most natural populations without intensive and extended field studies. Because of these difficulties in measuring different population criteria and predicting their consequences, it was of interest to investigate the extent to which the different criteria were used in the CAMP process and which were most influential in subsequent classification.

The criteria proposed by Mace and Lande (1991) did not include geographical range area although this has commonly been suggested as a surrogate for measures of endemism, and as an indicator of conservation status. Most notably, Bibby *et al.* (1992) have used restricted range area ($\leq 50\,000\,km^2$) to classify bird species, and to recommend areas where two or more restricted range birds are found as priority areas for conservation. Although it may be difficult to generalize across widely differing taxonomic groups, range area can be expected to correlate with population size and because it is relatively easily estimated for most species (for example from sites of known occurrence), it is a good candidate for inclusion in the criteria. The relationship with threatened status was therefore investigated here.

Methods

The CAMP reports include a column for listing population size estimates, subpopulation numbers, nature of population size change (increasing, stable or decreasing) and geographical range area. The population data are rarely precise, and generally an order of magnitude is given. In this analysis, therefore, population size was measured in orders of magnitude where any data that gave a greater or less than value were assumed to mean that the estimated population size was within one order of magnitude above or below the quoted value (thus, $\leq 10\,000$, was taken to mean 1001–10 000, etc.). Subpopulation data were rarely presented, and even when they were, migration rates among subpopulations could hardly ever be estimated. No further use was made of the subpopulation data. Population trends were generally presented as increasing, decreasing or stable. Geographical range areas were also mostly order of magnitude estimates, though island forms are identified separately. They were classified into seven classes (AA: islands $\leqslant$ 50 000 km^2; A $\leqslant$ 50 000 km^2; B $\leqslant$ 100 000 km^2, C $\leqslant$ 500 000 km^2; D $\leqslant$ 1 000 000 km^2; E $>$ 1 000 000 km^2).

Using data of this nature it was possible to analyse the basis for classification in the waterfowl CAMP (Anseriformes), where there was an exceptionally complete data set which had been through a comprehensive review subsequent to the workshop (Ellis-Joseph et al. 1992).

Results

Of the four variables listed in CAMP reports, three are applied in Mace and Lande (1991) criteria: population size, population trends and the number of subpopulations. Table 13.4 shows the number of Critical, Endangered and Vulnerable and Safe species and subspecies that were recorded as having increasing stable or declining populations. All Critical taxa are declining, but even among the Endangered taxa there are two taxa believed to be increasing, and nine (38%) were recorded as stable. In contrast, population size was much more closely correlated with threat category (Table 13.5). All Critical species and subspecies were estimated

TABLE 13.4. Relationship between trends in population size and category of threat assigned to (a) species and (b) subspecies in CAMP listings for waterfowl

Population trend	Category of threat				
	Critical	Endangered	Vulnerable	Safe	Total
(a) Declining	5	7	8	7	27
Stable	0	6	12	56	74
Increasing	0	1	1	20	22
Total	5	14	21	83	123
(b) Declining	5	6	14	14	39
Stable	0	3	7	49	59
Increasing	0	1	1	10	12
Total	5	10	22	73	110

TABLE 13.5. Relationship between population size and category of threat assigned to (a) species and (b) subspecies in CAMP listings for waterfowl

Population size estimate	Category of threat				
	Critical	Endangered	Vulnerable	Safe	Total
(a) $\leqslant 10^1$	2	0	0	0	2
$\leqslant 10^2$	1	0	0	0	1
$\leqslant 10^3$	2	2	0	0	4
$\leqslant 10^4$	0	12	19	1	32
$\leqslant 10^5$	0	0	2	38	40
$\leqslant 10^6$	0	0	0	31	31
$> 10^6$	0	0	0	14	14
Total	5	14	21	84	124
(b) $\leqslant 10^1$	1	0	0	0	1
$\leqslant 10^2$	2	0	0	0	2
$\leqslant 10^3$	2	2	0	0	4
$\leqslant 10^4$	0	8	12	0	20
$\leqslant 10^5$	0	0	9	28	37
$\leqslant 10^6$	0	0	1	28	29
$> 10^6$	0	0	0	17	17
Total	5	10	22	73	110

to number $\leqslant 10^3$, all Endangered taxa have estimated numbers in the classes $\leqslant 10^3$ and $\leqslant 10^4$, and all but one of the Vulnerable taxa had numbers estimated to be in the $\leqslant 10^4$ and $\leqslant 10^5$ classes; 31 (72%) of which were in the $\leqslant 10^4$ class. Taxa classified as Safe were all bar one in the class $> 10^4$. These relationships are illustrated for species of waterfowl in Figs 13.1 and 13.2.

Not surprisingly, there is a strong correlation between population size classes and range area size classes (Table 13.6). The partial correlation coefficients, removing the effects of both other variables (including trend) are reduced in all cases, but population size remains the most highly correlated variable. This is perhaps not surprising since range area does not, unlike the other two variables, formally have a role in the classification procedure, However, it remains significantly correlated, and for species at least, still accounts for 20% of the variance in the category classification.

Many Critical, Endangered and Vulnerable waterfowl species have geographical range areas in excess of $50\,000\,\text{km}^2$ (Table 13.7) and the spread of range area sizes among threatened species and subspecies can be seen to be quite extensive. In total, four of 10 Critical taxa, 15 of 24 Endangered taxa and 32 of 48 Vulnerable taxa have range areas in excess

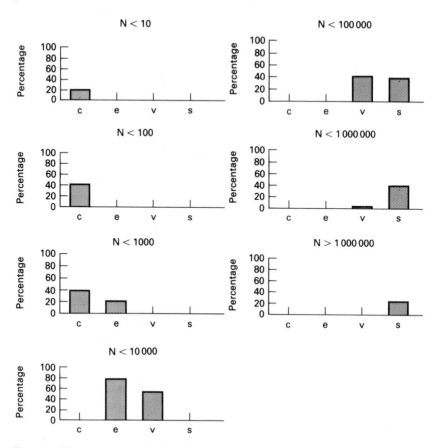

FIG. 13.1. The percentage of Critical (c), Endangered (e), Vulnerable (v) and Safe (s) species in each of the seven population size classes, compiled from the CAMP report for waterfowl.

of 50 000 km². Among Safe taxa, two of 156 have range areas less than 50 000 km².

The use of restricted range area (<50 000 km²) alone does not provide a good estimator of threatened status in this group. Waterfowl may be somewhat exceptional in this regard as they have quite extensive ranges, but it seems unlikely from inspection of the data in Table 13.7 that higher cut-off levels would enable range size alone to function as an efficient predictor of threatened status. However, in combination with observed or expected decline rates it may prove to be a useful measure for species where population size estimates are impossible to come by. With basic

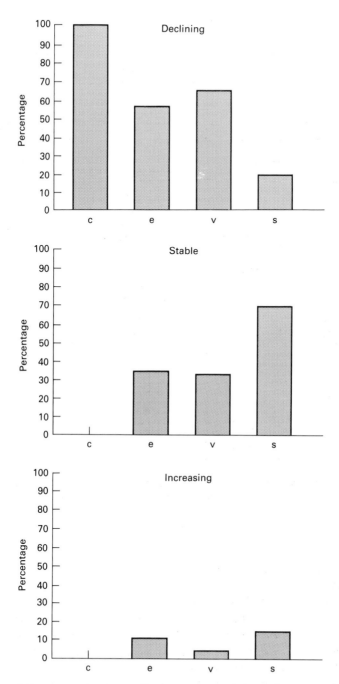

Fig. 13.2. The percentage of Critical (c), Endangered (e), Vulnerable (v) and Safe (s) species with declining, stable or increasing population numbers compiled from the CAMP report for waterfowl.

TABLE 13.6. Kendall rank correlation coefficients and partial rank correlation coefficients (in parentheses) among population size classes, population size trend classes and geographical range size area classes of species ($n = 110$) and subspecies ($n = 123$) of waterfowl. All coefficients are statistically significant at least at the $P < 0.001$ level

Taxonomic level	Variable	Correlation coefficients	
		Area	Threat category
Species	Population size	0.620	0.703 (0.560)
	Area		0.693 (0.448)
	Trend		0.415 (0.361)
Subspecies	Population size	0.594	0.741 (0.570)
	Area		0.539 (0.295)
	Trend		0.439 (0.272)

TABLE 13.7. Relationship between geographical range area and category of threat assigned to (a) species and (b) subspecies in CAMP listings for waterfowl

Geographical range area (km^2)	Category of threat				Total
	Critical	Endangered	Vulnerable	Safe	
(a) Island					
≤50 000	0	3	1	0	4
≤50 000	2	1	0	0	3
≤100 000	1	1	2	1	5
≤500 000	2	3	5	1	11
≤1 000 000	0	2	13	19	34
>1 000 000	0	0	1	52	53
Total	5	10	22	73	110
(b) Island					
≤50 000	2	5	4	2	13
≤50 000	2	0	0	0	2
≤100 000	1	2	4	4	11
≤500 000	0	5	4	6	15
≤1 000 000	0	2	5	23	30
>1 000 000	0	0	4	48	52
Total	5	14	21	83	123

information about the characteristic species densities, range area could be considered for inclusion in the criteria.

These results have important implications for the application of the criteria. The subpopulation criterion was almost never utilized, and the

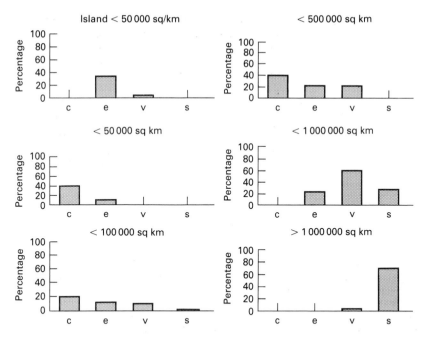

FIG. 13.3. The percentage of Critical (c), Endangered (e), Vulnerable (v) and Safe (s) species in each of the six geographical range area size classes compiled from the CAMP report for waterfowl.

population decline criterion only seems to come into play when there is good evidence (e.g. from exploitation for trade) that extremely high rates of decline are being experienced. In spite of the difficulties in predicting its consequences, population size was the dominant operational variable. Small geographical range area is not well correlated with high extinction risk in the waterfowl, though it is generally correlated with threat category. With further development it could provide an alternative criterion for species where population size data are not available.

EXTINCTION INDICATORS
IN NATURAL POPULATIONS

While population size dominates in much of the literature on threatened species (e.g. minimum viable populations, Soule 1987), other kinds of population parameters are more commonly employed in deterministic models of extinction, for example, r and the variance in r (Goodman 1987) and r and reproductive value (Lande & Orzack 1988). However, for most practical purposes where species are rare and declining, and field surveys are by necessity limited in time and scope, different kinds of variables are needed. Ideally, these should be variables that are relatively easy to census in field trips of limited duration, and which serve as reliable indicators of extinction. To this end they need to precede extinction with high probability, but also to give sufficient time between being observed and extinction for restorative intervention to be possible.

Using population simulation models some candidate population variables have been examined in studies based upon metapopulation models set to reflect observed demographic rates in wild populations of marine mammals, gorillas and ungulates (Durant 1991; Durant et al. 1992; Durant & Mace 1993). The variables include combinations of population size, adult, juvenile or population sex ratio, and years of no recruitment. Many of these variables, especially low recruitment and juvenile sex ratio, proved to be quite reliable both at predicting imminent extinction (within 60 years) as well as giving sufficient warning (at least 10 years) for management intervention to take place (Durant 1991; Durant & Mace 1993). However, the utility of these variables proved to be susceptible both to changes in population growth rate, in levels of environmental variation and to the extent of migration between breeding groups, all of which would require prohibitively extensive monitoring for such an approach to be effective. However, in a metapopulation model simple counts of the number of breeding groups proved to be an efficient indicator as long as the migration rates remained reasonably high.

It seems that relatively easily measured population parameters, such as years of no recruitment and biased juvenile sex ratios, in combination with small population size could be useful indicators for predicting high extinction risks, and this approach warrants further analysis to investigate alternative practical kinds of criteria for assessing extinction risk and for providing data upon which management actions to minimize extinction risk can be based.

CONCLUSIONS

In spite of the difficulties in assembling population data, the CAMP reports have shown that quantitative criteria can be applied at least to a range of vertebrate taxa. The results of this process led to the conclusion that threatened species are more abundant than is suggested by existing compilations, even for the most well-studied groups.

Population size estimates seem to dominate the decisions about threat categories, although there are both practical and theoretical problems with this measure. Geographical range area can contribute to an analysis of threatened status, but restricted area alone is not well correlated.

Simulation models suggest that other kinds of population parameters in combination with small population size may be more effective indicators of extinction risk, although these are case sensitive.

The development of new IUCN categories of threat is proceeding, with a new set of draft criteria (Mace *et al.* 1992) now under review and validation within the IUCN/SSC. These new criteria attempt to use a wider variety of population and distribution measure than did the Mace and Lande (1991) proposal, and also use population size more commonly in combination with other variables, especially with inferred or projected declines. Further examination of results from these new proposals and validation of the results will be likely to be a continuing process.

ACKNOWLEDGEMENTS

This work has been made possible by the efforts of many members of the IUCN/SSC network, and I am especially grateful to Dr Ulysses Seal for his work in applying the criteria and for making the data available for analysis. Simon Stuart has continued to guide and encourage the development of the new criteria. This work was made possible by the Pew Scholars Program in Conservation and the Environment.

REFERENCES

Berger, J. (1990). Persistence of different sized populations: an empirical assessment of rapid extinctions in bighorn sheep. *Conservation Biology*, **4**, 91–96.

Bibby, C.J., Collar, N.J., Crosby, M.J., Heath, M.F., Imboden, C., Johnson, T.H., Long, A.J., Stattersfield, A.J. & Thirgood, S.J. (1992). *Putting Biodiversity on the Map: Priority Areas for Conservation.* International Council for Bird Preservation, Cambridge.

CBSG (1992). *Conservation Assessment and Management Plan CAMP Summary Report.* IUCN/SSC/Captive Breeding Specialist Group, Minneapolis, MN.

Collar, N.G. & Andrew, P. (1988). *Birds to Watch: the ICBP Checklist of Threatened Birds.* International Council for Bird Preservation, Cambridge.

Diamond, J.M. (1987). Extant unless proven extinct? Or extinct unless proven extant? *Conservation Biology*, **1**(1), 77–79.

Durant, S.M. (1991). *Individual variation and dynamics of small populations: implications for conservation and management.* PhD Thesis, University of Cambridge.

Durant, S.M. & Mace, G.M. (1993). Species differences and population structure in population viability analysis. *Creative Conservation – the Interactive Management of Wild and Captive Animals* (Ed. by G.M. Mace, P.J. Olney & A.T.C. Feistner). Chapman & Hall, London.

Durant, S.M., Harwood, J. & Beudels, R. (1992). Monitoring and management strategies for endangered populations of marine mammals and ungulates. *Wildlife 2000* (Ed. by

D.R. McCullough & R.H. Barrett), pp. 252–261. Elsevier Applied Science, New York.

Ellis-Joseph, S., Hewston, N. & Green, A. (1992). *Global Waterfowl Conservation and Assessment Plan.* IUCN/SSC/Captive Breeding Specialist Group & The Wildfowl and Wetlands Trust, Minneapolis, MN.

Fitter, R. & Fitter, M. (Eds) (1987). *The Road to Extinction.* IUCN, Gland, Switzerland.

Goodman, D. (1987). The demography of chance extinction. *Viable Populations for Conservation* (Ed. by M.E. Soule), pp. 11–34. Cambridge University Press, Cambridge.

IUCN (1990). *1990 IUCN Red List of Threatened Animals.* IUCN, Gland, Switzerland.

Lande, R. & Orzack, S.H. (1988). Extinction dynamics of age structured populations in a fluctuating environment. *PNAS*, **85**, 7418–7421.

Mace, G.M. & Lande, R. (1991). Assessing extinction threats: toward a reevaluation of IUCN threatened species categories. *Conservation Biology*, **5**(2), 148–157.

Mace, G.M., Collar, N., Cooke, J., Gaston, K., Ginsberg, G., Leader-Williams, N., Maunder, M. & Milner-Gulland, E.J. (1992). The development of new criteria for listing species on the IUCN Red List. *Species*, **19**(Dec.), 16–22.

Munton, P. (1987). Concepts of threat to the survival of species used in Red Data Books and similar compilations. *The Road to Extinction* (Ed. by R. Fitter & M. Fitter), pp. 71–111. IUCN, Gland, Switzerland,

Pimm, S.L., Jones, H.L. & Diamond, J.M. (1988). On the risk of extinction. *American Naturalist*, **132**, 757–785.

Scott, P., Burton, J.A. & Fitter, R. (1987). Red Data Books: the historical background. *The Road to Extinction* (Ed by. R. Fitter & M. Fitter), IUCN, Gland, Switzerland.

Seal, U.S., Ellis, S.A., Foose, T.J. & Byers, A.P. (1993a). Conservation assessment and Management Plans (CAMPs) and Global Captive Action Plans (GCAPs). *IUCN/SSC/ Captive Breeding Specialist Group Newsletter*, **4**(2), 5–10.

Seal, U.S., Foose, T.J. & Ellis-Joseph, S. (1993b). Conservation assessment and management plans (CAMPs) and global captive action plans (GCAPs). *Creative Conservation – the Interactive Management of Wild and Captive Animals* (Ed. by G.M. Mace, P.J. Olney & A.T.C. Feistner). Chapman & Hall, London.

Soule, M. E. (1987). *Viable Populations for Conservation.* Cambridge University Press, Cambridge.

WCMC (1992). *Global Diversity: Status of the Earth's Living Resources.* Chapman & Hall, London.

14. TURNING CONSERVATION GOALS INTO TANGIBLE RESULTS: THE CASE OF THE SPOTTED OWL AND OLD-GROWTH FORESTS

DAVID S. WILCOVE

Environmental Defense Fund, 1875 Connecticut Avenue, NW, Washington, DC 20009, USA

SUMMARY

In the Pacific Northwest region of the United States, a bitter controversy over protection of the northern spotted owl (*Strix occidentalis caurina*) and the old-growth forests it inhabits illustrates how ecological research influences public policy. Three types of research have been used by environmentalists to bolster their contention that both the owl and the forests are imperilled by logging. Autecological studies have demonstrated the close association between spotted owls and old-growth forests, and they have supplied data on density, fecundity, and dispersal necessary to craft a conservation plan. Analyses of population viability have been used to highlight the inadequacy of various protection schemes proposed by the federal government. Inventories of old-growth forests have countered claims that such forests were abundant and in little need of protection. Victories (albeit temporary ones) have come by applying unrelenting pressure on legislators and land managers, via research studies, lawsuits, and media campaigns. This controversy has also rekindled a long-standing debate over the merits of ecosystem protection versus single-species protection. However, the distinction is a false dichotomy because the 'ecosystem' plans put forth for old-growth forests are based to a large extent on the habitat needs of individual species.

INTRODUCTION

On 2 April, 1993, the President of the United States, the vice-President, and three cabinet secretaries travelled to Portland, Oregon, for a public discussion of the long-standing, bitter controversy over logging of virgin, old-growth forests in the Pacific Northwest. The President's decision to convene a 'Forest Summit', as the event was quickly named by the press,

ıbiguous proof that the fate of these forests and the species
em had become one of the nation's paramount conservation
lthough the controversy is by no means resolved, the President's
involvement coupled with a recent decline in the amount of
logging are significant victories for environmentalists.

From an ecologist's perspective, the battle over the northern spotted
owl (*Strix occidentalis caurina* (Merriam)) and its old-growth forests is
especially noteworthy for two reasons. First, in contrast to previous
efforts to save a particular place or a particular species, this controversy
has centred around an entire ecosystem (the Pacific Northwest old-growth
forests) spread out over three states. Yet for scientific and legal reasons,
it is also very much a struggle to protect a single species, the outcome
of which may significantly alter our policies for the protection of all
endangered species. Second, more than any other natural resource controversy in recent years, this one has been shaped by scientific research.
I examine the type of research that has made this issue a national concern as well as the implications of protecting ecosystems as opposed to
individual species.

BACKGROUND

As northwestern timber companies cleared the virgin, old-growth forests
on their own property, they turned to federally owned lands, primarily
forests managed by the US Forest Service (and, to a lesser extent, the
Bureau of Land Management), to supply the big trees so valued for their
wood. Cutting intensities in the national forests of Washington and Oregon
climbed sharply after World War II, and hovered between four and five
billion board feet per year during the 1960s, 1970s, and 1980s (Fig. 14.1).
Most of the harvesting was done via clearcutting, in which all of the
trees on a site are removed. The cutting units were dispersed across the
landscape to minimize their visual and (presumably) ecological impact
and provide habitat for deer and elk. The cumulative loss of old-growth
forests alarmed and angered environmentalists.

The primary environmental laws governing management of the
national forests and other public lands contain no provisions specifically
mandating the protection of old-growth forests. There is, for example,
no 'Endangered Ecosystems Act' in the United States. However, the
National Forest Management Act of 1976 requires that management
plans for national forests 'provide for diversity of plant and animal
communities based on the suitability and capability of the specific land
area . . .'. To turn this vague concept into concrete administrative guide-

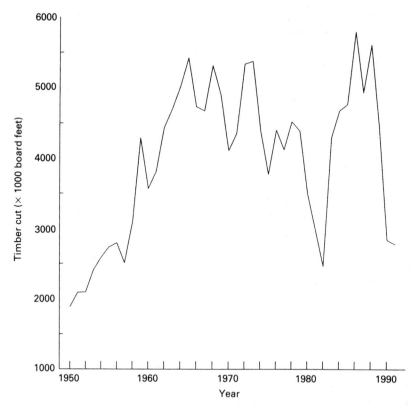

FIG. 14.1. Harvest levels in the national forests of Oregon and Washington, 1950–91. Data supplied by The Wilderness Society.

lines, the Forest Service developed a set of implementing regulations, one of which required the agency 'to maintain viable populations of existing native and desired non-native vertebrate species' in the national forests.

This mandate would set the stage for much of the ensuing controversy. Environmentalists would argue that the northern spotted owl was tied to old-growth forests and that only by protecting large amounts of these forests could its survival be ensured. Timber companies would question the extent to which the owls required old growth (versus younger forests) and the amount of protected land needed to ensure a viable owl population. Resolving either issue necessitated guidance from ecologists, who could determine the habitat requirements of the northern spotted owl and potentially answer the critical question 'How much is enough?'.

BUILDING THE SCIENTIFIC GROUNDS
FOR PROTECTION

Because so much literature has been generated on the topic of old-growth forests and spotted owls, it is unfair to single out any particular study or group of studies as most important. None the less, four studies – all quite different in substance and style – have greatly influenced the course of events.

Autecological studies

Intensive studies of northern spotted owls, largely sponsored or conducted by the Forest Service, began in the late 1960s and continue to the present. It became apparent quickly that northern spotted owls were closely associated with mature and old-growth forests (Forsman *et al.* 1984; Forsman 1988). For example, 95.5% of the sites in Oregon where spotted owls were found between 1969 and 1984 were dominated by old-growth forests or mixed stands of old growth and mature forest. In most cases, densities of spotted owls in younger (<80 years) forests are only one-tenth to one-seventh as high as densities in old growth (Thomas *et al.* 1990). The coastal redwood (*Sequoia* spp.) forests of northwestern California are an exception to this pattern. There, densities of owls in 50- to 80-year-old stands rival densities in old-growth forests. In this narrow region of the owl's range, a long growing season, highly productive soils, and abundant rainfall, combined with the stump-sprouting abilities of redwoods, cause forests to assume many of the structural characteristics of old growth at an early age.

Studies also revealed that spotted owls maintain large home ranges containing extensive amounts of old growth. Home-range size varies considerably, both within and among different geographical areas and forest types in the Pacific Northwest, making generalizations risky. None the less, median home ranges of 1200–2000 ha, including 250–1000 ha of old-growth forest, seem typical for pairs of spotted owls in much of the Pacific Northwest (Table 14.1). Home ranges in the State of Washington, approaching the northern limits of the bird's range, tend to be larger.

While the Forest Service could not ignore the linkage between spotted owls and old-growth forests, it was unwilling (or politically unable) to set aside significant amounts of old growth. Thus, throughout the 1970s and early 1980s, plans put forth by the agency protected an inadequate amount of forest for an inadequate number of owls. In 1977, for example, officials from the Forest Service and the Bureau of Land Management

TABLE 14.1. Home range size and amount of old growth and mature forest per home range for northern spotted owls. From Thomas *et al.* (1990, pp. 194–195)

	Annual home range (ha)				Old-growth and mature forest per home range (ha)			
	n	Median	Min.	Max.	*n*	Median	Min.	Max.
California								
Klamath Mountains								
Ukonom	9	1342	832	3167	9	1006	417	2289
Mad River	12	1204	730	1897	12	553	338	791
Willow Creek	2	685	509	861	2	324	148	499
Oregon								
South Coast								
Chetco	4	2273	2157	2509		Not available		
Klamath Mountains								
South Umpqua	3	571	419	609	3	249	228	311
Cow Creek	6	1662	1012	3034	6	627	587	803
Coast Ranges								
Tyee	5	1371	761	3349	5	822	666	1613
Peterson	4	2558	1410	4125	4	1056	520	1294
Eugene	4	2587	1504	3312	4	722	323	1449
Kellogg	5	1648	655	2543	5	412	282	803
Western Cascades	11	1196	584	3951	9	727	425	1533
Washington								
Western Cascades	13	2554	780	12 535	13	1328	780	8324
Olympic Peninsula	10	4020	1821	11 056	7	1854	1128	3420

agreed to protect 120 ha of old growth for each of 380 pairs of owls in Oregon (significantly less, it should be noted, than the amount of old growth in the home ranges of most pairs). By 1981, that plan had expanded to include the national forests in the State of Washington. The number of protected pairs was increased to approximately 500, and the amount of old growth per pair was increased to 400 ha. It was management by minima – choosing the smallest number of owls and smallest amount of old-growth per pair of owls in a weak effort to fulfil legal obligations.

If a significant area of old-growth forests were ever to be protected, (i) there would have to be a strong scientific justification for protecting more of these forests to fulfil legal obligations; and (ii) protecting old-growth forests would have to assume national importance, lest local interests override protection goals.

Analyses of population viability

In 1985, at the request of the National Wildlife Federation, Russell Lande produced a report on the demography and population viability of the northern spotted owl (R. Lande, unpublished report, subsequently published in modified form as Lande 1987, 1988). Applying data on survival rates, fecundity, and age of reproduction to a demographic model, Lande concluded that populations of the northern spotted owl were declining (i.e. lambda = 0.92). If correct, this would imply that even in the absence of further cutting, the owl was in trouble.

Lande did not believe the available demographic data were sufficient to accept fully his calculation of lambda. (When Lande published his results several years later, he incorporated newer data and concluded that lambda for the northern spotted owl was not significantly different from 1.0, indicating a stable population (Lande 1988).) Moreover, a standard demographic analysis could not be used to predict the impact of future losses of habitat on the birds. Therefore, assuming a stable population of owls in demographic equilibrium, Lande went on to develop a second model that gave the proportion of suitable habitat actually occupied by a species as a function of the proportion of suitable habitat (occupied and unoccupied) within a large region. Using this model, one can predict the effect of future habitat loss on the spotted owl by determining the proportion of suitable habitat occupied by birds when a new demographic equilibrium has been achieved. Applying data from various Forest Service studies to his model, Lande concluded that a spotted owl metapopulation cannot persist in a region where less than 20% of the landscape consists of suitable habitat. According to Lande, the plan then under consideration by the Forest Service would have kept only 6% of the landscape in old-growth cover, assuring the extinction of the northern spotted owl. The 6% value may have been an underestimate, since Lande presumed that additional old-growth forests would not be protected for reasons other than owl conservation, but it was clear the Forest Service did not intend to maintain anything approaching 20% of the landscape in old-growth cover.

Lande's paper was one of the first viability analyses of the spotted owl. Its stark conclusions helped the National Wildlife Federation to challenge successfully the Forest Service's owl management proposal, ultimately forcing the agency to begin a new round of more sophisticated research, analysis, and planning. Most subsequent management proposals would include viability analyses. In at least two cases (US Forest Service 1986, 1988) these analyses undercut the proposals by concluding that they provided no latitude in the event of natural catastrophes or in case the owls proved more susceptible to genetic, demographic, or environmental stochasticity than had been assumed originally.

Hidden in Lande's work was an explosive conclusion: the northern spotted owl could not be conserved unless a sizeable proportion of the landscape remained in old-growth forests (0.20 in the unpublished report; 0.21 in Lande 1988). A patchwork of small preserves would not suffice. It took 5 years for this idea to gain greater acceptance, in part because it was anathema to the timber companies.

Inventory and maps of old-growth forests

Few questions are more basic to resolving this controversy than 'how much old growth remains' and 'where is it located'?. Yet when environmentalists examined the Forest Service's inventory data, they discovered that no reliable estimates of the amount or distribution of old growth existed. The Forest Service's inventory was based on the timber characteristics of forests, rather than ecological characteristics. Each national forest employed its own definition of 'old growth', and in almost all cases, old-growth forests were not distinguished from much younger mature forests that lacked the structural complexity (snags, logs, multilayered canopy, etc.) of old growth.

In the late 1980s, The Wilderness Society, a private environmental organization, undertook an independent inventory of old-growth forests. The society used an old-growth definition developed by Forest Service scientists (Old-Growth Definition Task Group 1986) that emphasizes the structural characteristics of old-growth forests, including the presence of large, old trees, numerous snags and logs, and a deep, multilayered canopy. After surveying six national forests, The Wilderness Society found that only 45% of the old growth claimed by the agency met the ecological definition developed by its own scientists (Morrison 1988; see also Fig. 14.2). Subsequent research led the society to conclude that only about 0.85 million ha (2.1 million acres) of old growth remained in Washington and Oregon west of the Cascade Mountains, an 86–89% decrease from historical extent (Morrison *et al.* 1991). The Wilderness Society was also able to demonstrate that approximately 65% of the remaining old growth was unprotected and vulnerable to logging.

These data, coupled with maps showing historical losses of old-growth forests (Fig. 14.3), attracted considerable media attention and undermined claims by timber industry spokespersons that old-growth forests were abundant and in little need of protection. For the Forest Service, the

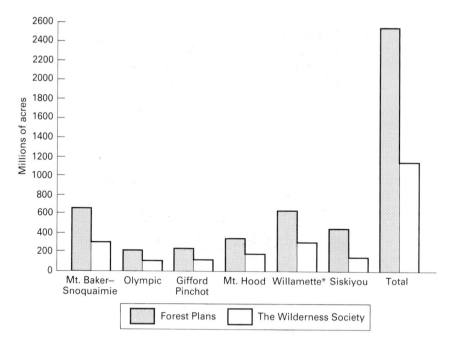

Fig. 14.2. Independent estimates of the amount of old-growth forest remaining in six national forests in western Washington and Oregon. The Forest Service combined mature and old-growth forests in its inventory; The Wilderness Society applied an ecological definition of old growth and came up with much lower estimates. From Morrison (1988).

studies were embarrassing evidence of the agency's inability to track accurately the natural resources under its domain.

A new approach

Convinced of the inadequacy of federal plans to protect the owl's habitat, federal courts issued injunctions barring new timber sales in old-growth forests in 1989. This action prompted federal land management agencies and the US Congress to convene an independent panel of experts to develop a conservation plan for the northern spotted owl. The panel's report (Thomas *et al.* 1990) was a milestone in efforts to protect old-growth forests.

Thomas *et al.* (1990) irrevocably changed the dynamics of the debate over forest management. Prior to this time, the Forest Service, Bureau of Land Management, and most scientists (e.g. Dawson *et al.* 1986) had focused on protecting individual pairs of owls or clusters of two to three

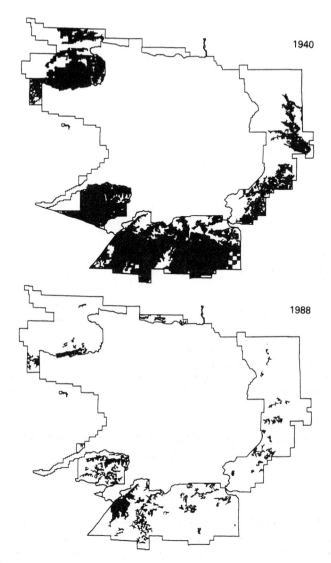

Fig. 14.3. Loss of old-growth forests in Olympic National Forest (Washington) between 1940 and 1988. The area delineated in the centre of the figure is Olympic National Park, whose forests are not shown but where logging is prohibited. Areas on the periphery of the park constitute Olympic National Forest, where logging is permitted. Figure provided by The Wilderness Society.

rrayed in a network. Thomas *et al.* (1990) concluded that such an
ach would inevitably lead to the owl's extinction because old-growth
es of a few hundred hectares would be vulnerable to small-scale
ral disturbances and edge effects. Also, protected areas capable of
supporting only one or two pairs of owls would be too small to support
internal dispersal of juvenile owls.

The Thomas Committee produced a set of maps delineating precisely
the locations for a recommended network of protected forests on federal
and state lands in Washington, Oregon, and northern California.
Wherever possible, each protected area contained a minimum of 20 pairs
of owls, and the maximum distance between protected areas was set at
19 km. To ensure adequate amounts of dispersal habitat between protected
areas, the panel restricted, but did not prohibit, logging of forest lands
surrounding protected areas. It required at least 50% of the landscape
surrounding protected areas to be covered by trees averaging $\geqslant 28$ cm (11
inches) in diameter with 40% canopy closure; this guideline became
known as the 50–11–40 rule. A total of approximately 1.5 million ha of
federal forestland would be set aside (i.e. protected from logging) for the
owls, mostly in large blocks of 20 000–30 000 ha, making it the most
ambitious plan ever developed for a vanishing species in the United
States.

The 20-pair criterion was based on both models and empirical studies.
Thomas *et al.* (1990) developed a simulation model to predict the
dynamics of a population of owls searching for unoccupied but suitable
territories arrayed in clusters of different sizes. Mean occupancy rates
reached an equilibrium when each cluster contained 15+ suitable territories. When the model assumed that only a fraction of the territories
within a cluster was suitable for immediate occupancy (e.g. if only 60% of
the habitat within a cluster was suitable, with the remaining 40% composed of forests too young to sustain owls), the mean occupancy rate did
not reach an equilibrium until the clusters each contained 20+ territories.

Support for the 20-pair criterion also came from several studies of bird
populations on islands. For example, after analysing census data spanning
nearly a century from the Channel Islands off the coast of southern
California, Jones & Diamond (1976) concluded that only about 15% of
the species represented by populations of about 20 pairs had become
extinct. Of the four large-bodied, non-migratory birds censused over a 29-
year period on the Farne Islands off the coast of Britain, only one (ringed
plover, *Charadrius hiaticula*) persisted throughout this time; its mean
population size was 13 pairs. Three other species, represented by one or
two pairs, did not persist (see Diamond & May 1977). Finally, using data
from Pimm *et al.* (1988) for land birds on 16 islands off the coast of

Britain, Thomas *et al.* (1990) constructed a linear regression model to describe the relationship between population size and persistence. For a large-bodied, non-migratory bird species, a population of 20 pairs had an average persistence time of about 50 years, versus only about 25 years for a population of 10 pairs.

Lacking any objective criteria for setting the maximum distance between protected areas, Thomas *et al.* (1990) chose 19 km because it was within the known dispersal distances of about two-thirds of radiomarked juvenile spotted owls. The committee arrived at this decision after consulting its advisers and outside authorities.

The recommendations of Thomas *et al.* (1990) have been incorporated into almost every subsequent proposal for protecting old-growth forests. Indeed, most of the parties to this dispute now accept the necessity of protecting large blocks of forest with numerous owl pairs, an idea many vigorously opposed prior to the panel's report.

What contributed to the success of Thomas *et al.* (1990)? First, the panel was inclusive. Although the core members of the panel consisted of federal scientists, it also included representatives from state wildlife agencies, environmental groups, forest industries, and academia. Second, the committee was able to command virtually unlimited resources to fulfil its mandate. It had access to all federal and state data on spotted owls and forest resources, and all its staffing requests were met. Third, the panel's deliberations were open to the public and free from political pressure, to the considerable credit of the federal agencies involved in this controversy. Fourth, the timing was right. Earlier studies, such as those by Russell Lande and The Wilderness Society, had damaged the credibility of the federal land agencies, and a growing number of legislators and administrators were now receptive to a seemingly impartial report.

Most important, Thomas *et al.* (1990) was a scientifically credible report. The panel's recommendations were derived from both population modelling and empirical studies; they could not be dismissed as either excessively theoretical or detached from natural history. And the panel cautiously applied both island biogeographic studies and population models to predict the fate of owl populations under different management scenarios. It was an expensive, labour-intensive, high-profile synthesis of years of field studies combined with a viability analysis.

Conclusions

One is struck by the pivotal role that non-glamorous, applied research has played in this controversy. Autecological studies demonstrated the dependence of spotted owls on older forests and supplied the data on demographics, home range, and dispersal needed to construct a reserve network. Most of the funding for this work came from the agencies themselves, and money allocated for studies of the northern spotted owl often came at the expense of other species. Unfortunately, the number of imperilled species in need of study vastly exceeds available funding – a situation that is unlikely to change in the near future.

The spotted owl/old growth controversy also made population viability analysis an indispensable part of endangered-species planning. Participants in future conflicts will almost certainly insist on subjecting proposed management plans to a state-of-the-art viability analysis. This will inevitably create an interesting moral dilemma. Questions of population viability reduce to questions of risk: what is the probability that a particular population will persist for a given period of time? It then becomes a public policy issue to choose both the appropriate survival odds and time frame. For example, Thomas *et al.* (1990) developed a plan whose objective was to 'assure the viability of an owl population well-distributed throughout its range . . . for at least 100 years' (Thomas *et al.* 1990, p. 9). But who is to say that 100 years is the appropriate time frame? Over much of the owl's range, two centuries or more are probably necessary to regenerate logged habitat.

The evolutionary forces that create and extinguish species typically operate over a span of thousands or millions of years. In managing for viable populations, a logical approach would be to use a similar time scale. Soulé (1987), for example, has defined a viable population as one that 'maintains its vigour and its potential for evolutionary adaptation'. Yet it is inconceivable that political leaders or the public-at-large would adopt such a long-term perspective, especially when it entails short-term costs. One suspects the best we can do in most circumstances is look a century or two ahead and hope that our skills and resources for protecting bio-diversity grow in the meantime.

BUILDING NATIONAL INTEREST

Were old-growth forests to remain solely a regional concern, there would be little prospect of significant reductions in logging. Most political leaders from the Northwest have been unwilling to incur the wrath of the timber companies, and in the rural communities whose economies are tied to logging, sentiment runs strongly against preservation of old-growth forests. National support is critical to protecting these forests, yet it cannot be built solely on the basis of a threatened owl.

Environmental organizations have therefore stressed that the northern spotted owl embodies a larger issue: the loss and fragmentation of old-growth forests. Two of their most powerful campaigning tools have been the Pacific yew (*Taxus brevifolia* Nuttall), a small, slow-growing tree found in old-growth forests, and the numerous declining stocks of salmon (*Oncorhynchus* spp.) in the region.

The yew acquired fame when medical researchers discovered that a compound (Taxol) extracted from its bark was surprisingly effective in reducing some human cancers, including ovarian cancer (US Forest Service 1993). Before this discovery, the yew was considered by most foresters to be a 'weed tree'. Yews felled during commercial timber harvests were simply burned along with the other woody debris. Environmentalists seized upon the yew as an example of the undiscovered benefits of old-growth forests.

Salmon provide an especially powerful argument for old-growth protection because they are both economically and symbolically important to people in the Northwest. The commercial harvest of salmon from spawning grounds in federal forests in Oregon and Washington is conservatively valued at $40 million annually (C. Frissell, unpublished report to The Wilderness Society); more important, that harvest creates numerous jobs in the catching and processing of fish. Rights to catch salmon are also included in various treaties between the federal government and Native American tribes.

Fish populations are vulnerable to the environmental changes associated with logging. Logging roads alter natural drainage patterns and accelerate erosion, thereby causing harmful changes in stream flow, channel configuration, sediment transport, and stream-bottom structure. The removal of trees can increase sedimentation rates and water temperature, change seasonal flow patterns, and alter the quantity and distribution of woody debris in streams, all of which redound to the detriment of anadromous fish populations.

At least 106 major populations of salmonid fish on the West Coast of the United States have been extirpated, and an additional 214 are considered to be at risk of extinction or of special concern (Nehlsen *et al.* 1991). Areas with the greatest numbers of imperilled salmonid stocks tend to lie in regions of steep, forested uplands where extensive logging and road construction have occurred. In similar terrain where logging is prohibited (i.e. in parks and wilderness), fewer stocks are imperilled (Anderson & Olson 1991).

By stressing the benefits of old-growth forest protection for salmon or medicinal compounds, environmentalists have gained national support for their side. Of course, the grandeur of the forests and the visual shock of a freshly logged site are as important in arousing public ire as any study or endangered species.

SPECIES AND ECOSYSTEMS

Having argued that the plight of the spotted owl is indicative of a broader threat to an ecosystem, one is left with the difficult task of deciding what it means to protect an ecosystem. The stakes are very high: scientists, environmentalists, and representatives from the extractive industries have all spoken of the need to manage on the level of entire ecosystems, rather than single species (see Noss 1991; Scott *et al.* 1991). When all parties to a bitter controversy begin using the same terms, they are unlikely to be referring to the same things. One environmentalist quipped after the President's Forest Summit, 'I hear ecosystem management, I hear "ecosystem". The industry hears ecosystem management, they hear "management". To them that means cutting lots of trees' (McCoy 1993).

In May 1991, two committees of the US House of Representatives asked a panel of scientists to evaluate 'different approaches for protecting ecologically significant old growth and late successional ecosystems, species, and processes, including, but not confined to, spotted owls' (Johnson *et al.* 1991, p. 44). The panel's report, released in October 1991, outlined a series of management scenarios providing successively more protection for the old-growth forests and attendant species (Johnson *et al.* 1991). It is a useful departure point for discussing ecosystem management.

In the opinion of the panel, the spotted-owl plan proposed by Thomas *et al.* (1990) would provide a 'high' probability of sustaining the owls over the course of a century but only a 'medium low' probability of maintaining breeding populations of marbled murrelets (*Brachyramphus marmoratus'*, a threatened alcid that nests in old-growth forests), and a 'very low' probability of sustaining sensitive fish stocks. The panel gave the plan an overall rating of 'medium low' for sustaining a functional old-growth ecosystem. The ratings are subjective, but they illustrate a significant point: what suffices for the owl may not suffice for other old-growth species or for the more nebulous ecosystem as a whole. (This point was vividly demonstrated by a recent report (Thomas *et al.* 1993) that identified over 380 species of plants, animals, and fungi associated with old-growth forests that would not be adequately protected under the plan proposed by Thomas *et al.* (1990).)

To provide a high probability of sustaining a functional old-growth ecosystem with all its species, including salmon, the panel recommended the following steps:

1 Protection of virtually all of the remaining late-successional and old-growth forests.

2 Protected buffer zones around all streams and rivers.

3 A prohibition on the construction of logging roads in all watersheds containing high-quality fisheries, rare fish populations, or especially clean water.

4 Reduction in the density of existing roads in some of these key watersheds.

5 A harvest rotation length of 180 years in areas outside protected zones (current rotation lengths are much shorter).

6 Retention of snags, logs, and some living trees in all areas where logging is permitted.

Under this scenario the sustainable harvest level on federal lands in western Washington, Oregon, and northern California would be 814 million board feet per year. In contrast, from 1985 to 1989, the annual harvest from these lands averaged 5.3 billion board feet.

Thus, moving from a spotted-owl plan to a more comprehensive 'ecosystem plan' entailed three steps: preserving more old growth, providing additional protection for aquatic ecosystems, and retaining more structural diversity within harvest areas. The aquatic guidelines were developed with the needs of salmon and other fish in mind; they are, in essence, species-specific guidelines. The structural diversity guidelines were deemed critical to sustaining healthy forests because snags and logs provide habitats for numerous plant and animal species, replenish soil nutrients, prevent erosion, etc. – a mixture of species requirements and 'ecosystem services' (themselves produced by species). In short, ecosystem protection reduces to species protection. This is hardly surprising, given that ecosystems are defined by their species.

An 'ecosystem plan' that fails to protect all of the species associated with a particular natural community can only be characterized as deficient. On the other hand, a management plan for the spotted owl or any single species is likely to fall short of protecting the full diversity of species inhabiting the old-growth forests of the Pacific Northwest. The solution is to incorporate the needs of more species into the plan, as Johnson *et al.* (1991) began to do, rather than abandoning this approach in favour of a far more nebulous one that proclaims 'ecosystem' but means all things to all people.

ACKNOWLEDGEMENTS

I thank Dr Jack Ward Thomas for his thoughtful comments on an earlier draft of this chapter and Dr Mark Shaffer for his helpful discussion of population viability models. Jeffrey Olson kindly provided the graph of timber harvest levels in the national forests of Oregon and Washington.

REFERENCES

Anderson, H.M. & Olson, J.T. (1991). *Federal Forests and the Economic Base of the Pacific Northwest. A Study of Regional Transitions.* The Wilderness Society, Washington, DC.

Dawson, W.R., Ligon, J.D., Murphy, J.R., Myers, J.P., Simberloff, D. & Verner, J. (1986). *Report of the Advisory Panel on the Spotted Owl.* Audubon Conservation Report No. 7. National Audubon Society, New York.

Diamond, J.M. & May, R.M. (1977). Species turnover rates on islands: Dependence on census interval. *Science*, 197, 266–270.

Forsman, E.D. (1988). The spotted owl: literature review. *Final Supplement to the Environmental Impact Statement for an Amendment to the Pacific Northwest Regional Guide. Spotted Owl Guidelines* (U.S. Forest Service). pp. C1–C35. US Forest Service, Pacific Northwest Region, Portland, Oregon.

Forsman, E.D., Meslow, E.C. & Wight, H.M. (1984). Distribution and biology of the spotted owl in Oregon. *Wildlife Monographs*, 87, 1–64.

Johnson, K.N., Franklin, J.F., Thomas, J.W. & Gordon, J. (1991). Alternatives for Management of Late-Successional Forests of the Pacific Northwest. A Report to the Agriculture Committee and the Merchant Marine and Fisheries Committee of the US House of Representatives. Privately published.

Jones, H.L. & Diamond, J.M. (1976). Short-time-base studies of turnover in breeding bird populations on the California Channel Islands. *Condor*, 78, 526–549.

Lande, R. (1987). Extinction thresholds in demographic models of territorial populations. *American Naturalist*, 130, 624–635.

Lande, R. (1988). Demographic models of the northern spotted owl (*Strix occidentalis caurina*), *Oecologia*, 75, 601–607.

McCoy, C. (1993). Environmentalists appear to score in timber battle. *The Wall Street Journal*, April 5.

Morrison, P.H. (1988). *Old Growth in the Pacific Northwest. A Status Report.* The Wilderness Society, Washington, DC.

Morrison, P.H. (1990). *Ancient Forests on the Olympic Peninsula. Analysis From a Historical Perspective.* The Wilderness Society, Washington, DC.

Morrison, P.H., Kloepfer, D., Leversee, D.A., Socha, C.M. & Ferber, D.L. (1991). *Ancient Forests in the Pacific Northwest. Analysis and Maps of Twelve National Forests.* The Wilderness Society, Washington, DC.

Nehlsen, W., Williams, J.E. & Lichatowich, J.A. (1991). Pacific salmon at the crossroads: Stocks at risk from California, Oregon, Idaho, and Washington. *Fisheries*, 16(2), 4–21.

Noss, R.F. (1991). From endangered species to biodiversity. *Balancing on the Brink of Extinction. The Endangered Species Act and Lessons for the Future* (Ed. by K.A. Kohm), pp. 227–246. Island Press, Washington, DC.

Old-Growth Definition Task Group (1986). *Interim Definitions for Old-Growth Douglas-fir and Mixed-Conifer Forests in the Pacific Northwest and California.* US Forest Service, Pacific Northwest Research Station Research Note PNW-447, Portland, OR.

Pimm, S.L., Jones, H.L. & Diamond, J. (1988). On the risk of extinction. *American Naturalist*, 132, 757–785.

Scott, J.M., Csuti, B., Smith, K., Estes, J.E. & Caicco, S. (1991). Gap analysis of species richness and vegetation cover: an integrated biodiversity conservation strategy. *Balancing on the Brink of Extinction. The Endangered Species Act and Lessons for the Future* (Ed. by K.A. Kohm), pp. 282–297. Island Press, Washington, DC.

Soulé, M.E. (1987). Introduction. *Viable Populations for Conservation* (Ed. by M.E. Soulé), pp. 1–10. Cambridge University Press, Cambridge.

Thomas, J.W., Forsman, E.D., Lint, J.B., Meslow, E.C., Noon, B.R. & Verner, J. (1990). *A Conservation Strategy for the Northern Spotted Owl.* US Forest Service, Bureau of Land Management, US Fish and Wildlife Service, and National Park Service, Portland, Oregon.

Thomas, J.W., Raphael, M.G., Anthony, R.G., Forsman, E.D., Gunderson, A.G., Holthausen, R.S., Marcot, B.G., Reeves, G.H., Sedell, J.R. & Solis, D.M. (1993). *Viability Assessments and Management Considerations for Species Associated With Late-Successional and Old-Growth Forests of the Pacific Northwest.* US Forest Service, Washington, DC.

US Forest Service (1986). *Draft Supplement to the Environmental Impact Statement for an Amendment to the Pacific Northwest Regional Guide. Spotted Owl Guidelines.* US Forest Service, Portland, Oregon.

US Forest Service (1988). *Final Supplement to the Environmental Impact Statement for an Amendment to the Pacific Northwest Regional Guide. Spotted Owl Guidelines.* US Forest Service, Portland, Oregon.

US Forest Service (1993). *Pacific Yew Draft Environmental Impact Statement.* US Forest Service, Bureau of Land Management, Food and Drug Administration, Portland, Oregon.

15. THE ECOLOGICAL COMPONENT OF ECONOMIC POLICY

JOHN S. MARSH

Department of Agricultural Economics, University of Reading,
4 Earley Gate, Reading RG6 2AR, UK

SUMMARY

Economics is concerned with the choices people make. They are forced to choose because it is impossible to satisfy all demands, given the resources at our disposal and the technologies we possess. Improvement, in economic terms, means being able to increase the range of choice.

Every economic decision has an ecological implication – it affects the opportunities open to other species, it changes the form and may render useless some natural resource and it creates new opportunities for some organisms. There is here a balance of effects, positive and negative in terms of current and future choices.

The economist has to attribute values to these ecological changes. Some can be derived from the market place but many can not. There is thus an important theoretical task in giving comparable weights to such values and an empirical job of actually determining values for specific resource-using alternatives.

Only when satisfactory estimates of these values are made and generally accepted, can policy give proper emphasis to ecological considerations in the determination of national and international resource use. This has become a very sensitive issue for agriculture where environmental issues attract growing concern.

Ecological considerations require a more adequate method of evaluating economic activity. Conventional economic measures do so on an 'annual flow' basis. This can allow for changes in the value of man-made resources, 'capital'. It does not incorporate considerations relating to stocks of natural resources. To do so involves assumptions about costs and benefits to future generations which are difficult to validate. To ignore such issues means getting the economics wrong.

WHAT DO ECONOMISTS DO?

Introductory economic textbooks usually start with an attempt to define the subject. The length and sophistication of such an account varies from full-length studies such as Lionel Robbins *The Nature and Significance of Economic Science* (Robbins 1932) to a few breezy paragraphs which tend to sound like common sense. Such explanations generally point to four main ideas:

1 That there is an unlimited list of 'wants' (of goods and services which people wish to enjoy).

2 That the capacity to meet these wants is constrained by our understanding of how resources may be used.

3 Because the stock of resources is finite – this forces people to make choices.

4 That, according to our understanding of the implications of alternative sets of choices, we may be able to satisfy more or fewer of these wants.

The task of economics is to study these choices and their implications. It is not to prescribe some particular set of resource uses for society but to inform people of the options that are open to them and the consequences of different decisions by individuals or by public authorities.

This is a very different agenda from that which seems to exist in the public mind when economics is mentioned. There, economics is often thought only to be about those relationships that involve money. It is associated with profitability and business success or failure. Thus speakers are often invited to address such questions as the 'economic future of British agriculture' and companies that go bankrupt are said to be victims of 'economic forces'. Quite understandably, mixing in such company economists are often thought to be 'hard men', concerned with impersonal magnitudes and indifferent to anything that stands in the way of maximizing some money magnitude.

Money is important. It matters because it is through spending money that people exercise many of their choices over the use of resources. It provides the most convenient means by which such behaviour can be aggregated so that we can talk at the same time about choices relating to apples and pears. It is, however, an imperfect measure. Partly it is imperfect because of technical defects. It is elastic – the amount of money in relation to other things is not constant. Partly it is unsatisfactory because our 'monies' relate only to specific places and so make the aggregation of choices that cross frontiers very difficult. Most of all it fails because there are many values that are real but which do not enter into the market place. These values and the choices they determine are just as 'economic' as those that do but they are much more difficult to handle. Since many of the concerns of ecologists fall into this 'non-market' category this is an especially important consideration for those seeking to understand the way in which they may influence policy.

We have policies because we believe that we can improve on the outcome of autonomous decisions. In formulating such a view it is important to understand what is meant by 'we'. In a fully participative democratic system the 'we' concerned might include all citizens. In practice the 'we' concerned is more narrowly defined. It includes the government and those who formulate and carry out policies in the bureaucracies. It includes pressure groups who identify particular aspects of the autonomous working of society which they feel need to be changed. The 'we' may be an altruistic group or it may be those who quite consciously seek the benefit of their own group at the expense of, or at least without regard for, the welfare of others. In all cases policy is in fact determined by a subset of the whole population.

Economic policies are expressions of non-market intervention in relation to the choices people make or are allowed to make. They quite naturally are a cause of considerable debate both about their goals and about their efficiency in bringing about such goals. Because our understanding is weak we may often subscribe to goals that prove incompatible and frequently apply methods that do not achieve their intended purposes. Sometimes we attempt to express these non-market values by attaching money numbers to them. This is a difficult process and the methods used cannot be relied upon to give an impartial view of the values society actually has. This is a criticism of economics as such and a good reason why economists should be modest about their contribution to society. At the same time, it underlines the need for more and better research if in this important area society is to be empowered to steer choices in the preferred direction.

Economics does make its own value judgements. It assumes that an economic improvement occurs when the choices open to people are enlarged. This is in fact quite a restrictive concept of improvement. In so far as we find it impossible to evaluate the welfare of one individual in relation to others, it means not just that overall choice is enhanced but that, in the process, no individual must find himself faced by fewer choices than before. In reality this is a very difficult condition to fulfil or to monitor so that economic policies are often, in these terms, likely to be ambiguous in their effects. The practical outcome is that we make decisions about comparative welfare and apply them in the form of the economic policies we actually adopt. Most people do not find it too difficult to distinguish situations which we regard as 'economically' preferable. Thus a society in which there is abundant food, in which people can live in good houses, wear attractive clothes and enjoy more leisure, is regarded as better than one in which those choices are not open to its residents.

It is on such broad generalizations that much modern economic policy is founded. In particular the idea of economic growth has become an important element in economic policy making. By ensuring that each year the bundle of goods and services produced is larger than in previous years, the needs of a growing population and the expectations of many people for rising 'living standards' can be met. Further, without economic growth many of those important economic resources we call 'people' may not be needed in the productive process. In a purely market system such people would receive no income. Although we make value judgements which may mean that some income is transferred to such people this is usually unpopular with both the recipients and the taxpayers. Thus a government which fails to deliver 'growth' is unlikely to remain a government for long.

THE ECOLOGICAL IMPLICATIONS OF ECONOMIC ACTIONS

It is at this point that the ecological significance of economic actions becomes inescapable. Given that the stock of natural resources is limited, if economic growth speeds up the rate at which they are exploited, it shortens the period during which it will be possible to enjoy the endowments of nature. In this sense it deprives future generations of choices that they would otherwise enjoy. There is an implicit conflict between present and future generations. Given the interdependence of many of the biological and physical processes that underpin economic growth, there is a conflict between the short-term survival of the present system and the longer term capacity of the planet to cope with its effects.

It is important not to overstate this conflict. There are a number of considerations that qualify the apparent conflict between economic growth and the welfare of future generations:

1 Not all growth depends on heavier use of finite natural resources. Many of the choices that people may make as they become richer may be for non-material benefits such as poetry and music, leisure and conversation rather than additional resource-consuming goods. People may even choose to conserve the natural inheritance as a form of wealth in itself.

2 As economies grow they invent new technologies which make more efficient use of resources. As a result, from a given stock of natural resources a larger volume of outputs can be produced. An important aspect of growth is to increase the stock of resources by finding new ways of using previously useless assets. The silica chip provides one example. Without economic growth such developments would not occur – the poor

countries of the world are not the places from which new technologies emerge. In this way the value of the assets passed to succeeding generations may be enhanced.

3 It is, in practice, very difficult to attach sensible values today to alternative bundles of resource in the future. Even over so short a period as the past 100 years, technologies have changed, tastes have changed and entirely new products have come to dominate the market place. Past experience, if anything, suggests that modern society has a greater capacity to solve purely technical problems of relative resource scarcity than to cope with the human pressures that result from the pace of economic change. There is more risk of humanity blowing itself up than of running out of resources. It would be as unreasonable to impose the values of the 1990s on the rates of use of alternative resources for the next century as it would have been to apply those of the 1890s to the twentieth century.

Despite these qualifications this does not mean that we can be complacent about the ecological impact of economic activity. Every human activity, including those we think of as economic, has some impact on the environment as a whole. Natural resources are used. From the viewpoint of some species they may be rendered valueless. For others new opportunities may arise. There is no way in which the economy can be ecologically neutral. It will have some balance of effects which may be judged either positive or negative.

Much of the current debate about the environment implies a judgement that this balance has become negative in a number of important respects. The list of concerns is long, it includes for example:

1 Urban congestion, the decay of traditional urban areas, noise, the inadequacies of transport, water systems and sewage disposal.

2 The loss of traditional landscapes, the decay of rural buildings or their conversion to uncongenial but 'profitable' purposes.

3 Issues relating to global warming, pollution of water and air, the loss of the ozone layer.

4 Loss of biodiversity, the over exploitation of marine resources, the loss of rainforest and the extinction of rare species, the exhaustion of natural resources.

Such examples illustrate the wide range of interactions between economic activity and ecological systems. They also draw attention to two important but rather different set of problems.

First, we need to understand better the implications of the economic choices we make for the world in which we live. This is a continuous need because our economic activity is always changing. It also needs to be frequently reviewed as our scientific understanding grows. Some current anxieties exist because we are now capable of measuring things that were previously invisible to our science. This does not mean they are trivial or unimportant but it does stress the need for continued vigilance. It also underlines one of the important difficulties facing scientists in this dialogue. It is the nature of science to make progress by invalidating yesterday's received wisdom. However, for the layman, who looks to the expert for guidance and makes his decisions based on what is generally thought to be true, this changing of 'truth' is disconcerting. Politically it may be deeply embarrassing, since for some reason politicians are thought to have a responsibility to 'know everything'. Practically, it may lead to sudden changes in demand, sudden shifts in regulations applied to industrial processes and therefore to windfall profits or losses for business. When such changes occur the tendency may be for the public to believe that it is the scientist that is discredited not just yesterday's ideas.

Second, society has to make its choices not only in the light of the 'knowledge' it has but also to reflect its values. The values of different people or communities are unlikely to be similar. Thus some may value the fox as an integral part of the natural balance of the British countryside, other may esteem it in terms of the opportunity it provides for the 'unspeakable to chase the uneatable'; a poultry farmer who favours free range may see it as a deadly enemy content not just to feed its family but intent on destroying his stock. Each of these views may be held with equal sincerity by individuals and there is no objective, scientific way of proving one right and the other wrong. The outcome so far as the fox is concerned is thus likely to depend on who holds power at the time. At a global level these differences of perception and interest are no less sharp. The governments of countries confronted by hunger, poverty and disease as a result of poor water and sanitation are unlikely to take the same view about preserving wildlife or preventing global warming as those whose agricultural problems are of 'surplus' and whose populations are regularly entertained by nicely packaged televized wildlife programmes. Both approaches are valid but they are not readily compatible.

THE CONTRIBUTION OF THE ECONOMIST

It was Oscar Wilde who put into the mouth of one of his characters the definition of a cynic as 'A man who knows the price of everything and the value of nothing' (Oscar Wilde, *Lady Windermere's Fan*, Act III). Economists are sometimes put into the same category. In order to illuminate the implications of the choices that people make they have to be subjected to some common system of measurement. The measure used is money.

The authenticity of market prices as a measure of value should not be underestimated. It is at the point when we decide to buy that we simultaneously abandon the capacity to turn that purchasing power into other goods or services. No-one is better placed to know what we really want than ourselves, even if some of the things we want we may not need and others we may be ashamed of. The phrase 'put your money where your mouth is' has an uncomfortable validity for many of us.

At the same time, by making these decisions about what we buy we also determine what is produced. Producers whose products are not bought lose the resources they embody and eventually will be forced out of business. Producers who meet a need will be able to expand and so attract a larger share of the available resources to supply what people choose to buy. Since changes in demand may change the relative scarcity of factors of production, they will also revalue these resources on a continuous basis, making sure that they are used to produce those goods and services that are most highly valued.

Despite these important characteristics of markets they may not always give rise to authentic values. Three sorts of defect are of particular importance in the context of this chapter:

1 Markets may be distorted if monopolistic power enables suppliers of some factor or product to create an artificial scarcity. Where this is the case the 'monopolist' may be able to attract a larger share of the consumer's expenditure to himself than the value of the resources he uses and the products he supplies justifies. The supply of the product will be smaller than is feasible and prices higher.

2 Economic activity may involve costs and returns that affect society but do not appear in the accounts of the businesses concerned. For example, the construction of a supermarket may impose extra costs in the form of noise, dirt and traffic congestion for those who live in its immediate environment. Such costs do not figure in the accounts of the supermarket. At the same time, by attracting a concentration of people to a specific place it may create opportunities for other enterprises, for example, for leisure facilities, which could not exist without the numbers of people and car parks created for the supermarket. Any full account of the value of economic activity has to capture these externalities which are not represented in the market price.

3 The pleasure of an attractive landscape is not reduced because it is shared by others. We call such benefits 'public goods'. We cannot buy them individually nor do the providers of such goods receive any direct

reward for making them available. Should they disappear society would certainly be poorer but the market as such has no way of ensuring their survival.

In each of these areas we look to policy to supplement or correct the working of markets. However, whilst it is easy to identify some of these issues it is hard to attribute appropriate weights or to describe them in terms that permit their assessment in comparison with market-provided goods and services.

Ecological values usually fall into this non-market character. At any one time any economic system must work within boundaries set by ecological considerations. However, these boundaries may not be visible to the consumer either individually or collectively. If such boundaries are transgressed the economic system is forced to change. More likely it will change as boundaries, which are effective in the short term, begin to bite. Thus, for example, as the demand for water in a particular place begins to outstrip the local renewable supply additional costs will be incurred bringing in water from outside or learning to manage with less water. In a sense the market is working to reflect ecological realities. In most cases it does so, however, after changes have taken place or when they are imminent rather than in anticipation. By the time a response is made by this route, irreversible damage is likely to have been done. Thus although, at any one time, there may be an equilibrium between the economy and ecological considerations this is unlikely to be a satisfactory outcome in the light of our knowledge of ecology. We have therefore to modify the market by policies that imply a different set of values than those revealed by the current choices of producers and consumers.

The basis for doing this has to be information that is not present in the market system. Economists have devised a number of strategies to acquire such information. For example they may invite people to indicate their preferences between different possible scenarios for the future use of some resource, they may seek to estimate how much people value a view or a site by establishing how much they spend in travel to see it, or they may examine property values of similar houses to discover if the presence of trees in a road makes it more attractive. Very considerable work has gone into the development and refinement of such methods. Surveys of opinion of this type are expensive and depend heavily on careful design and interpretation. Where several studies point in the same direction they suggest that there is a strong possibility that they have identified a genuine non-market value and some indication of its strength (see e.g. ESRC 1993).

Such procedures are, however, dependent upon the understanding of the public, the extent to which they take seriously a hypothetical type question and especially the degree to which choices in relation to the issue under discussion are really related to the totality of choice that may be involved. This is less of a problem for some site-specific issue, such as the decision to build a new housing estate. It is extremely difficult where the concern is diffused, extends over a long period of years and rests on very imperfect knowledge. Many ecological concerns, for example, those summed up in the term 'biodiversity' fall into this category. It is in practice extremely difficult to put non-scientific members of the public in a position in which they could make informed judgements on such matters.

Policies, at least in democratic societies, are dependent upon the consent of the governed. Thus, although it is very difficult to attach much credence to uninformed views, they do actually count in the determination of what is possible. This brings the ecologist and the economist together in a task of explanation and information. Ecologists have not so much to educate the public in the details of their science as to command the public's respect for their authority. Once this exists the economist can then begin to establish the likely implications of sets of choices about resource use which take greater or less account of the concerns which stem from ecological studies. In turn, this leads to a presentation to the public of choices in terms that are likely to be well understood: prices of products, levels of taxes and the existence of more or less convenient public services.

THE PRACTICE OF POLICY

Much of the success of the 'green movement' in recent years has been its ability to communicate ideas, many of which stem from science, to an interested but non-scientific public. Nowhere has this been more evident than in the area of agricultural policy. Pressure groups such as the Royal Society for the Protection of Birds, Friends of the Earth and Greenpeace have become well known through the media. Each of them has an agenda for agricultural policy. Within government, environmental issues have assumed much higher prominence involving tensions between departments of state, commitments to international treaties and a flow of papers on planning, on agriculture, on the disposal of sewage, etc., all of which have an environmental origin. These to a considerable extent determine the area of practicable policy in the immediate environment. (See for example the commitments entered into at the UNCED conference in Rio in 1992. These include conventions on biodiversity, climate change, undertakings relating to forests and a long list of requirements under Agenda 21. A summary of these commitments is given in Johnson 1993).

In taking account of ecological issues in economic policy many conflicting policy goals have to be rebalanced. These include:

1　The need to ensure that an industry remains able to compete – or to accept the consequences of its disappearance. The coal industry provides a dramatic example. In agriculture the process of change has already involved dramatic reductions in the numbers of people employed and a revolution in the way in which farming is managed.

2　The need to identify ecological threats that undermine the ability of the system to continue to produce food. Here the rate of change is important – very long-run threats may have no political price. Short-run alarms, for example, the spread of a new disease, may call for dramatic and possibly draconian action in order to ensure that the government is seen to 'be doing something'.

3　The reality that interests conflict and policies cannot satisfy all parties. Nowhere is this conflict more evident than in current debates concerning land use. Here those anxious to foster the prosperity of the rural economy call for relaxation of planning rules whilst those anxious about the insidious expansion of urban life seek to constrain development.

4　The extent to which a particular concern is amenable to policy manipulation – will it require public expenditure, could it raise a tax, might it be left to regulation and policing? If action is to be taken it must be affordable and effective. Of course, if the purpose is only to be seen to be doing' something' it may not matter if the policy actually affects the 'real world' – it will keep the 'chattering classes' quiet.

Recent changes in the Common Agricultural Policy (CAP) have reflected some of these concerns. For the most part they have been driven by the need to limit budget expenditure and by a wish to present a more accommodating face to the world in the General Agreement on Tariffs and Trade (GATT) negotiations. For full details of the 1992 CAP reform settlement see EC (1992). They have also been concerned to protect the 'family farm' as perceived on the continent and so to avoid any very abrupt reduction in income to the farmers concerned. The route taken has been that of supply control. In essence, prices have been cut, measures adopted which should result in lower output and compensation payments made to farmers on a basis that reflects the historic levels of stocking or areas of crops grown. In doing so, however, some attention has been given to environmental matters. Thus, in the policies for beef, stocking density rules have been adopted that will gradually be tightened up over the next few years. Ceilings have been set on the maximum number of sheep or beef animals that may qualify for compensation in the form of premia payments. Higher rates of compensation are to be paid to those who apply an extensification scheme. An agri-environmental package is to be applied within each of the member states. At the time of writing, some 9 months after the agreement, it remains unclear precisely what steps the UK government is to adopt in relation to this requirement but a package is 'on the way'.

This approach to reform makes it clear that, in this case, environmental considerations have been perceived as an 'add on' to agricultural policy. Welcome as their presence may be, this is still not an adequate approach to the economic question: how can we best use our resources to ensure that the maximum range of choices is open to the community? To do this we have to start from the opposite end of the debate – what do people really want and what are the environmental constraints within which our economic system, including agriculture, has to work? These constraints are seldom absolute. They demand trade-offs which may be between difference resources, between various products which can meet consumer demands and between today and tomorrow. They are nevertheless fundamental. Since economic activity can never exceed its ecological boundaries, it needs to know where those boundaries lie and how they may be moved by different choices that society may make. The concern is not peculiar to agriculture or to land use. Since there is an element of both substitutability and complementarity between resources and products in agriculture and other sectors, the choices that have to be made are choices for the economy as a whole.

The implication of this discussion is that, in the practical world of policy making, it is no longer sufficient to be concerned only with one sector. The economist needs to hear and apply messages that come from ecology.

ECONOMISTS AND ECOLOGISTS: DO THEY DIFFER IN THE LONG TERM?

Both economics and ecology have their root in the same Greek word *oikos* which is concerned with household management. The final section of this chapter stresses some of the fundamental similarities between the concerns of those who today call themselves 'ecologists' and those who sport the label 'economist'.

One of the dictionaries I use has a double-page picture spread to illustrate ecology (*Oxford Reference Dictionary*, 1986). The diagrams display a series of flows in energy and in the food chain. They also demonstrate the effect of a change in ecological balance as a result of the outbreak in 1954–55 of myxomatosis. Essentially the diagrams stress the circular nature of the systems on which we depend and of which we form part. Economics does not qualify for such graphic treatment. Instead it has to rely on two definitions:

1 The science of the production and distribution of wealth: the application of this to a particular subject.
2 The financial aspects of something.

It is the first of these concepts that comes closest to the use of the term in this chapter. Using this approach, it is evident that the production and distribution of wealth, too, forms part of a circular system. Production can only continue as long as there is a corresponding flow of payments for what is produced. In that continuous flow the distribution of incomes and the level of output is determined. The issues that bother economists and those who seek their advice are largely concerned with what is produced, for whom, and what incomes will result.

There is, however, a sense in which such an approach can miss an issue that is of concern to both economists and ecologists, that is the level of 'stocks' which we carry forward to future years. In economic terms these stocks may be either raw materials, capital equipment or accumulations of unsold products. We can only assess adequately what has been produced and what resources have been used if we charge the opening stocks to our system and credit those that remain at the end of the time period in which we are interested.

Ecologists are less likely to draw up annual accounts in the same way as economists but the need to take account not only of changes in the current flow of activity but also of changes in the level of stocks is common to both disciplines. It focuses attention on one of the areas in which there is a shared concern.

Sustaining a thing, by itself, is not necessarily a good thing. For example, sustaining an oppressive political regime would seem to be decidedly bad. Nor is it simply a defence to say we have left a resource untouched for the future. It may turn out that the future has no need for the unused resource or that it may choose to apply it to some purpose much less useful than current consumption. There is no virtue in leaving a stock of coal or oil in the ground if it is never to be used. These ideas are well developed in a paper by Professor C.R.W. Spedding given to a conference of the Royal Agricultural Society of England (RASE) in March 1993.

The justification for our concern about the level of stocks implies judgements about future generations and the values they will hold. Clearly the present set of parents is likely to attach some significance to the future of its children – if not why would they pay for school fees or put up with the considerable discomforts of living with a teenager? We might also imagine that part of the welfare of our children will depend upon their ability to meet the needs of their children. In this way a chain of concern builds up that could make us take seriously the notion of foregoing consumption now in order to enable future generations to enjoy the resources we have not used. This is a not unreasonable approach but it seems hard to press it too far. Whilst I might even feel some real concern for my great, great grandchildren – who will certainly not know me – I find it incredible to be asked to forego my motor car now in order to ensure that someone can drive to work in AD 3090. In other words whilst the future is not without value to present generations, its value is not infinite.

This draws economists and ecologists together. We need to have clear ideas about the rates and directions of change implicit in present economic practices. We need to consider what alternative packages exist which might vary these rates of change in specific directions. We have to be able to inform the debate so that society itself can make better choices about the way it behaves today. To ignore such considerations is not just to be self indulgent. It is to get economic policy wrong for this generation as well as for the future.

REFERENCES AND BIBLIOGRAPHY

DOE (**1991**). *Policy Appraisal and the Environment*, Department of the Environment, HMSO, London.

EC (**1992**). Press Release No. 6539 (21.5.92), EC Council, General Secretariat Agricultural Council, Brussels.

ESRC (**1993**). *Assessing Methodologies to Value the Benefits of Environmentally Sensitive Areas*. ESRC Countryside Change Initiative, Working Paper 39, University of Newcastle upon Tyne.

Hoggart, K. (Ed.) (**1992**). *Agricultural Change, Environment and Economy. Essays in Honour of W.B. Morgan*, pp. 25–88. Mansell, London.

Furness, G.W., Russell, N.P. & Colman, D.R. (**1990**). *Developing Proposals for Cross Compliance*. Department of Agricultural Economics, University of Manchester, Manchester.

Johnson, S.P. (**1993**). *The Earth Summit: The United Nations Conference on Environment and Development (UNCED)*. Graham & Troman/Martinus Nijhoff, The Netherlands.

Oxford Reference Dictionary (**1986**). Clarendon Press, Oxford.

Pearce, D.W. & Turner, R.K. (**1990**). *Economics of Natural Resources and the Environment*. Harvester Wheatsheaf, Hemel Hempstead.

Robbins, L. (**1932**)., *An Essay on the Nature and Significance of Economic Science*. Macmillan.

RSPB (1991). *A Future for Environmentally Sensitive Farming.* RSPB Submission to the UK Review of Environmentally Sensitive Areas 1991.

This Common Inheritance (1990). A Summary of the White Paper of the Environment, HMSO, London.

16. TRANSLATING ECOLOGICAL SCIENCE INTO PRACTICAL POLICY

JONATHON PORRITT

Thornbury House, 18 High Street, Cheltenham GL50 1DZ, UK

SUMMARY

In contrast to many areas of economic and social policy-making (in which the notionally empirical data of social scientists are ruthlessly adulterated by a variety of heavy ideological inputs), environmental policy-making remains more or less genuinely science-based. Ideological interpretations may shape the interpretation of scientific evidence, but there is wide consensus that decisions must be based on the best available evidence at the time.

Theoretically this should promote good governance. Unfortunately, the processes by which the results of ecological research are converted into specific policy remain extremely opaque. Scientists concentrate on the elucidation of increasingly specialized parts of the whole, whereas civil servants and politicians prefer to work in terms of simplified wholes! Scientists are understandably reluctant to infer too much in policy terms from their partial (sometimes hotly contested) evidence, and politicians are for the most part quite incapable of grappling with anything other than generalizations and half-truths.

The media's role in attempting to bridge this gap is ambivalent.

This chapter considers the complexities and implications of this gap with regard to the historical debate about 'acid rain', and the contemporary dilemma of dealing with global warming.

The most recent injection into this debate is the much-vaunted 'precautionary principle'. An attempt is made to analyse exactly what this means in policy development terms: is it a genuinely useful aid to the interpretation of complex scientific evidence or a rhetorical device for politicians in the war of green words?

Consideration is also given to the adequacy of existing 'half-way houses' between scientists and policy makers in this area: for example, the royal commission, parliamentary select committees, the research councils themselves.

I address this issue neither as a scientist, nor as a policy maker, but rather as an environmental activist with more than a working knowledge of both! As such, I am conscious of the British Ecological Society's slight departure from normal practice in inviting a non-specialist to give a paper to this Conference, and I was very keen to do it, notwithstanding my first, somewhat daunting experience of the Society back in 1975. Standing then for the first time as a candidate for the newly formed Ecology Party, I had the misfortune to canvass one of your more irascible members who made it abundantly clear to me that ecology had no business in politics and politicians had even less business meddling in ecology!

It took the Ecology Party roughly 10 years to change its name to the Green Party, though its members remain unapologetic in seeking to persuade politicians to make room in their ideological kitbag for the basic principles of ecology. And while there has been very little politicization of the science of ecology over the intervening 18 years, it is more than possible to detect a certain ecologization of politics.

Indeed, advances in ecological science have provided the intellectual authority for a whole array of policy changes during that time. There are many who would vigorously assert that environmental policy-making has been driven far more by empirical evidence, and tarred far less with various ideological brushes than any other area of contemporary social or economic policy.

That may well be the case, but of itself it tells us little about the ease with which ecological science is being converted into practical policy. Success in accelerating this essential process depends primarily on overcoming three persistent impediments:

1   Continuing uncertainty across a huge range of so-called 'environmental issues', all the way through from very large-scale complex problems like global warming to very small-scale, notionally simple problems like dog nuisance.

2   Even when there is agreement as to the validity of the empirical evidence available at any one time, that still leaves almost unlimited scope for differential and often conflicting interpretation of that evidence.

3   The regrettable fact that much of that interpretative debate (let alone the science itself) remains inaccessible to the vast majority of people, with the consequence that there is often very little genuine public participation in the policy development process. It should also be pointed out that much of this debate remains inaccessible to the decision-makers themselves!

There is a growing sense of consternation amongst activists and scientists alike at the growing gap between available scientific evidence and policy responses to it. In terms of the three impediments referred to above, the challenge obviously has to be to reduce progressively (where possible to eliminate) residual uncertainty, to establish some kind of interpretive consensus (as happened over global warming through the efforts of the Inter-Governmental Panel on Climate Change), and to use that emerging consensus as the main lever in achieving *policy* consensus, preferably through increased public participation.

Achieving that kind of consensus can take years, if not decades, and may well entail painful compromises on the part of environmental activists desperate to see *some* progress as the world seems to be falling to pieces around them.

Consider the whole desperate issue of species extinction and loss of biodiversity. The last decade has witnessed what might be described as a gathering consensus about the scale of the problem (assessed by Edward Wilson at an alarming 27 000 species lost every year, or roughly three an hour), and the grave seriousness of dismantling whole ecosystems before we have begun to understand how they work or how they might actually benefit us.

And that is where the difficulties begin in establishing any kind of policy consensus as to what we should actually be doing. The prevailing rationale is both pragmatic and simple: there has to be a way of deriving an income from that biological diversity in order to justify any interventionist decisions to protect it. 'Chemical prospecting' (and even chemical hijacking) are all the rage as pharmaceutical transnational companies scour the forests of the south for future profit-spinning drugs or agricultural products.

Pragmatic it may be, but such rampant, unapologetic utilitarianism has laid a heavy ethical burden on many environmental organizations. Given the reality of financial hardship and poverty in the Third World, 'the sustainable utilization of wildlife' may well be the only game in town, and playing a skilful hand in it may well be the only means of dragging tens of thousands of life forms back from the abyss of extinction. But such an approach has a corrosive debilitating effect on the human spirit itself, reinforcing our arrogant assumptions that the only value any other creature might have is its use value directly to us.

The commoditization of biological diversity is now an unstoppable process. Even those countries whose legislatures have permitted the saving of creatures purely for their own sake rather than for ours (such as in the United States) are increasingly influenced by the insidious delights of getting a payback on achieving survival.

In that context, I was intrigued to see in David Wilcove's paper on the northern spotted owl that even so non-utilitarian a campaign as that (based essentially on the intrinsic value of the owl itself and the old-growth forests in which they live) had been greatly strengthened by profoundly utilitarian arguments based on the commercial value of both the salmon spawning grounds in Oregon and Washington's federal forests, and of the extraction of taxol (a compound that has proved very effective in reducing certain human cancers) from the bark of the Pacific yew found in large numbers of the old-growth forests.

To many of the hard-nosed lawyers and scientists that now make up the staff of leading environmental organizations, talk of 'intrinsic value' is just much metaphysical moonshine. Though it is true to say that we humans have *always* made use of other creatures for our own specific purposes, the current dominance of the 'sustainable utilization' concept accurately reflects the rather arid spirit of the last 15 years or so. *Our* sustainable development depends wholly on the sustainable exploitation of everything else.

Including, quite possibly, the whale. No creature has won over more human hearts than the great whale. And there is no more powerful symbol of a collective desire on our part to expiate our ecological sins and to forge a new relationship with the rest of life on Earth. It is for such reasons that a very large number of environmentalists believe that whales should be protected absolutely. Not only do they have 'intrinsic value' regardless of their use value to us, but their mammalian intelligence and sensitivity, and their extraordinary communication systems, puts them in a quite different league from any other creature.

Ethically speaking, of course, such league tables are dangerously subjective territory. And it is territory on which the International Whaling Commission (IWC) has never felt at home. Behind the headline confrontations, the business of IWC meetings has been all about quotas, maximum sustainable yields, population age structures, depletion levels, and catch-limit algorithmic calculations.

Clinical and alienating though it undoubtedly is for a very large number of people, this is the language of sustainable utilization. And sustainable utilization (harvesting nature's wealth without depleting its capital stocks) is an all-embracing concept that governments and most non-governmental organizations are now signed up to. But are we about to be hoisted by our own sustainability petard?

Scientists are now agreed that the number of minke whales (the smallest of the great whales) in the Antarctic is anywhere from 600 000 upwards. It is hard to deny that a certain number could theoretically be 'harvested' every year on an absolutely sustainable basis. In this context, the whaling nations argue that there is no difference whatsoever between the whale and the many species that are hunted for food by humankind the world over.

Many environmentalists have come to eschew the ethical case based on intrinsic value, acknowledging that there is no objection in principle to the resumption of whaling whilst simultaneously insisting on certain conditions. For one thing, the IWC is legally bound to establish beyond reasonable doubt the effects of the 1982 moratorium on whale populations before the moratorium can be lifted, and some have argued that this will take decades to achieve.

Beyond that, there must clearly be full scientific proof that a whale population has recovered sufficiently to sustain certain catch levels (the 'precautionary principle' makes it extremely difficult to adduce that degree of proof), and there must be an effective management and control system to ensure that the catch quotas are not violated. At present, these conditions have not been met for any whale species anywhere in the world, including the Antarctic minke.

But just imagine they were met, what then? The next line of defence is the argument that the method of killing whales (with exploding grenades) is cruel and inhumane. This is essentially the position of the British Government and several other anti-whaling nations. It is reckoned that exploding harpoons kill 'cleanly' in only 70% of strikes, and it can sometimes take up to half an hour for a whale to die. But the whaling nations argue that this already represents a much better clean-kill rate than deer-stalking, and that they can improve considerably on the 70% figure.

Just for the sake of the argument, imagine that they can. What then? We come back full circle to the ethical argument that we should not be killing whales commercially (i.e. for sale) in *any* circumstances – however humanely or how sustainably. (This would still allow a small number of whales to be killed by indigenous peoples such as the Inuits for their own use rather than for commercial sale.)

The logic of the sustainable utilization case is unforgiving: if the conditions for achieving it can be met, then there is nowhere to hide. No more for the whale than for the broiler chicken.

But as we all know, there are very many people who simply cannot accept that logic, seeing in it not so much reconciliation between us and the rest of life on Earth as deepening of the divide and a further belittling of our own humanity. And for what it is worth, I think they are right. Sustainable utilization is a useful but not sufficient basis for a conservation ethic today. Indeed, we must challenge those who wield it as some green panacea, and simultaneously encourage a far broader debate about totally non-exploitative relationships between ourselves and the natural world. One indication of this emerges from the report recently carried out by the Whale and Dolphin Conservation Society indicating that at least as much money is currently being made from whale-watching as is from whale-killing.

But on what grounds would we argue the non-empirical, ethical defence of the whale. Max Weber, the renowned sociologist, talked of three basic forms of social authority that activists and politicians tend to use:

1 The traditional (predominantly ideological),
2 The charismatic (predominantly spiritual),
3 The legal/rational (predominantly scientific and technological).

Of all today's progressive social movements, the environment movement is perhaps most committed to the legal/rational route to achieving influence and 'authority'. That may well be a sound approach, but it is one that has been arrived at accidentally rather than purposefully, and one which is now defended almost as an article of faith.

Tom Wilkie put his finger on it in an article for *The Independent* on 28 July 1992:

> Today's concentration on technical questions to which there are scientific answers is all-pervasive. It has become difficult to articulate any other concerns because the language in which they might be expressed lacks legitimacy in the modern world. Only questions which can be formulated in technical terms appear to be respectable.

This creates a peculiar but inevitable tension within the environment movement. Many individual environmentalists are equivocal about the role of modern science, seeing scientists at their worst as active collaborators in the destruction of the Earth, and at their best as almost irrelevant bystanders locked into utterly pathetic illusions about the 'objectivity' of their craft.

They are worried that on issue after issue the debate has become primarily if not exclusively science-led, often to the point where a more value oriented approach is deemed totally invalid. What Mary Midgley has often referred to as 'the omnicompetence of science' and its hegemonistic authority in contemporary industrial society are sources of enormous concern to environmentalists.

By the same token, many environmentalists are drawn more intuitively to non-legal/rational forms of social authority, and particularly to religious and spiritual forms. This, incidentally, may well explain the astonishing appeal of the Gaia hypothesis, formulated by Jim Lovelock and Lyn Margulis, for the idea of a self-regulating biological super organism may be interpreted either scientifically or spiritually – or both!

But most still acknowledge that they have no option but to continue working within the dominant legal/rational paradigm.

For most politicians, such concerns might appear just a little too abstract! It is not omnicompetence that politicians complain of in scientists today, but rather a lack of competence: how do they keep getting it wrong so often? And it's not an excessive certainty that they complain of, but an irritating level of residual uncertainty!

Scientists and politicians are very different organisms, and as our Minister of Science, William Waldegrave, must surely have discovered by now in his review process, symbiosis is not easily arrived at. Scientists are often involved in a very long haul, working something through over many years: politicians need to get things processed and policies formulated in months if not weeks.

Scientists are constantly hypothesizing, testing, rejecting, amending, positively thriving in between the 'grey areas'; politicians like it black and white, or not at all. Scientists these days tend to concentrate on the elucidation of increasingly specialized and often minute parts of the whole; politicians and their civil servants cannot afford to get bogged down too much in the detail, and feel much happier working with simplified wholes. Scientists are reluctant to infer too much from partial evidence; politicians will selectively infer as much as they need to in order to reinforce their existing prejudices. Lastly, scientists tend to be rather more independent of the vagaries of public opinion (what is in or out of fashion), whereas politicians find their agenda fashioned in no small part by those very vagaries.

The debate about global warming exemplifies many of the differences that arise out of these rather different profiles. Vast amounts of new research are now emerging both about the causes and the political impacts of global warming. None of that research is conclusive, some is inconsistent if not contradictory. Little of it is capable of instant translation into policy prescription. The very short period of time in which it has been possible to make year-on-year studies of potential impacts on special populations, distribution or adaptation, makes it extremely difficult for scientists to use the past authoritatively to predict the future.

Each of those separate studies can be seen as parts of a vast jigsaw spread out in research departments and institutions across the entire world. By any standards, putting those pieces together is not easy, but politicians are still making far heavier weather of it than might be expected given that the edges of the jigsaw were all pretty much joined together by the reports of the Inter-Governmental Panel on Climate Change back in 1990 and 1992.

Even at the time, these reports crystallized the 'balance of probabilities' approach to global warming, with a 'best gestimate' of a temperature increase of around 3°C by the end of the next century, and an upper estimate of around 4.5°C. There is a lot of circumstantial evidence that is consistent with those estimates. 1990 was the hottest year since records began. 1991 was the second hottest year despite the so-called Pinatubo effect, referring to the way the debris from the Pinatubo explosion must certainly have added some cooling effect. The seven hottest years have all occurred over the course of the last 10 years. Coral reefs are beginning to bleach out and die. Hurricanes and cyclones of record strength are popping up all over the place. Anomalously intense droughts are disrupting the lives of millions of people the world over. Average snowfall has dropped by roughly 8% since 1973. Mountain glaciers are in retreat almost everywhere.

All of that certainly adds up to a little bit more than some of the sceptics continue to claim. But it is true to say that no one has yet died specifically of global warming! No one has yet lost a job as a *direct* consequence of global warming – or so the sceptics will argue. No species has yet become extinct as a direct proven consequence of global warming – also the sceptics will argue. It is, therefore, tomorrow's problem, ideally to be grappled with by tomorrow's politicians, especially given that we now have so many problems of our own to grapple with. The massive uncertainties associated with global warming research merely strengthen the hand of the prevaricators.

And in such circumstances, just where do politicians get their advice from? The science/policy interface here in the UK seems often to be as muddied and grey as some of the evidence that politicians are being asked to process.

I started by suggesting that the business of environmental policy-making was probably less tainted by political abuse than many other areas. But ultimately it all comes down to how the facts themselves are interpreted, and there are just as many vested interests playing the interpretation game here as in any other area of government. Indeed, there will always be some scientist at hand to justify whatever has been decided by the politicians!

It is true today that scientists are forever complaining of the inadequacies of our arts-based political culture, and it is certainly true that survey after survey reveals startling and extraordinarily disturbing levels of ignorance amongst the general public on scientific and environmental issues. And it is not so much better with our MPs, only a handful of whom have scientific degrees and the vast majority of whom are overwhelmed both by their reading load and by the sheer complexity of it all.

But I have to say that this cuts both ways. As a dispassionate if not wholly objective observer of scientists at work within the political community, I cannot help but point out that most scientists are fantastically bad communicators! Some just cannot do it at all, others seem positively to disapprove of the business of clear, accessible popularization. Heinz Wolff talks of scientists' 'masonic attitudes', with them defending the purity of their caste and its secrets against all comers.

And that, I venture to suggest, is as much a problem for ecologists today as for any other branch of modern science. It is impossible to overestimate the importance of the work that is being done today, for it is only on the basis of that research, slowly but authoritatively built up over so many years, that campaigners and activists can develop any kind of persuasive case for change. The research findings of ecologists make up the raw material that feeds the conveyor belt of policy change, that progressively allows areas of policy uncertainty (be they legitimate or illegitimate) to be dissolved, and which empowers politicians to get on and do what needs to be done.

To argue such a case is of course to remain firmly within the matrix of the legal/rational! To say anything else to such an audience would be extremely presumptuous, but having first reconfirmed that approach as the most appropriate way for ecologists to achieve 'social authority' there are two further, interrelated issues still to be addressed. The first is *communication* and the second is *engagement*.

It is not just good science that we need today, but passionate advocacy, and the best people to be passionate are precisely those whose passion is empirically based! People today do not just need your knowledge and astonishing expertise: they need your enthusiasm, that sense of wonder that can be tracked down even in the most academic of papers, that uncorrupted joy in the workings of the natural world and our relationship with it.

And why hide such lights under such blighted bushels? Eighteen years ago I was entirely at a loss as to how to respond to that irascible British Ecological Society member on the electoral doorstep. Today I would tell him that there is still politics in ecology (whether he likes it or not!) but more important still, there is poetry in ecology, there is reverence in ecology, there is illumination and inspiration in ecology. And it would be as well for ecologists themselves to realize that as many politicians and non-scientists are likely to be moved by those attributes as by the science itself.

AUTHOR INDEX

Figures in *italics* refer to pages where full references appear.

SUBJECT INDEX

Page references in *italics* refer to figures or tables.